Student Solutions Manual

for

Waner and Costenoble's
Applied Calculus

Third Edition

Stefan Waner
Hofstra University

Steven R. Costenoble
Hofstra University

Australia • Canada • Mexico • Singapore • Spain • United Kingdom • United States

COPYRIGHT © 2004 Brooks/Cole, a division of Thomson Learning, Inc. Thomson Learning™ is a trademark used herein under license.

ALL RIGHTS RESERVED. No part of this work covered by the copyright hereon may be reproduced or used in any form or by any means—graphic, electronic, or mechanical, including but not limited to photocopying, recording, taping, Web distribution, information networks, or information storage and retrieval systems—without the written permission of the publisher.

Printed in Canada
1 2 3 4 5 6 7   07 06 05 04 03

Printer: Webcom Limited

ISBN: 0-534-41960-7

For more information about our products, contact us at:
Thomson Learning Academic Resource Center
1-800-423-0563

For permission to use material from this text, contact us by:
Phone: 1-800-730-2214
Fax: 1-800-731-2215
Web: http://www.thomsonrights.com

Brooks/Cole—Thomson Learning
10 Davis Drive
Belmont, CA 94002-3098
USA

Asia
Thomson Learning
5 Shenton Way #01-01
UIC Building
Singapore 068808

Australia/New Zealand
Thomson Learning
102 Dodds Street
Southbank, Victoria 3006
Australia

Canada
Nelson
1120 Birchmount Road
Toronto, Ontario M1K 5G4
Canada

Europe/Middle East/South Africa
Thomson Learning
High Holborn House
50/51 Bedford Row
London WC1R 4LR
United Kingdom

Latin America
Thomson Learning
Seneca, 53
Colonia Polanco
11560 Mexico D.F.
Mexico

Spain/Portugal
Paraninfo
Calle/Magallanes, 25
28015 Madrid, Spain

Table of Contents

Chapter 1	1
Chapter 2	35
Chapter 3	51
Chapter 4	83
Chapter 5	99
Chapter 6	127
Chapter 7	147
Chapter 8	179
Chapter 9	209
Appendix A	223

Chapter 1
1.1

1. Using the table,
(a) $f(0) = 2$
(b) $f(2) = 0.5$.

3. Using the table,
(a) $f(2) - f(-2) = 0.5 - 2 = -1.5$
(b) $f(-1)f(-2) = (4)(2) = 8$
(c) $-2f(-1) = -2(4) = -8$

5. $f(x) = 4x - 3$
(a) $f(-1) = 4(-1) - 3 = -4 - 3 = -7$
(b) $f(0) = 4(0) - 3 = 0 - 3 = -3$
(c) $f(1) = 4(1) - 3 = 4 - 3 = 1$
(d) Substitute y for x to obtain
$$f(y) = 4y - 3$$
(e) Substitute $(a+b)$ for x to obtain
$$f(a+b) = 4(a+b) - 3$$

7. $f(x) = x^2 + 2x + 3$
(a) $f(0) = (0)^2 + 2(0) + 3$
$= 0 + 0 + 3 = 3$
(b) $f(1) = 1^2 + 2(1) + 3$
$= 1 + 2 + 3 = 6$
(c) $f(-1) = (-1)^2 + 2(-1) + 3$
$= 1 - 2 + 3 = 2$
(d) $f(-3) = (-3)^2 + 2(-3) + 3$
$= 9 - 6 + 3 = 6$
(e) Substitute a for x to obtain
$f(a) = a^2 + 2a + 3$
(f) Substitute $(x+h)$ for x to obtain
$f(x+h) = (x+h)^2 + 2(x+h) + 3$

9. $g(s) = s^2 + \dfrac{1}{s}$
(a) $g(1) = 1^2 + \dfrac{1}{1} = 1 + 1 = 2$
(b) $g(-1) = (-1)^2 + \dfrac{1}{(-1)} = 1 - 1 = 0$

Section 1.1

(c) $g(4) = 4^2 + \dfrac{1}{4} = 16 + \dfrac{1}{4}$
$= 16\tfrac{1}{4}$ or $\dfrac{65}{4}$ or 16.25

(d) Substitute x for s to obtain
$$g(x) = x^2 + \dfrac{1}{x}$$

(e) Substitute $(s+h)$ for s to obtain
$$g(s+h) = (s+h)^2 + \dfrac{1}{s+h}$$

(f) $g(s+h) - g(s)$
= Answer to part (e) − Original function
$= \left((s+h)^2 + \dfrac{1}{s+h}\right) - \left(s^2 + \dfrac{1}{s}\right)$

11. $f(t) = \begin{cases} -t & \text{if } t < 0 \\ t^2 & \text{if } 0 \le t < 4 \\ t & \text{if } t \ge 4 \end{cases}$

(a) $f(-1) = -(-1) = 1$ (using the first formula, since $-1 < 0$).
(b) $f(1) = 1^2 = 1$ (using the second formula, since $0 \le 1 < 4$).
(c) $f(4) = 4$ (using the third formula, since $4 \ge 4$).
$f(2) = 2^2 = 4$ (using the second formula, since $0 \le 2 < 4$).
Therefore,
$f(4) - f(2) = 4 - 4 = 0$
(d) $f(3) = 3^2 = 9$ (using the second formula, since $0 \le 3 < 4$).
$f(-3) = -(-3) = 3$ (using the first formula, since $-3 < 0$).
Therefore,
$f(3)f(-3) = (9)(3) = 27$.

13. $f(x) = x - \dfrac{1}{x^2}$, with domain $(0, +\infty)$
(a) Since 4 is in $(0, +\infty)$, $f(4)$ is defined.
$f(4) = 4 - \dfrac{1}{4^2} = 4 - \dfrac{1}{16} = \dfrac{63}{16}$

1

Section 1.1

(b) Since 0 is not in $(0, +\infty)$, $f(0)$ is not defined.
(c) Since -1 is not in $(0, +\infty)$, $f(-1)$ is not defined.

15. $f(x) = \sqrt{x+10}$, with domain $[-10, 0)$
(a) Since 0 is not in $[-10, 0)$, $f(0)$ is not defined.
(b) Since 9 is not in $[-10, 0)$, $f(9)$ is not defined.
(c) Since -10 is in $[-10, 0)$, $f(-10)$ is defined.
$f(-10) = \sqrt{-10+10} = \sqrt{0} = 0$

17. $f(x) = x^2$
(a) $f(x+h) = (x+h)^2$
Therefore,
$$\begin{aligned} f(x+h) - f(x) &= (x+h)^2 - x^2 \\ &= x^2 + 2xh + h^2 - x^2 \\ &= 2xh + h^2 \\ &= h(2x + h) \end{aligned}$$
(b) Using the answer to part (a)
$$\frac{f(x+h) - f(x)}{h} = \frac{h(2x+h)}{h} = 2x + h$$

19. $f(x) = 2 - x^2$
(a) $f(x+h) = 2 - (x+h)^2$
Therefore,
$f(x+h) - f(x)$
$= 2 - (x+h)^2 - (2-x^2)$
$= 2 - x^2 - 2xh - h^2 - 2 + x^2$
$= -2xh - h^2$
$= -h(2x + h)$

(b) Using the answer to part (a)
$$\frac{f(x+h) - f(x)}{h} = -\frac{h(2x+h)}{h} = -(2x+h)$$

21. $f(x) = 0.1x^2 - 4x + 5$
Technology formula: `0.1*x^2-4*x+5`
Table of Values:

x	0	1	2	3
f(x)	5	1.1	−2.6	−6.1
x	4	5	6	7
f(x)	−9.4	−12.5	−15.4	−18.1
x	8	9	10	
f(x)	−20.6	−22.9	−25	

23. $h(x) = \dfrac{x^2-1}{x^2+1}$

Technology formula: `(x^2-1)/(x^2+1)`
Table of Values (rounded to four decimal places):

x	0.5	1.5	2.5	3.5
h(x)	−0.6000	0.3846	0.7241	0.8491
x	4.5	5.5	6.5	7.5
h(x)	0.9059	0.9360	0.9538	0.9651
x	8.5	9.5	10.5	
h(x)	0.9727	0.9781	0.9820	

25.
(a) Using the table,
$P(5) = 117 \qquad P(10) = 132$
Since $P(9) = 130$ and $P(10) = 132$, we estimate $P(9.5)$ to be midway between these 130 and 132:
$$P(9.5) \approx \frac{130+132}{2} = 131.$$
Interpretation:
Since $P(t)$ gives the number of million people employed in July 1 of year t, we have:
$P(5) = 117$: Approximately 117 million people were employed in the US on July 1, 1995.
$P(10) = 132$: Approximately 132 million people were employed in the US on July 1, 2000.
$P(9.5) \approx 131$: Approximately 131 million people were employed in the US on January 1, 2000. (Note that January 1, 2000 is midway between July 1, 1999 and July 1, 2000.)
(b) The table gives $P(t)$ for $5 \le t \le 11$. Hence, the domain of P is the set of numbers t with $5 \le t \le 11$, that is, $[5, 11]$.

2

Section 1.1

27. $C(t) = \begin{cases} 500t + 800 & \text{if } 0 \le t \le 4 \\ 1300t - 2400 & \text{if } 4 < t \le 10 \end{cases}$

(a) $C(0) = 500(0) + 800 = 800$ We used the first formula, since 0 is in [0, 4].
$C(4) = 500(4) + 800 = 2800$ We used the first formula, since 4 is in [0, 4].
$C(5) = 1300(5) - 2400 = 6500 - 2400 = 4100$ We used the second formula, since 5 is in (4, 10].
$C(4) = 2800$, $C(5) = 4100$.
Interpretation:
Since $C(t)$ gives the number of coffee shops in year 1990+t, we have:
$C(0) = 800$: There were 800 coffee shops in 1990.
$C(4) = 2800$: There were 2800 coffee shops in 1994.
$C(5) = 4100$: There were 4100 coffee shops in 1995.

(b) We want $C(t) = 5400$. The first formula gives no more than 2800 coffee shops (see part (a)), so we use the second formula instead:
$$C(t) = 5400$$
$$1300t - 2400 = 5400$$
$$1300t = 7800$$
$$t = 7800/1300 = 6$$
which corresponds to 1996.

(c) A technology formula for C is
`(500*t+800)*(t<=4)+(1300*t-2400)*(4<t)`
(For a graphing calculator, use x instead of t, and $\le$ instead of <=)
Table of Values:

t	0	1	2	3	4	5
C(t)	800	1300	1800	2300	2800	4100
t	6	7	8	9	10	
C(t)	5400	6700	8000	9300	10,600	

29. $C(t) = \begin{cases} 0.08t + 0.6 & \text{if } 0 \le t < 8 \\ 0.355t - 1.6 & \text{if } 8 \le t \le 11 \end{cases}$

(a) A technology formula for C is
`(0.08*t+0.6)*(t<8)+(0.355*t-1.6)*(t>=8)`
(For a graphing calculator, use x instead of t, and $\ge$ instead of >=)
Table of Values:

t	0	1	2	3	4	5
C(t)	0.6	0.68	0.76	0.84	0.92	1
t	6	7	8	9	10	11
C(t)	1.08	1.16	1.24	1.595	1.95	2.305

(b) The costs of a Superbowl ad in 1998, 1999, and 2000 were:
1998: $C(8) = \$1.24$ million
1999: $C(9) = \$1.595$ million
2000: $C(10) = \$1.95$ million
From 1998 to 1999, the cost increased by $\$1.595 - \$1.24 = \$0.355$ million dollars.
From 1999 to 2000, the cost increased by $\$1.95 - \$1.595 = \$0.355$ million dollars.
Thus, the cost increased at a rate of $0.355 million dollars (or $355,000) per year between 1998 and 2000.

31. A taxable income of $26,000 falls in the category "Over $6000 but not over $27,050" and so we use the formula "$600.00 + 15% of the amount over $6000."
$T(26,000) = \$600 + 0.15(26,000 - 6000)$
$= \$3600.00$
A taxable income of $65,000 falls in the category "Over $27,050 but not over $65,550" and so we use the formula "$3757.50 + 27% of the amount over $27,050."
$T(65,000) = \$3757.50 + 0.27(65,000 - 27,050)$
$= \$14,004$

Section 1.1

33. $q(p) = 361{,}201 - (p+1)^2$

(a) Note p is expressed in cents, not dollars, so 50¢ is represented by $p = 50$, not 0.50.
$$q(50) = 361{,}201 - (50+1)^2$$
$$= 361{,}201 - 2601$$
$$= 358{,}600 \text{ brownie dishes}$$

(b) If they give them away, $p = 0$, so
$$q(0) = 361{,}201 - (0+1)^2$$
$$= 361{,}201 - 1$$
$$= 361{,}200 \text{ brownie dishes}$$

(c) If they sell no dishes, $q = 0$, and so
$$0 = 361{,}201 - (p+1)^2$$
$$(p+1)^2 = 361{,}201$$
$$p+1 = \sqrt{361{,}201} = 601$$
$$p = 601 - 1 = 600\text{¢, or \$6.00.}$$

35. $n(t) = 5t^2 - 49t + 232$

(a) Since the model is specified as being valid from 1986 ($t = 0$) to 1994 ($t = 8$) the appropriate domain for $n(t)$ is [0, 8], or $0 \le t \le 8$.

(b) $t \ge 0$ is not an appropriate domain for two reasons:

(1) It would predict investments in South Africa into the indefinite future with no basis.

(2) It would also lead to preposterous results for large values of t; for example, $n(400) = 780{,}632$ US companies investing in South Africa in the year 2386!

37. The technology formulas are:
(1): `-0.2*t^2+t+16`
(2) `0.2*t^2+t+16`
   (3) `t+16`

The following table shows the values predicted by the three models.

t	0	1	2	3	4	5	6	7
S	16	18	18	20	22	26	28	30
(1)	16	16.8	17.2	17.2	16.8	16	14.8	13.2
(2)	16	17.2	18.8	20.8	23.2	26	29.2	32.8
(3)	16	17	18	19	20	21	22	23

As shown in the table, the values predicted by model (2) are much closer to the observed values S than those predicted by the other models.

(b) Since 1998 corresponds to $t = 8$,
$$S(t) = 0.2\,t^2 + t + 16$$
$$S(8) = 0.2(8)^2 + 8 + 16 = 36.8$$
So the spending on corrections in 1998 was predicted to be \$36.8 billion.

39. $C(q) = 2000 + 100q^2$

(a) $C(10) = 2000 + 100(10)^2 = 2000 + 10{,}000$
$$= \$12{,}000$$

(b) Net Cost = Total Cost − Subsidy
$$N(q) = C(q) - S(q)$$
$$= 2000 + 100q^2 - 500q$$
$$N(20) = 2000 + 100(20)^2 - 500(20)$$
$$= 2000 + 40{,}000 - 10{,}000$$
$$= \$32{,}000$$

41. $p(t) = 100\left(1 - \dfrac{12{,}200}{t^{4.48}}\right)$ $\qquad (t \ge 8.5)$

(a) Technology formula:
`100*(1-12200/t^4.48)`

(b) Table of values:

t	9	10	11	12	13	14
$p(t)$	35.2	59.6	73.6	82.2	87.5	91.1
t	15	16	17	18	19	20
$p(t)$	93.4	95.1	96.3	97.1	97.7	98.2

(c) From the table, $p(12) = 82.2$, so that 82.2% of children are able to speak in at least single words by the age of 12 months.

(b) We seek the first value of t such that $p(t)$ is at least 90. Since $t = 14$ has this property ($p(14) = 91.1$) we conclude that, at 14 months, 90% or

Section 1.1

more children are able to speak in at least single words.

43. The dependent variable is a function of the independent variable. Here, the market price of gold m is a function of time t. Thus, the independent variable is t and the dependent variable is m.

45. To obtain the function notation, write the dependent as a function of the independent variable. Thus $y = 4x^2 - 2$ can be written as
$f(x) = 4x^2 - 2$ or $y(x) = 4x^2 - 2$

47. Number of sound files
= Starting number + New files
= $200 + 10 \times$ Number of days

So, $N(t) = 200 + 10t$
($N =$ number of sound files, $t =$ time in days)

49. As the text reminds us: to evaluate f of a quantity (such as $x+h$) replace x everywhere by the *whole quantity* $x+h$:
$$f(x) = x^2 - 1$$
$$f(x+h) = (x+h)^2 - 1.$$

52. A numerical model supplies only the values of a function at specific values of the independent variable, whereas an algebraic model supplies the value of a function at every point in its domain. Thus, an algebraic model supplies more information.

1.2

1. From the graph, we find
(a) $f(1) = 20$ (b) $f(2) = 30$

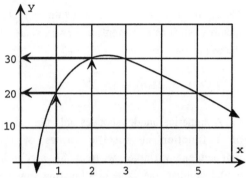

In a similar way, we find

(c) $f(3) = 30$ (d) $f(5) = 20$

(e) $f(3) - f(2) = 30 - 30 = 0$

3. From the graph,

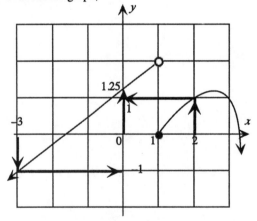

(a) $f(-3) = -1$ (b) $f(0) = 1.25$

(c) $f(1) = 0$ since the solid dot is on $(1, 0)$
($f(1) \neq 2$ since the hollow dot at $(1, 2)$ indicates that $(1, 2)$ is not a point of the graph.)

(d) $f(2) = 1$

(e) Since $f(3) = 1$ and $f(2) = 1$,
$$\frac{f(3) - f(2)}{3 - 2} = \frac{1 - 1}{3 - 2} = 0$$

5.
(a) $f(x) = x$ $(-1 \le x \le 1)$
Since the graph of $f(x) = x$ is a diagonal 45° line through the origin inclining up from left to right, the correct graph is (I)

(b) $f(x) = -x$ $(-1 \le x \le 1)$
Since the graph of $f(x) = -x$ is a diagonal 45° line through the origin inclining down from left to right, the correct graph is (IV)

(c) $f(x) = \sqrt{x}$ $(0 < x < 4)$
Since the graph of $f(x) = \sqrt{x}$ is the top half of a sideways parabola, the correct graph is (V)

(d) $f(x) = x + \frac{1}{x} - 2$ $(0 < x < 4)$
If we plot a few points like $x = 0.1, 1, 2,$ and 3 we find that the correct graph is (VI).

(e) $f(x) = |x|$ $(-1 \le x \le 1)$
Since the graph of $f(x) = |x|$ is a "V"-shape with its vertex at the origin, the correct graph is (III).

(f) $f(x) = x - 1$ $(-1 \le x \le 1)$
Since the graph of $f(x) = x-1$ is a straight line through $(0, -1)$ and $(1, 0)$, the correct graph is (II).

7. $f(x) = -x^3$ (domain $(-\infty, +\infty)$)
Technology formula: $-(x^3)$

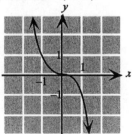

Section 1.2

9. $f(x) = x^4$ (domain $(-\infty, +\infty)$)
Technology formula: x^4

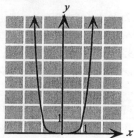

11. $f(x) = \frac{1}{x^2}$ $(x \neq 0)$

Technology formula: 1/x^2

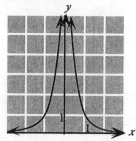

13. $f(x) = \begin{cases} x & \text{if } -4 \leq x < 0 \\ 2 & \text{if } 0 \leq x \leq 4 \end{cases}$

Technology formula: x*(x<0)+2*(x>=0)
(For a graphing calculator, use $\geq$ instead of >=.)

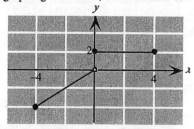

(a) $f(-1) = -1$. We used the first formula, since -1 is in $[-4, 0)$.
(b) $f(0) = 2$. We used the second formula, since 0 is in $[0, 4]$.
(c) $f(1) = 2$. We used the second formula, since 1 is in $[0, 4]$.

15. $f(x) = \begin{cases} x^2 & \text{if } -2 < x \leq 0 \\ 1/x & \text{if } 0 < x \leq 4 \end{cases}$

Technology formula:
(x^2)*(x<=0)+(1/x)*(0<x)
(For a graphing calculator, use $\leq$ instead of <=.)

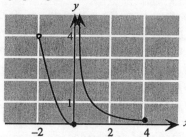

(a) $f(-1) = 1^2 = 1$. We used the first formula, since -1 is in $(-2, 0]$.
(b) $f(0) = 0^2 = 0$. We used the first formula, since 0 is in $(-2, 0]$.
(c) $f(1) = 1/1 = 1$. We used the second formula, since 1 is in $(0, 4]$.

17. $f(x) = \begin{cases} x & \text{if } -1 < x \leq 0 \\ x+1 & \text{if } 0 < x \leq 2 \\ x & \text{if } 2 < x \leq 4 \end{cases}$

Technology formula:
x*(x<=0)+(x+1)*(0<x)*(x<=2)+
x*(2<x)
(For a graphing calculator, use $\leq$ instead of <=.)

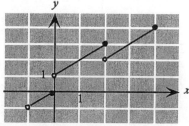

(a) $f(0) = 0$. We used the first formula, since 0 is in $(-1, 0]$.

Section 1.2

(b) $f(1) = 1+1 = 2$. We used the second formula, since 1 is in (0, 2].
(c) $f(2) = 2+1 = 3$. We used the second formula, since 2 is in (0, 2].
(d) $f(3) = 3$. We used the third formula, since 3 is in (2, 4].

19. Reading from the graph, $f(6) \approx 2000$, $f(9) \approx 2800$, $f(7.5) \approx 2500$. Interpretation: Since $f(t)$ is the is the number of SUVs (in thousands) sold in the U.S. in the year starting $t+1990$, we interpret the results as follows:
$f(6) \approx 2000$: In 1996 (1990+6), 2,000,000 SUVs were sold.
$f(9) \approx 2800$: In 1999 (1990+9), 2,800,000 were sold.
$f(7.5) \approx 2500$: Since 1990+7.5 = 1997.5, or half way through 1997, we interpret this by saying that, in the year beginning July, 1997, 2,500,000 were sold.

21. From the graph,
$f(6)-f(5) \approx 2000-1750 = 250$
$f(10)-f(9) \approx 2900-2750 = 150$
Therefore, $f(6)-f(5)$ is larger.
Interpretation: In general, the difference $f(b)-f(b)$ measures the change in sales from year a to year b. So, SUV sales increased more from 1995 to 1996 than from 1999 to 2000.

23.
(a) From the graph, we see that $N(t)$ is defined for $-1.5 \le t \le 1.5$. Thus, the domain of N is $[-1.5, 1.5]$.
(b) From the graph, $N(-0.5) \approx 131$, $N(0) \approx 132$, $N(1) \approx 132$. Since $N(t)$ gives the number of million people employed in the U.S. in year Jan. 2000+t, we interpret the results as follows:

$N(-0.5) \approx 131$: In July, 1999 (Jan. 2000 − 0.5), approximately 131 million people were employed.
$N(0) \approx 132$: In January 2000, approximately 132 million people were employed.
$N(1) \approx 132$: In January 2001 Jan. 2000 + 1), approximately 132 million people were employed.
(c) The graph is descending from around $t = 0.5$ to $t = 1.5$. Thus, $N(t)$ is falling on the interval [0.5, 1.5], showing that employment was falling during the period July, 2000–July, 2001.

25.
(a) The technology formulas are:
(A): `0.005*x+2.75`
(B): `0.01*x+20+25/x`
(C): `0.0005*x^2-0.07*x+23.25`
(D): `25.5*1.08^(x-5)`
The following table shows the values predicted by the four models.

x	5	25	40	100	125
A(x)	22.91	21.81	21.25	21.25	22.31
(A)	2.775	2.875	2.95	3.25	3.375
(B)	25.05	21.25	21.025	21.25	21.45
(C)	22.913	21.813	21.25	21.25	22.313
(D)	25.5	118.85	377.03	38177	261451

Model (C) fits the data almost perfectly—more closely that any of the other models.

(b) Graph of model (C):

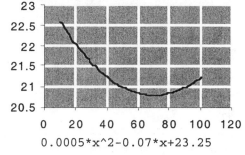

`0.0005*x^2-0.07*x+23.25`

The lowest point on the graph occurs at $x = 70$ with a y-coordinate of 20.8. Thus, the lowest cost per shirt is $20.80, which the team can obtain by buying 70 shirts.

27. A plot of the given data suggests a curve that is concave down and peaks somewhere between 1985 and 1994. A linear model would predict perpetually increasing or decreasing number of visitors (depending on whether the slope is positive or negative) and an exponential model $n(t) = Ab^t$ would also be perpetually increasing or decreasing (depending whether b is larger than 1 or less than 1). This leaves a quadratic model as the only possible choice. In fact, a quadratic can always be found that passes through any three points not on the same straight line with different x-coordinates. Therefore, a quadratic model would give an exact fit.

29. $p(t) = 100\left(1 - \dfrac{12{,}200}{t^{4.48}}\right)$ $(t \geq 8.5)$

(a) Technology formula:
   100*(1-12200/t^4.48)

(b) Graph:

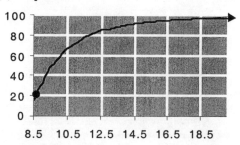

(c) If we zoom in to the graph near $x = 12$ we find that $p(12) \approx 82$. Thus, 82% of children are able to speak in at least single words by the age of 12 months.

(d) We need to find the approximate value of t so that $p(t) = 90$. From the graph above, it occurs somewhere between $t = 12.5$ and 14.5. Here is a close-up of the graph near that point:

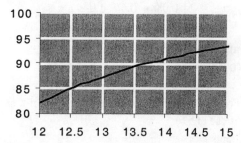

This close-up shows that $f(13.7) \approx 90$. Therefore, 90% of children speaking in at least single words at approximately 13.7 months of age, or 14 months if we round to the nearest month.

31. $C(t) = \begin{cases} 500t + 800 & \text{if } 0 \leq t \leq 4 \\ 1300t - 2400 & \text{if } 4 < t \leq 10 \end{cases}$

Technology formula:
   (500*t+800)*(t<=4)+(1300*t-2400)*(4<t)
(For a graphing calculator, use x instead of t, and $\leq$ instead of <=)
To graph this function, sketch the lines
   $y = 500t + 800$
and $y = 1300t - 2400$
and then switch from the first to the second at $t = 4$:

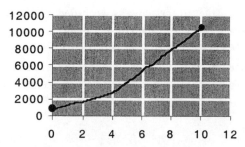

To answer the question, we need to find t such that $C(t) = 9000$. From the graph, $C(9) \approx 9000$. Since $t = 9$ corresponds to 1999, we conclude

Section 1.2

that there were 9000 coffee shops in the U.S. in 1999 (to the nearest year).

33. $C(t) = \begin{cases} 0.08t + 0.6 & \text{if } 0 \le t < 8 \\ 0.355t - 1.6 & \text{if } 8 \le t \le 11 \end{cases}$

(a) Technology formula:
(0.08*t+0.6)*(t<8)+(0.355*t-1.6)*(t>=8)
(For a graphing calculator, use x instead of t, and ≥ instead of >=)
Graph:

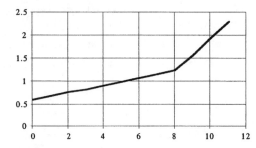

(b) To answer the question, we need to find the first integer value of t such that $C(t)$ exceeds 2. Although $C(10) \approx 2$, it is a little less than 2. On the other hand, $C(11) \ge 2$, and $t = 11$ represents the first year such that $C(t) \ge 2$. Thus, a Superbowl ad first exceeded $2 million in 1990+11 = 2001.

35. True. A graphically specified function is specified by a graph. Given a graph, we can read off a set of values to construct a table, and hence specify the function numerically. (The more accurate the graph is, the more accurate the numerical values are.)

37. False. A numerically specified function with domain [0, 10] is specified by a table of some values between 0 and 10. Since only certain values of the function are specified, we can obtain only certain points on the graph.

39. If two functions are specified by the same formula $f(x)$ say, their graphs must follow the same curve $y = f(x)$. However, it is the domain of the function that specifies what portion of the curve appears on the graph. Thus, if the functions have different domains, their graphs will be different portions of the curve $y = f(x)$.

41. Suppose we already have the graph of f and want to construct the graph of g. We can plot a point of the graph of g as follows: Choose a value for x ($x = 7$, say) and then "look back" 5 units to read off $f(x-5)$ ($f(2)$ in this instance). This value gives the y-coordinate we want. In other words, points on the graph of g are obtained by "looking back 5 units" to the graph of f and then copying that portion of the curve. Put another way, the graph of g is the same as the graph of f, but shifted 5 units to the right.

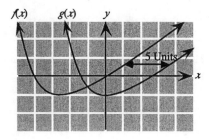

1.3

1.

x	−1	0	1
y	5	8	

We calculate the slope m first. The first two points shown give a changes in x and y of
$$\Delta x = 0 - (-1) = 1$$
$$\Delta y = 8 - 5 = 3$$
This gives a slope of
$$m = \frac{\Delta y}{\Delta x} = \frac{3}{1} = 3.$$
Now look at the second and third points: The change in x is again
$$\Delta x = 1 - 0 = 1$$
and so Δy must be given by the formula
$$\Delta y = m\Delta x$$
$$\Delta y = 3(1) = 3.$$
This means that the missing value of y is
$$8 + \Delta y = 8 + 3 = 11.$$

3.

x	2	3	5
y	−1	−2	

We calculate the slope m first. The first two points shown give a changes in x and y of
$$\Delta x = 3 - 2 = 1$$
$$\Delta y = -2 - (-1) = -1$$
This gives a slope of
$$m = \frac{\Delta y}{\Delta x} = \frac{-1}{1} = -1.$$
Now look at the second and third points: The change in x is
$$\Delta x = 5 - 3 = 2$$
and so Δy must be given by the formula
$$\Delta y = m\Delta x$$
$$\Delta y = (-1)(2) = -2.$$
This means that the missing value of y is
$$-2 + \Delta y = -2 + (-2) = -4.$$

5.

x	−2	0	2
y	4		10

We calculate the slope m first. The first and third points shown give a changes in x and y of
$$\Delta x = 2 - (-2) = 4$$
$$\Delta y = 10 - 4 = 6$$
This gives a slope of
$$m = \frac{\Delta y}{\Delta x} = \frac{6}{4} = \frac{3}{2}.$$
Now look at the first and second points: The change in x is
$$\Delta x = 0 - (-2) = 2$$
and so Δy must be given by the formula
$$\Delta y = m\Delta x$$
$$\Delta y = (\frac{3}{2})(2) = 3.$$
This means that the missing value of y is
$$4 + \Delta y = 4 + 3 = 7.$$

7. From the table,
$$b = f(0) = -2.$$
The slope (using the first two points) is
$$m = \frac{y_2 - y_2}{x_2 - x_1} = \frac{-2 - (-1)}{0 - (-2)} = \frac{-1}{2} = -\frac{1}{2}.$$
Thus, the linear equation is
$$f(x) = mx + b = -\frac{1}{2}x - 2,$$
or $f(x) = -\frac{x}{2} - 2.$

9. The slope (using the first two points) is
$$m = \frac{y_2 - y_2}{x_2 - x_1} = \frac{-2 - (-1)}{-3 - (-4)} = \frac{-1}{1} = -1.$$
To obtain $f(0) = b$, notice that, since the slope is −1, y decreases by 1 for every one-unit increase in x. Thus,
$$f(0) = f(-1) + m = -4 - 1 = -5.$$
This gives

Section 1.3

$f(x) = mx + b = -x - 5$.

11. In the table, x increases in steps of 1 and f increases in steps of 4, showing that f is linear with slope
$$m = \frac{\Delta y}{\Delta x} = \frac{4}{1} = 4$$
and intercept
$$b = f(0) = 6$$
giving
$$f(x) = mx + b = 4x + 6.$$
The function g does not increase in equal steps, so g is not linear.

13. In the first three points listed in the table, x increases in steps of 3, but f does not increase in equal steps, whereas g increases in steps of 6. Thus, based on the first three points, only g could possibly be linear, with slope
$$m = \frac{\Delta y}{\Delta x} = \frac{6}{3} = 2$$
and intercept
$$b = g(0) = -1$$
giving
$$g(x) = mx + b = 2x - 1.$$
We can now check that the remaining points in the table fit the formula $g(x) = 2x-1$, showing that g is indeed linear.

15. Slope = coefficient of $x = -\frac{3}{2}$

17. Write the equation as
$$y = \frac{x}{6} + \frac{1}{6}$$
Slope = coefficient of $x = \frac{1}{6}$

19. If we solve for x we find that the given equation represents the vertical line $x = -1/3$, and so its slope is infinite (undefined).

21. $3y + 1 = 0$
Solving for y:
$$3y = -1$$
$$y = -\frac{1}{3}$$
Slope = coefficient of $x = 0$

23. $4x + 3y = 7$
Solve for y:
$$3y = -4x + 7$$
$$y = -\frac{4}{3}x + \frac{7}{3}$$
Slope = coefficient of $x = -\frac{4}{3}$

25. $y = 2x - 1$
y-intercept $= -1$, slope $= 2$

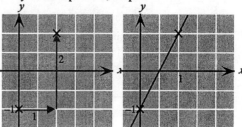

27. $y = -\frac{2}{3}x + 2$
y-intercept $= 2$, slope $= -\frac{2}{3}$

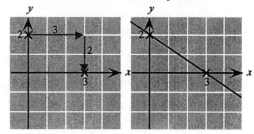

Section 1.3

29. $y + \frac{1}{4}x = -4$

Solve for y to obtain $y = -\frac{1}{4}x - 4$

y-intercept = -4, slope = $-\frac{1}{4}$

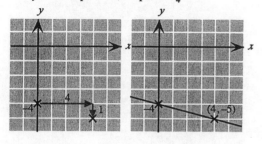

31. $7x - 2y = 7$

Solve for y:

$-2y = -7x + 7$

$y = \frac{7}{2}x - \frac{7}{2}$

y-intercept = $-\frac{7}{2} = -3.5$, slope = $\frac{7}{2} = 3.5$

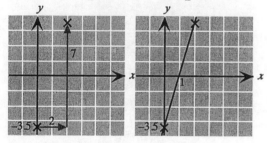

33. $3x = 8$

Solve for x to obtain $x = \frac{8}{3}$.

The graph is a vertical line:

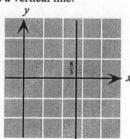

35. $6y = 9$

Solve for y to obtain $y = \frac{9}{6} = \frac{3}{2} = 1.5$

y-intercept = $\frac{3}{2} = 1.5$, slope = 0

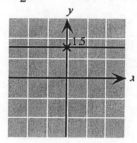

37. $2x = 3y$

Solve for y to obtain $y = \frac{2}{3}x$

y-intercept = 0, slope = $\frac{2}{3}$

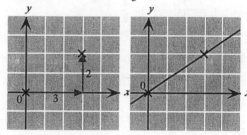

39. $(0, 0)$ and $(1, 2)$

$m = \dfrac{y_2 - y_1}{x_2 - x_1} = \dfrac{2 - 0}{1 - 0} = 2$

41. $(-1, -2)$ and $(0, 0)$

$m = \dfrac{y_2 - y_1}{x_2 - x_1} = \dfrac{0 - (-2)}{0 - (-1)} = \dfrac{2}{1} = 2$

43. $(4, 3)$ and $(5, 1)$

$m = \dfrac{y_2 - y_1}{x_2 - x_1} = \dfrac{1 - 3}{5 - 4} = \dfrac{-2}{1} = -2$

Section 1.3

45. (1, −1) and (1, −2)

$m = \dfrac{y_2 - y_1}{x_2 - x_1} = \dfrac{-2 - (-1)}{1 - 1}$ Undefined

47. (2, 3.5) and (4, 6.5)

$m = \dfrac{y_2 - y_1}{x_2 - x_1} = \dfrac{6.5 - 3.5}{4 - 2} = \dfrac{3}{2} = 1.5$

49. (300, 20.2) and (400, 11.2)

$m = \dfrac{y_2 - y_1}{x_2 - x_1} = \dfrac{11.2 - 20.2}{400 - 300}$

$= \dfrac{-9}{100} = -0.09$

51. $(0, 1)$ and $(-\tfrac{1}{2}, \tfrac{3}{4})$

$m = \dfrac{y_2 - y_1}{x_2 - x_1} = \dfrac{\tfrac{3}{4} - 1}{-\tfrac{1}{2} - 0}$

$= \dfrac{-\tfrac{1}{4}}{-\tfrac{1}{2}} = \dfrac{1}{4} \cdot 2 = \dfrac{1}{2}$

53. (a, b) and (c, d) $(a \neq c)$

$m = \dfrac{y_2 - y_1}{x_2 - x_1} = \dfrac{d - b}{c - a}$

55.
(a)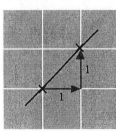

$m = \dfrac{\Delta y}{\Delta x} = \dfrac{1}{1} = 1$

(b)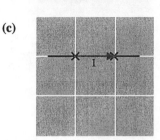

$m = \dfrac{\Delta y}{\Delta x} = \dfrac{1}{2}$

(c)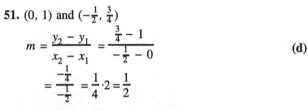

$m = \dfrac{\Delta y}{\Delta x} = \dfrac{0}{1} = 0$

(d)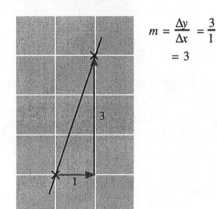

$m = \dfrac{\Delta y}{\Delta x} = \dfrac{3}{1} = 3$

(e)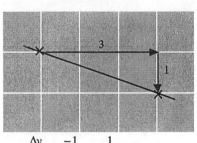

$m = \dfrac{\Delta y}{\Delta x} = \dfrac{-1}{3} = -\dfrac{1}{3}$

Section 1.3

(f)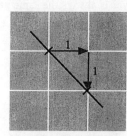

$$m = \frac{\Delta y}{\Delta x} = \frac{-1}{1}$$
$$= -1$$

(g)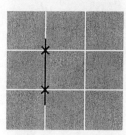

Vertical line; undefined slope

(h)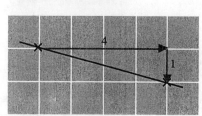

$$m = \frac{\Delta y}{\Delta x} = \frac{-1}{4} = -\frac{1}{4}$$

(i)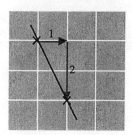

$$m = \frac{\Delta y}{\Delta x} = \frac{-2}{1}$$
$$= -2$$

57. Through $(1, 3)$ with slope 3
Point: $(1, 3)$
Slope: $m = 3$
$$\begin{aligned}b &= y_1 - mx_1\\&= 3 - 3 \cdot 1 = 0\end{aligned}$$
Thus, the equation is
$$y = mx + b$$
$$y = 3x + 0$$
$$y = 3x$$

59. Through $(1, -\frac{3}{4})$ with slope $\frac{1}{4}$
Point: $(1, -\frac{3}{4})$ **Slope:** $m = \frac{1}{4}$
$$\begin{aligned}b &= y_1 - mx_1\\&= -\tfrac{3}{4} - \tfrac{1}{4} \cdot 1 = -1\end{aligned}$$
Thus, the equation is
$$y = mx + b$$
$$y = \tfrac{1}{4}x - 1$$

61. Through $(20, -3.5)$ and increasing at a rate of 10 units of y per unit of x
Point: $(20, -3.5)$
Slope: $m = \dfrac{\Delta y}{\Delta x} = \dfrac{10}{1} = 10$
$$\begin{aligned}b &= y_1 - mx_1\\&= -3.5 - (10)(20)\\&= -3.5 - 200 = -203.5\end{aligned}$$
Thus, the equation is
$$y = mx + b$$
$$y = 10x - 203.5$$

63. Through $(2, -4)$ and $(1, 1)$
Point: $(2, -4)$
Slope: $m = \dfrac{y_2 - y_1}{x_2 - x_1} = \dfrac{1 - (-4)}{1 - 2} = \dfrac{5}{-1} = -5$
$$\begin{aligned}b &= y_1 - mx_1\\&= -4 - (-5)(2) = 6\end{aligned}$$
Thus, the equation is
$$y = mx + b$$
$$y = -5x + 6$$

65. Through $(1, -0.75)$ and $(0.5, 0.75)$
Point: $(1, -0.75)$
Slope: $m = \dfrac{y_2 - y_1}{x_2 - x_1} = \dfrac{0.75 - (-0.75)}{0.5 - 1}$
$$= \dfrac{1.5}{-0.5} = -3$$

Section 1.3

$b = y_1 - mx_1$
$= -0.75 - (-3)(1) = -0.75 + 3 = 2.25$
Thus, the equation is
$y = mx + b$
$y = -3x + 2.25$

67. Through (6, 6) and parallel to the line $x + y = 4$
Point: (6, 6)
Slope: Same as slope of $x + y = 4$. To find the slope, solve for y, getting
$y = -x + 4$
Thus, $m = -1$.
$b = y_1 - mx_1$
$= 6 - (-1)(6) = 6 + 6 = 12$
Thus, the equation is
$y = mx + b$
$y = -x + 12$

69. Through (0.5, 5) and parallel to the line $4x - 2y = 11$
Point: (0.5, 5)
Slope: Same as slope of $4x - 2y = 11$. To find the slope, solve for y, getting
$2y = 4x - 11$
$y = 2x - \frac{11}{2}$
Thus, $m = 2$.
$b = y_1 - mx_1$
$= 5 - (2)(0.5) = 5 - 1 = 4$
Thus, the equation is
$y = mx + b$
$y = 2x + 4$

71. A table of values of x and y comes from a linear equation precisely when the successive ratios $\Delta y / \Delta x$ are all the same. Thus, to test such a table of values, compute the corresponding successive changes Δx in x and Δy in y, and compute the ratios $\Delta y / \Delta x$. If the answer is always the same number, then the values in the table come from a linear function.

73. To find the linear function, solve the equation $ax + by = c$ for y:
$by = -ax + c$
$y = -\frac{a}{b}x + \frac{c}{b}$
Thus, the desired function is $f(x) = -\frac{a}{b}x + \frac{c}{b}$.
If $b = 0$, then $\frac{a}{b}$ and $\frac{c}{b}$ are undefined, and y cannot be specified as a function of x. (The graph of the resulting equation would be a vertical line.)

75. The slope of the line is $m = \frac{\Delta y}{\Delta x} = \frac{3}{1} = 3$.
Therefore, if, in a straight line, y is increasing three times as fast as x, then its slope is $\underline{3}$.

77. If m is positive then y will increase as x increases; if m is negative then y will decrease as x increases; if m is zero then y will not change as x changes.

79.

	A	B	C	D
1	x	y	m	b
2	1	2	=(B3-B2)/(A3-A2)	=B2-C2*A2
3	3	-1		Slope	Intercept

The slope computed in cell C2 is given by
$m = \frac{y_2 - y_1}{x_2 - x_1} = \frac{-1 - 2}{3 - 1} = -1.5$
If we increase the y-coordinate in cell B3, this increases y_2, and thus increases the numerator $\Delta y = y_2 - y_1$ without effecting the denominator Δx. Thus the slope will increase. .

1.4

1. For a linear cost function,
$$C(x) = mx + b$$
m = marginal cost = \$1500 per piano
b = fixed cost = \$1200
Thus, the daily cost function is
$$C(x) = 1500x + 1200.$$
(a) The cost of manufacturing 3 pianos is
$$C(3) = 1500(3) + 1200$$
$$= 4500 + 1200 = \$5700$$
(b) The cost of manufacturing each additional piano (such as the third one or the 11th one) is the marginal cost, m = \$1500.
(c) Same answer as (b).

3. We are given two points on the graph of the linear cost function: (100, 10,500) and (120, 11,0000) (x is the number of items, and y is the cost C).
Marginal cost:
$$m = \frac{C_2 - C_1}{x_2 - x_1} = \frac{11{,}000 - 10{,}500}{120 - 100}$$
$$= \frac{500}{20} = \$25 \text{ per bicycle.}$$
Fixed cost:
$$b = C_1 - mx_1 = 10{,}500 - (25)(100)$$
$$= 10{,}500 - 2500 = \$8000$$

5.
(a) For a linear cost function,
$$C(x) = mx + b.$$
m = marginal cost = \$0.40 per copy
b = fixed cost = \$70
Thus, the cost function is
$$C(x) = 0.4x + 70.$$
The revenue function is
$$R(x) = 0.50x \quad (x \text{ copies @ } 50¢ \text{ per copy})$$
The profit function is
$$P(x) = R(x) - C(x)$$
$$= 0.5x - (0.4x + 70)$$
$$= 0.5x - 0.4x - 70$$
$$= 0.1x - 70$$
(b) $P(500) = 0.1(500) - 70$
$$= 50 - 70 = -20$$
Since P is negative, this represents a loss of \$20.
(c) For break-even,
$$P(x) = 0$$
$$0.1x - 70 = 0$$
$$0.1x = 70$$
$$x = \frac{70}{0.1} = 700 \text{ copies}$$

7. A linear demand function has the form
$$q = mp + b.$$
(x is the price p, and y is the demand q). We are given two points on its graph: (1, 1960) and (5, 1800).
Slope:
$$m = \frac{q_2 - q_1}{p_2 - p_1} = \frac{1800 - 1960}{5 - 1} = \frac{-160}{4} = -40$$
Intercept:
$$b = q_1 - mp_1 = 1960 - (-40)(1)$$
$$= 1960 + 40 = 2000$$
Thus, the demand equation is
$$q = mp + b$$
$$q = -40p + 2000$$

9. A linear demand function has the form
$$q = mp + b.$$
(x is the price p, and y is the demand q). We are given two points on its graph:
1997 second quarter data: (235, 21)
1997 third quarter data: (215, 24)
Slope:
$$m = \frac{q_2 - q_1}{p_2 - p_1} = \frac{24 - 21}{215 - 235} = \frac{3}{-20} = -0.15$$
Intercept:
$$b = q_1 - mp_1 = 21 - (-0.15)(235)$$
$$= 21 + 35.25 = 56.25$$

Thus, the demand equation is
$$q = mp + b$$
$$q = -0.15p + 56.25$$
In the second part of the question, we are asked if the demand equation fits the 1998 first quarter data: (210, 23). There are two methods we can use to see that this data does *not* fit the demand equation.

Method 1: Reasoning

The demand equation has a negative slope, and thus predicts that demand increases as the price goes down. However, the 1998 first quarter price was lower than the 1997 third quarter price, and the demand was also lower. Thus we know that (210, 23) cannot fit the demand equation.

Method 2: Direct Computation

Substituting the 1998 first quarter data into the demand equation gives
$$23 = -0.15(210) + 56.25$$
$$23 = -31.5 + 56.25 = 24.75 \; ✗$$
The demand equation predicts sales of 24.5 million microprocessors—greater than the observed sales of 23 million microprocessors.

11.

(a) Demand Function: The given points are
$$(p, q) = (1, 90) \text{ and } (2, 30)$$
Slope:
$$m = \frac{q_2 - q_1}{p_2 - p_1} = \frac{30 - 90}{2 - 1} = -60$$
Intercept:
$$b = q_1 - mp_1 = 90 - (-60)(1) = 150$$
Thus, the demand equation is
$$q = mp + b$$
$$q = -60p + 150$$
Supply Function: The given points are
$$(p, q) = (1, 20) \text{ and } (2, 100)$$
Slope:
$$m = \frac{q_2 - q_1}{p_2 - p_1} = \frac{100 - 20}{2 - 1} = 80$$

Intercept:
$$b = q_1 - mp_1 = 20 - (80)1 = -60$$
Thus, the supply equation is
$$q = mp + b$$
$$q = 80p - 60$$

(b) For equilibrium,
Supply = Demand
$$80p - 60 = -60p + 150$$
$$140p = 210$$
$$p = \frac{210}{140} = 1.5$$
Thus, the chias should be marked at $1.50 each.

13.

(a) Here is the given graph with successive points joined by line segments.

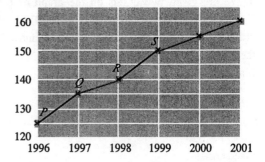

The segments PQ and RS are the steepest (they have the largest Δy ($\Delta y = 10$) for the fixed $\Delta x = 1$), and hence have the largest slope ($m = \Delta y/\Delta x = 10$). (You can check that all line segments joining non-successive pairs of points—for example P and R—have smaller slopes.) Thus, we have two possible answers:

PQ: Points $P(1996, 125)$ and $Q(1997, 135)$
RS: Points $R(1998, 140)$ and $(1999, 150)$.

(b) Consulting the textbook, we find that the slope m measures the rate of change of the number of in-ground swimming pools, and is measured in thousands of pools per year (units of

Section 1.4

y per unit of x). Thus, the number of new in-ground pools increased most rapidly during the periods 1996–1997 and 1998–1999, when it rose by 10,000 new pools in a year.

15. We are asked for a linear model of N as a function of t. That is, $N = mt + b$. (N is playing the role of y and t is playing the role of x.) The two points we are given are

$(t, N) = (-1, 350)$ $t = -1$ represents 1999
and
$(t, N) = (1, 450)$ $t = 1$ represents 2001

(Notice that we are also thinking of N in millions of transactions.)

Slope:
$$m = \frac{N_2 - N_1}{t_2 - t_1} = \frac{450 - 350}{1 - (-1)} = \frac{100}{2} = 50$$

Intercept:
$b = N_1 - mt_1 = 350 - (50)(-1) = 400$

Thus, the linear model is
$N = mt + b$
$N = 50t + 400$ million transactions

Consulting the textbook, we find that the slope m measures the rate of change of the number of on-line shopping transactions, and is measured in millions of transactions per year (units of N per unit of t).

17.

(a) We are asked for a linear model of s as a function of t. That is, $s = mt + b$. The two points we are given are

$(t, s) = (0, 240)$ $t = 0$ represents 2000
and
$(t, s) = (25, 600)$ $t = 1$ represents 2001

(Notice that we are also thinking of s in billions of dollars.)

Slope:

$$m = \frac{s_2 - s_1}{t_2 - t_1} = \frac{600 - 240}{25 - 0} = \frac{360}{25} = 14.4$$

Intercept:
$b = 240$ (The point $(0, 240)$ gives us the y-intercept.)

Thus, the linear model is
$s = mt + b$
$s = 14.4t + 240$ billion dollars.

The quantity s increases by $m = 14.4$ for every one-unit increase in t. Thus, Medicare spending is predicted to rise at a rate of $14.4 billion per year.

(b) In 2040, $t = 40$, so
$s(40) = 14.4(40) + 240 = 576+240 = 816$.

Thus, Medicare spending in 2040 will be $816 billion.

19. $s(t) = 2.5t + 10$

(a) Velocity = slope = 2.5 feet/sec.

(b) After 4 seconds, $t = 4$, so
$s(4) = 2.5(4) + 10 = 10+10 = 20$
Thus the model train has moved 20 feet along the track.

(c) The train will be 25 feet along the track when $s = 25$. Substituting gives
$25 = 2.5t + 10$
Solving for time t gives
$2.5t = 25-10 = 15$
$$t = \frac{15}{2.5} = 6 \text{ seconds}$$

21.

(a) Take s to be displacement from Jones Beach, and t to be time in hours. We are given two points

$(t, s) = (10, 0)$ $s = 0$ for Jones Beach
$(t, s) = (10.1, 13)$ 6 minutes = 0.1 hours

We are asked for the speed, which equals the magnitude of the slope.

Section 1.4

$$m = \frac{s_2 - s_1}{t_2 - t_1} = \frac{13 - 0}{10.1 - 10} = \frac{13}{0.1} = 130$$

Units of slope = units of s per unit of t
= miles per hour

Thus, the police car was traveling at 130 mph.

(b) For the displacement from Jones Beach at time t, we want to express s as a linear function of t; namely, $s = mt + b$. We already know $m = 130$ from part (a). For the intercept, use

$$b = s_1 - mt_1 = 0 - 130(10) = -1300$$

Therefore, the displacement at time t is

$$s = mt + b$$
$$s = 130t - 1300$$

23. F = Fahrenheit temperature,
C = Celsius temperature,
and we want F as a linear function of C. That is,

$$F = mC + b$$

(F plays the role of y and C plays the role of x.)
We are given two points:

$(C, F) = (0, 32)$ Freezing point
$(C, F) = (100, 212)$ Boiling point

Slope:

$$m = \frac{F_2 - F_1}{C_2 - C_1} = \frac{212 - 32}{100 - 0} = \frac{180}{100} = 1.8$$

Intercept:

$$b = F_1 - mC_1 = 32 - 1.8(0) = 32.$$

Thus, the linear relation is

$$F = mC + b$$
$$F = 1.8C + 32$$

When $C = 30°$

$$F = 1.8(30) + 32 = 54 + 32 = 86°$$

When $C = 22°$

$$F = 1.8(22) + 32 = 39.6 + 32 = 71.6°$$

Rounding to the nearest degree gives 72°F.

When $C = -10°$

$$F = 1.8(-10) + 32 = -18 + 32 = 14°$$

When $C = -14°$

$$F = 1.8(-14) + 32 = -25.2 + 32 = 6.8°$$

Rounding to the nearest degree gives 7°F.

20

25. Income = royalties + screen rights

I = 5% of net profits + 50,000
$I = 0.05N + 50,000$ Equation notation
$I(N) = 0.05N + 50,000$ Function notation

For an income of $100,000,

$$100,000 = 0.05N + 50,000$$
$$0.05N = 50,000$$
$$N = \frac{50,000}{0.05} = \$1,000,000$$

Her marginal income is her increase in income per $1 increase in net profit. This is the slope, m = 0.05 dollars of income per dollar of net profit, or 5¢ per dollar of net profit.

27.

n	82	26
c	5.8	3.0

(a) We want circulation c as a linear function of the number n of newspapers published. That is, $c = mn + b$.

Slope:

$$m = \frac{c_2 - c_1}{n_2 - n_1} = \frac{3.0 - 5.8}{26 - 82} = \frac{-2.8}{-56} = 0.05$$

Intercept:

$$b = c_1 - mn_1 = 5.8 - (0.05)(82) = 1.7$$

Thus, the linear function is

$$c = mn + b$$
$$c = 0.05n + 1.7$$

(b) In general, the slope gives the number of additional units of y (circulation c) per additional unit of x (newspapers published n). Thus, the slope represents the increase in daily circulation per additional newspaper that a company publishes.

(c) We are told that Gannett was adding 11 new papers to its existing 82, to give $n = 93$. Thus, the circulation would be

Section 1.4

$c(11) = 0.05(93) + 1.7$
$= 4.65 + 1.7 = 6.35$ million newspapers

29. We want w as a linear function of n:
$w = mn + b$
Thus, w plays the role of y and n plays the role of x. Here is the (milk) data listed in the customary way ($x = n$ first, and $y = w$ second).

n	57	59
w	56	60

Slope:
$$m = \frac{w_2 - w_1}{n_2 - n_1} = \frac{60-56}{59-57} = \frac{4}{2} = 2$$

Intercept:
$b = w_1 - mn_1 = 56 - (2)(57)$
$= 56 - 114 = -58$

Thus, the linear function is
$w = 2n - 58$.

For the second part of the question, we are told that $n = 50$. Thus,
$w = 2(50) - 58 = 100 - 58$
$= 42$ billion pounds of milk

31. We want c (cheese production) as a linear function of m (milk production) in the western states.

$c = km + b$ We are using k for the slope

Thus, c plays the role of y and m plays the role of x. Here is the (western states) data listed in the customary way ($x = m$ first, and $y = c$ second).

m	56	60
c	2.7	3.0

Slope:
$$k = \frac{c_2 - c_1}{m_2 - m_1} = \frac{3.0 - 2.7}{60 - 56} = \frac{0.3}{4} = 0.075$$

$b = c_1 - km_1 = 2.7 - (0.075)(56)$
$= 2.7 - 4.2 = -1.5$

Thus, the linear equation is
$c = km + b$
$c = 0.075m - 1.5$

For the second part of the question, we want the number of pounds of cheese produced for every 10 pounds of milk. The slope $k = 0.075$ gives us the number of pounds of cheese produced per (additional) *one* pound of milk. Multiplying this by 10 gives 0.75 pounds of cheese per 10 pounds of milk.

33. We want the temperature T as a linear function of the rate r of chirping. That is,
$T(r) = mr + b$.
Thus, T plays the role of y and r plays the role of x. We are given two points:
$(r, T) = (140, 80)$ and $(120, 75)$

Slope:
$$m = \frac{T_2 - T_1}{r_2 - r_1} = \frac{75-80}{120-140} = \frac{-5}{-20} = \frac{1}{4}$$

Intercept:
$b = T_1 - mr_1 = 80 - \frac{1}{4}(140) = 80 - 35 = 45$

Thus, the linear function is
$T(r) = mr + b$
$T(r) = \frac{1}{4}r + 45$

When the chirping rate is 100 chirps per minute, $r = 100$, and so the temperature is
$T(100) = \frac{1}{4}(100) + 45 = 25 + 45 = 70°F$

35. The hourly profit function is given by
Profit = Revenue − Cost
$P(x) = R(x) − C(x)$
(Hourly) cost function: This is a fixed cost of $5132 only:
$C(x) = 5132$
(Hourly) revenue function: This is a variable of $100 per passenger cost only:
$R(x) = 100x$

21

Section 1.4

Thus, the profit function is
$$P(x) = R(x) - C(x)$$
$$P(x) = 100x - 5132$$

For the domain of $P(x)$, the number of passengers x cannot exceed the capacity: 405. Also, x cannot be negative. Thus, the domain is given by $0 \le x \le 405$, or $[0, 405]$.

For break-even, $P(x) = 0$
$$100x - 5132 = 0$$
$$100x = 5132, \text{ or } x = \frac{5132}{100} = 51.32$$

If x is larger than this, then the profit function is positive, and so there should be at least 52 passengers; $x \ge 52$, for a profit.

37. To compute the break-even point, we use the profit function:

Profit = Revenue − Cost
$$P(x) = R(x) - C(x)$$
$$R(x) = 2x \quad \$2 \text{ per unit}$$
$$\begin{aligned} C(x) &= \text{Variable Cost} + \text{Fixed Cost} \\ &= 40\% \text{ of Revenue} + 6000 \\ &= 0.4(2x) + 6000 \\ &= 0.8x + 6000 \end{aligned}$$

Thus,
$$P(x) = R(x) - C(x)$$
$$P(x) = 2x - (0.8x + 6000)$$
$$P(x) = 1.2x - 6000$$

For break-even, $P(x) = 0$
$$1.2x - 6000 = 0$$
$$1.2x = 6000$$
$$x = \frac{6000}{1.2} = 5000$$

Therefore, 5000 units should be made to break even.

39. To compute the break-even point, we use the revenue and cost functions:
$$R(x) = \text{Selling price} \times \text{Number of units}$$
$$= SPx$$

$$\begin{aligned} C(x) &= \text{Variable Cost} + \text{Fixed Cost} \\ &= VCx + FC \end{aligned}$$

(Note that "variable cost per unit" is marginal cost.) For break-even
$$R(x) = C(x)$$
$$SPx = VCx + FC$$
$$SPx - VCx = FC$$
$$x(SP - VC) = FC$$
$$x = \frac{FC}{SP - VC}$$

41. Take x to be the number of grams of perfume he buys and sells. The profit function is given by

Profit = Revenue − Cost
$$P(x) = R(x) - C(x)$$

Cost function $C(x)$:

Fixed costs:	20,000
Cheap perfume @ \$20 per g:	$20x$
Transportation @ \$30 per 100 g:	$0.3x$

Thus the cost function is
$$C(x) = 20x + 0.3x + 20{,}000$$
$$C(x) = 20.3x + 20{,}000$$

Revenue function $R(x)$
$$R(x) = 600x \quad \$600 \text{ per gram}$$

Thus, the profit function is
$$P(x) = R(x) - C(x)$$
$$P(x) = 600x - (20.3x + 20{,}000)$$
$$P(x) = 579.7x - 20{,}000,$$
with domain $x \ge 0$.

For break-even, $P(x) = 0$
$$579.7x - 20{,}000 = 0$$
$$579.7x = 20{,}000$$
$$x = \frac{20{,}000}{579.7} \approx 34.50$$

Thus, he should buy and sell 34.50 grams of perfume per day to break even.

43. $C(t) = \begin{cases} 0.08t + 0.6 & \text{if } 0 \le t < 8 \\ 0.355t - 1.6 & \text{if } 8 \le t \le 11 \end{cases}$

million dollars.

Since 1999 is represented by $t = 9$, we use the second formula,

$C(t) = 0.355t - 1.6.$

Since $m = 0.355$ million dollars per year, or \$355,000 per year, the cost of an ad was increasing by \$355,000 per year.

45. The data is

t	0	1	3
p	15	25	55

(a) 1997–1998 (first two data points):

Slope: $m = \dfrac{p_2 - p_1}{t_2 - t_1} = \dfrac{25-15}{1-0} = 10$

Intercept: $b = 15$ Specified in first data point

Thus, the linear model is

$p = mt + b$

$p = 10t + 15$

(b) 1998–2000 (second and third data points):

Slope: $m = \dfrac{p_2 - p_1}{t_2 - t_1} = \dfrac{55-25}{3-1} = \dfrac{30}{2} = 15$

Intercept: $b = p_1 - mt_1 = 25 - 15(1) = 10$

Thus, the linear model is

$p = mt + b$

$p = 15t + 10$

(c) Since the first model is valid for $0 \le t \le 1$ and the second one for $1 \le t \le 2$, we put them together as

$$p = \begin{cases} 10t + 15 & \text{if } 0 \le t \le 1 \\ 15t + 10 & \text{if } 1 \le t \le 3 \end{cases}$$

Notice that, since both formulas agree at $t = 1$, we can also say

$$p = \begin{cases} 10t + 15 & \text{if } 0 \le t < 1 \\ 15t + 10 & \text{if } 1 \le t \le 3 \end{cases}.$$

(d) Since 1999 is represented by $t = 2$, we use the second formula to obtain

$p(2) = 15(2) + 10 = 30+10 = 40\%$

47.

1989–1994 ($0 \le t \le 5$)

Points: $(t, C) = (0, 30{,}000)$ 1989 data

$(t, C) = (5, 23{,}000)$ 1994 data

Slope: $m = \dfrac{C_2 - C_1}{t_2 - t_1} = \dfrac{23{,}000-30{,}000}{5 - 0}$

$= \dfrac{-7000}{5} = -1400$

Intercept: $b = 30{,}000$

Thus, the linear model is

$C = mt + b$

$C = -1400t + 30{,}000$

1994–1999 ($5 \le t \le 10$)

Points: $(t, C) = (5, 23{,}000)$ 1994 data

$(t, C) = (10, 60{,}000)$ 1999 data

Slope: $m = \dfrac{C_2 - C_1}{t_2 - t_1} = \dfrac{60{,}000-23{,}000}{10 - 5}$

$= \dfrac{37{,}000}{5} = 7400$

Intercept: $b = C_1 - mt_1 = 23{,}000 - 7400(5)$

$= 23{,}000 - 37{,}000 = 14{,}000$

Thus, the linear model is

$C = mt + b$

$C = 7400t + 14{,}000$

Putting them together gives

$$C(t) = \begin{cases} -1400t + 30{,}000 & \text{if } 0 \le t \le 5 \\ 7400t - 14{,}000 & \text{if } 5 < t \le 10 \end{cases}$$

Since 1992 corresponds to $t = 3$, we use the first formula to obtain

$C(3) = -1400(3) + 30{,}000$

$= -4200 + 30{,}000 = 25{,}800$ students

49. We want d as a piecewise-linear function of r, using the three points (r, d) given:

$(r, d) = (1.3, 22), (1.6, 35)$, and $(1.1, 30)$

Section 1.4

Caution: These points are not given in ascending order of r. We rearrange them in increasing order of r:
$(r, d) = (1.1, 30), (1.3, 22)$, and $(1.6, 35)$

First pair of points:
$(r, d) = (1.1, 30)$ and $(1.3, 22)$
Slope $m = \dfrac{d_2 - d_1}{r_2 - r_1} = \dfrac{22 - 30}{1.3 - 1.1} = \dfrac{-8}{0.2} = -40$
Intercept: $b = d_1 - mr_1 = 30 - (-40)1.1$
$= 30 + 44 = 74$

Thus, when $1.1 \le r \le 1.3$, the linear model is
$d = mr + b$
$d = -40 + 74$

Second pair of points:
$(r, d) = (1.3, 22)$ and $(1.6, 35)$
Slope $m = \dfrac{d_2 - d_1}{r_2 - r_1} = \dfrac{35 - 22}{1.6 - 1.3} = \dfrac{13}{0.3} = \dfrac{130}{3}$
Intercept: $b = d_1 - mr_1 = 22 - \dfrac{130}{3}(1.3)$
$= 22 - \dfrac{169}{3} = -\dfrac{103}{3}$

Thus, when $1.3 \le r \le 1.6$, the linear model is
$d = mr + b$
$d = \dfrac{130}{3}r - \dfrac{103}{3}$

Putting the two linear models together gives
$d(r) = \begin{cases} -40r + 74 & \text{if } 1.1 \le r \le 1.3 \\ \dfrac{130r}{3} - \dfrac{103}{3} & \text{if } 1.3 < r \le 1.6 \end{cases}$

When there are the same number of available men as women, the ratio r of available men to women is 1, so we are asked to calculate $d(1)$. Since $1 \le 1.3$, we extrapolate the first formula:
$d(1) = -40(1) + 74 = 34\%$

Thus, the divorce rate is 34%.

51. The units of the slope m are units of y (bootlags) per unit of x (Martian yen). The intercept b is on the y-axis, and is thus measured in units of y (bootlags). Thus, m is measured in <u>bootlags per Martian yen</u> and b is measured in <u>bootlags</u>.

53. $v = 0.1t + 20$ m/sec
Since the slope is 0.1, the velocity is increasing at a rate of 0.1 m/sec every second. Since the velocity is increasing, the object is accelerating (choice B).

55. If a quantity changes linearly with time, it must change by the same amount for every unit change in time. Thus, since it increases by 10 units in the first day, it must increase by 10 units each day, including the third.

57. Increasing the number of items from the break-even results in a profit: Since the slope of the revenue graph is larger than the slope of the cost graph, it is higher than the cost graph to the right of the point of intersection, and hence corresponds to a profit.

1.5

1. (1, 1), (2, 2), (3, 4) ; $y = x-1$

x	y	Predicted $\hat{y} = x-1$	Residue $y - \hat{y}$	Residue^2 $(y - \hat{y})^2$
1	1	0	1	1
2	2	1	1	1
3	4	2	2	4

SSE = Sum of squares of residues
= 4+1+1 = 6

3. (0,−1), (1,3), (4,6), (5,0); $y = -x+2$

x	y	Predicted $\hat{y} = -x+2$	Residue $y - \hat{y}$	Residue^2 $(y - \hat{y})^2$
0	−1	2	−3	9
1	3	1	2	4
4	6	−2	8	64
5	0	−3	3	9

SSE = Sum of squares of residues
= 9 + 4 + 64 + 9 = 86

5. (1, 1), (2, 2), (3, 4)
(a) $y = 1.5x-1$

x	y	$\hat{y}$	$y - \hat{y}$	$(y - \hat{y})^2$
1	1	0.5	0.5	0.25
2	2	2	0	0
3	4	3.5	0.5	0.25

SSE = Sum of squares of residues = 0.5

(b) $y = 2x - 1.5$

x	y	$\hat{y}$	$y - \hat{y}$	$(y - \hat{y})^2$
1	1	0.5	0.5	0.25
2	2	2.5	−0.5	0.25
3	4	4.5	−0.5	0.25

SSE = Sum of squares of residues = 0.75
The model that gives the better fit is (a) because it gives the smaller value of SSE.

7. (0, −1), (1, 3), (4, 6), (5, 0)
(a) $y = 0.3x + 1.1$

x	y	$\hat{y}$	$y - \hat{y}$	$(y - \hat{y})^2$
0	−1	1.1	−2.1	4.41
1	3	1.4	1.6	2.56
4	6	2.3	3.7	13.69
5	0	2.6	−2.6	6.76

SSE = Sum of squares of residues = 27.42

(b) $y = 0.4x+0.9$

x	y	$\hat{y}$	$y - \hat{y}$	$(y - \hat{y})^2$
0	−1	0.9	−1.9	3.61
1	3	1.3	1.7	2.89
4	6	2.5	3.5	12.25
5	0	2.9	−2.9	8.41

SSE = Sum of squares of residues = 27.16
The model that gives the better fit is (b) because it gives the smaller value of SSE.

9. (1,1), (2,2), (3,4)

	x	y	xy	x^2
	1	1	1	1
	2	2	4	4
	3	4	12	9
Σ (Sum)	6	7	17	14

n = 3 (number of data points)

Slope: $m = \dfrac{n(\Sigma xy) - (\Sigma x)(\Sigma y)}{n(\Sigma x^2) - (\Sigma x)^2}$

$= \dfrac{3(17) - (6)(7)}{3(14) - 6^2} = \dfrac{9}{6} = 1.5$

Intercept: $b = \dfrac{\Sigma y - m(\Sigma x)}{n}$

$= \dfrac{7 - 1.5(6)}{3} = -\dfrac{2}{3} \approx -0.6667$

Thus, the regression line is
$y = mx + b$
$y = 1.5x - 0.6667$

Section 1.5

Graph:

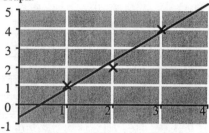

11. $(0, -1), (1, 3), (4, 6), (5, 0)$

x	y	xy	x^2
0	-1	0	0
1	3	3	1
4	6	24	16
5	0	0	25
Σ (Sum) 10	8	27	42

$n = 4$ (number of data points)

Slope: $m = \dfrac{n(\Sigma xy) - (\Sigma x)(\Sigma y)}{n(\Sigma x^2) - (\Sigma x)^2}$

$= \dfrac{4(27) - (10)(8)}{4(42) - 10^2} = \dfrac{28}{68} \approx 0.4118$

Intercept: $b = \dfrac{\Sigma y - m(\Sigma x)}{n}$

$= \dfrac{8 - \left(\dfrac{28}{68}\right)(10)}{4} \approx 0.9706$

Thus, the regression line is

$y = mx + b$

$y = 0.4118x + 0.9706$

Graph:

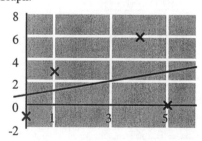

13. (a) $\{(1, 3), (2, 4), (5, 6)\}$

x	y	xy	x^2	y^2
1	3	3	1	9
2	4	8	4	16
5	6	30	25	36
Σ 8	13	41	30	61

$n = 3$ (number of data points)

$r = \dfrac{n(\Sigma xy) - (\Sigma x)(\Sigma y)}{\sqrt{n(\Sigma x^2) - (\Sigma x)^2} \cdot \sqrt{n(\Sigma y^2) - (\Sigma y)^2}}$

$= \dfrac{3(41) - (8)(13)}{\sqrt{3(30)-(8)^2} \sqrt{3(61)-(13)^2}}$

$\approx \dfrac{19}{19.078784} \approx 0.9959$

(b) $\{(0, -1), (2, 1), (3, 4)\}$

x	y	xy	x^2	y^2
0	-1	0	0	1
2	1	2	4	1
3	4	12	9	16
Σ 5	4	14	13	18

$n = 3$ (number of data points)

$r = \dfrac{n(\Sigma xy) - (\Sigma x)(\Sigma y)}{\sqrt{n(\Sigma x^2) - (\Sigma x)^2} \cdot \sqrt{n(\Sigma y^2) - (\Sigma y)^2}}$

$= \dfrac{3(14) - (5)(4)}{\sqrt{3(13)-(5)^2} \sqrt{3(18)-(4)^2}}$

$\approx \dfrac{22}{23.0651252} \approx 0.9538$

(c) $\{(4, -3), (5, 5), (0, 0)\}$

x	y	xy	x^2	y^2
4	-3	-12	16	9
5	5	25	25	25
0	0	0	0	0
Σ 9	2	13	41	34

$n = 3$ (number of data points)

$r = \dfrac{n(\Sigma xy) - (\Sigma x)(\Sigma y)}{\sqrt{n(\Sigma x^2) - (\Sigma x)^2} \cdot \sqrt{n(\Sigma y^2) - (\Sigma y)^2}}$

$= \dfrac{3(13) - (9)(2)}{\sqrt{3(41)-(9)^2} \sqrt{3(34)-(2)^2}}$

Section 1.5

$$\approx \frac{21}{64.1560597} \approx 0.3273$$

The value of r in part (a) has the largest absolute value. Therefore, the regression line for the data in part (a) is the best fit.

The value of r in part (c) has the smallest absolute value. Therefore, the regression line for the data in part (c) is the worst fit.

Since r is not ± 1 for any of these lines, none of them is a perfect fit.

15. Calculation of the regression line:

x	y	xy	x^2
0	7.4	0	0
5	5.6	28	25
10	5.1	51	100
15	4.7	70.5	225
19	4.5	85.5	361
Σ (Sum) 49	27.3	235	711

$n = 5$ (number of data points)

Slope: $m = \dfrac{n(\Sigma xy) - (\Sigma x)(\Sigma y)}{n(\Sigma x^2) - (\Sigma x)^2}$

$= \dfrac{5(235) - (49)(27.3)}{5(711) - (49)^2}$

$= \dfrac{-162.7}{1154} \approx -0.141$

(Rounded to 3 significant digits)

Intercept: $b = \dfrac{\Sigma y - m(\Sigma x)}{n}$

$\approx \dfrac{27.3 - (-0.140988)(49)}{5}$

≈ 6.84 (to 3 significant digits)

Thus, the regression line is
$y = mt + b$ Independent variable is called t
$y = -0.141t + 6.84$

In 2005, $t = 2005 - 1980 = 25$, so government expenditures are
$y(25) = -0.141(25) + 6.84$

$= 3.315 \approx 3.3$ (to 2 significant digits)
Thus, government expenditures are predicted to be \$3.3 billion in 2005.

17. Calculation of the regression line:

x	y	xy	x^2
12	21	252	144
14	30	420	196
22	42	924	484
30	49	1470	900
Σ 78	142	3066	1724

$n = 4$ (number of data points)

Slope: $m = \dfrac{n(\Sigma xy) - (\Sigma x)(\Sigma y)}{n(\Sigma x^2) - (\Sigma x)^2}$

$= \dfrac{4(3066) - (78)(142)}{4(1724) - (78)^2}$

$= \dfrac{1188}{812} \approx 1.46$

(Rounded to 3 significant digits)

Intercept: $b = \dfrac{\Sigma y - m(\Sigma x)}{n}$

$\approx \dfrac{142 - (1.463054)(78)}{4}$

≈ 6.97 (to 3 significant digits)

Thus, the regression line is
$y = mx + b$
$y = 1.46x + 6.97$

If $x = \$10$, then the recovery is given by
$y(10) = 1.46(10) + 6.97 = 14.6 + 6.97$
$= 21.57 \approx 22$ billion barrels
(Rounded to 2 significant digits)

19.
(a) Since the independent variable (t) is years since 1994, the given data are as follows

t	0	1	2	3
y	6.3	5.2	5.0	5.0

Section 1.5

Calculation of the regression line:

x	y	xy	x^2
0	6.3	0	0
1	5.2	5.2	1
2	5	10	4
3	5	15	9
4	4.7	18.8	16
Σ 10	26.2	49	30

$n = 5$ (number of data points)

Slope: $m = \dfrac{n(\Sigma xy) - (\Sigma x)(\Sigma y)}{n(\Sigma x^2) - (\Sigma x)^2}$

$= \dfrac{5(49) - (10)(26.2)}{5(30) - (10)^2}$

$= \dfrac{-17}{50} = -0.34$

Intercept: $b = \dfrac{\Sigma y - m(\Sigma x)}{n}$

$= \dfrac{26.2 - (-0.34)(10)}{5}$

$= 5.92$

Thus, the regression line is

$y = mt + b$ Independent variable is called t
$y = -0.34t + 5.92$

Graph:

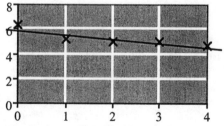

(b) The slope, $m = -034$, gives the change in y (cameras sold) per unit increase in t (per year). Thus, camera sales were declining at a rate of 0.34 million per year.

(c) Use of a such a linear model to make predictions is not reasonable, since the possible factors that may influence sales are not taken into account.

In 1999, $t = 5$, and so predicted camera sales are given by
$y(5) = -0.34(5) + 5.92 = 4.22$
≈ 4.2 million cameras
(Rounded to the nearest 0.1 million cameras)

(d) y(Observed) is given as 10 million, and y(Predicted) was calculated in part (c) as 4.2 million. Thus, the residue is

Residue $= y - \hat{y}$ y(Observed)−y(Predicted)
$= 10 - 4.2 = 5.8$ million cameras
This is large compared with the other residues, confirming the point raised in part (c).

21. Here is the given data, with $t = 0$ representing 1987:

t	0	1	2	3	4	5	6
y	2900	3400	4000	4200	4400	5500	7000

Using technology, we obtain the following regression line (coefficients rounded to 5 significant digits):
$y = 603.57t + 2675$
In 1995, $t = 8$, so
$y(8) = 603.57(8) + 2675$
≈ 7500 phone lines
(rounded to the nearest 100 phone lines).

23. Here is the given data, with $t = 0$ representing 1900:

t	20	30	40	50	60
y	54.1	59.7	62.9	68.2	69.7
t	70	80	90	98	
y	70.8	73.7	75.4	76.7	

Using technology, we obtain the following regression line (coefficients rounded to 4 significant digits):
$y = 0.2737t + 51.55$
The rate at which life expectancy (y) was changing is the slope,

Section 1.5

$m \approx 0.27$ years (of life) per year

25. Here is the data given:

x	14.73	29.31	19.76	21.76	36.49
y	76.4	70.6	69.3	62.8	61.3
x	28.86	48.64	59.36	32.91	
y	59.3	52.4	27.6	22.3	

(a) Graph:

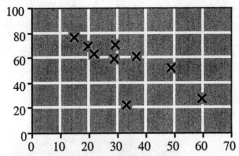

The graph suggests (except for one extraneous "outlier" point) that the percentage rate (y) decreases as the size of the 1991 homicide rate (x) increases.

(b) Using technology, we obtain the following regression line (coefficients rounded to 4 significant digits):
$$y \approx -0.9148x + 85.44$$
$$r^2 \approx 0.4748$$
$$r = -\sqrt{0.4748} \approx -0.689$$
(We used the negative square root because the slope is negative.) The value of r suggests a reasonable fit.

(c) The slope gives the number of units of y (percentage drop in the homicide rate) per unit of x (homicide rate). Since the slope is negative, it suggests that the percentage drop in the homicide rate decreases with increasing 1991 homicide rate.

(d) For a city experiencing 40 homicides per 100,000 people in 1991, $x = 40$, the percentage drop in the homicide rate is
$$y(40) \approx -0.9148(40) + 85.44 \approx 48.8.$$

27. The regression line is defined to be the line that gives the lowest sum-of-squares error, SSE. If we are given two points, (a, b) and (c, d) with $a \neq c$ then there is a line that passes through these two points, giving SSE = 0. Since 0 is the smallest value possible, this line must be the regression line.

29. Calculation of the regression line:

x	y	xy	x^2
0	0	0	0
$-a$	a	$-a^2$	a^2
a	a	a^2	a^2
Σ 0	$2a$	0	$2a^2$

$n = 3$ (number of data points)

Slope: $m = \dfrac{n(\Sigma xy) - (\Sigma x)(\Sigma y)}{n(\Sigma x^2) - (\Sigma x)^2}$

$= \dfrac{3(0) - (0)(2a)}{3(2a^2) - 0^2} = 0$

Regression Coefficient:
$$r = \dfrac{n(\Sigma xy) - (\Sigma x)(\Sigma y)}{\sqrt{n(\Sigma x^2) - (\Sigma x)^2} \cdot \sqrt{n(\Sigma y^2) - (\Sigma y)^2}}$$

has the same numerator as m, and we have just seen that this numerator is zero, Hence, $r = 0$.

31. If the points $(x_1, y_1), (x_2, y_2), \ldots, (x_n, y_n)$ lie on a straight line, then the sum-of-squares error, SSE, for this line is zero. Since 0 is the smallest value possible, this line must be the regression line.

33. No. The regression line through $(-1, 1)$, $(0, 0)$, and $(1, 1)$ passes through none of these points

Chapter 1 Review Test

1.
(a) $y = -2x + 5$
y-intercept = 5   slope = -2

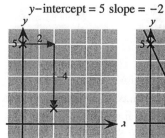

(b) $2x - 3y = 12$
Solving for y gives
$$-3y = -2x + 12$$
$$y = \frac{2}{3}x - 4$$
y-intercept = -4 slope = $\frac{2}{3}$

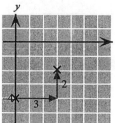 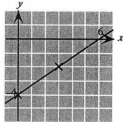

(c) $y = \begin{cases} \frac{1}{2}x & \text{if } -1 \leq x \leq 1 \\ x - 1 & \text{if } 1 < x \leq 3 \end{cases}$

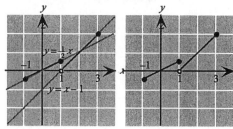

(d) $f(x) = 4x - x^2$ with domain $[0, 4]$
Technology formula: `4*x-x^2`

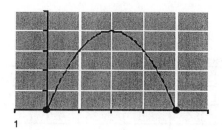

2.

x	-2	0	1	2	4
$f(x)$	4	2	1	0	2
$g(x)$	-5	-3	-2	-1	1
$h(x)$	1.5	1	0.75	0.5	0
$k(x)$	0.25	1	2	4	16
$u(x)$	0	4	3	0	-12

Here are the graphs of the five functions, with the points connected:

$f(x)$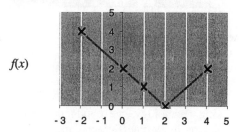

V-shape indicates **absolute value** model.

$g(x)$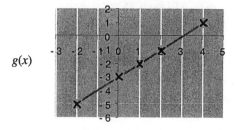

Linear

30

$h(x)$

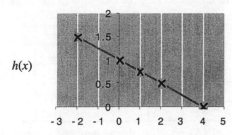

Linear

$k(x)$

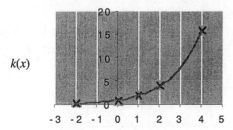

y doubles for each 1-unit increase in x, indicating an **exponential** model.

$u(x)$

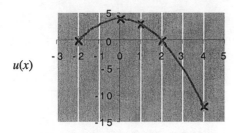

Parabolic shape of graph indicates a **quadratic** model.

3.

(a) Line through (3, 2) with slope -3
Point: (3, 2)
Slope: $m = -3$
Intercept: $b = y_1 - mx_1$
$= 2-(-3)(3) = 2+9 = 11$

Thus, the equation is
$y = mx + b$
$y = -3x + 11$

(b) Through $(-1, 2)$ and $(1, 0)$
Point: (1, 0)
Slope: $m = \dfrac{y_2 - y_1}{x_2 - x_1} = \dfrac{0 - 2}{1 - (-1)} = \dfrac{-2}{2} = -1$
Intercept: $b = y_1 - mx_1 = 0-(-1)(1) = 1$

Thus, the equation is
$y = mx + b$
$y = -x + 1$

(c) Through (1, 2) parallel to $x - 2y = 2$
Point: (1, 2)
Slope: Parallel to the line $x - 2y = 2$, and so has the same slope. Solve for y to obtain
$-2y = -x - 2$
$y = \tfrac{1}{2}x + 1$
Thus the slope is $m = \tfrac{1}{2}$

Intercept: $b = y_1 - mx_1 = 2-\tfrac{1}{2}(1) = \dfrac{3}{2}$

Thus, the equation is
$y = mx + b$
$y = \tfrac{1}{2}x + \tfrac{3}{2}$

(d) With slope 1/2 crossing $3x + y = 6$ at its x-intercept
Point: We need the x-intercept of $3x + y = 6$. This is given by setting $y = 0$ and solving for x:
$3x + 0 = 6$
$x = 2$.
Thus, the point is (2, 0) On the x-axis, so $y=0$
Slope: $m = \tfrac{1}{2}$
Intercept: $b = y_1 - mx_1 = 0 - \tfrac{1}{2}(2) = -1$

Thus, the equation is
$y = mx + b$
$y = \tfrac{1}{2}x - 1$

4. $n(x) = \begin{cases} 0.02x & \text{if } 0 \le x \le 1000 \\ 0.025x - 5 & \text{if } 1000 < x \le 2000 \end{cases}$

(a) $n(500) = 0.02(500) = 10$ books per day
We used the first formula, since 500 is in [0, 1000].
$n(1000) = 0.02(1000) = 20$ books per day
We used the first formula, since 1000 is in [0, 1000].
$n(1500) = 0.025(1500) - 5$
$= 37.6 - 5 = 32.5$ books per day
We used the second formula, since 1500 is in (1000, 2000].

(b) The coefficient 0.025 is the slope of the second formula, indicating that, for web site traffic of more than 1000 hits per day and up to 2000 hits per day ($1000 < x \le 2000$) book sales are increasing by 0.025 books per additional hit.

(c) To sell an average of 30 books per day, we desire $n(x) = 30$. Of we try the first formula, we get
$$0.02x = 30$$
giving $x = 1500$, which is not in the domain of the first formula. So, we try the second formula:
$$0.025x - 5 = 30$$
$$0.025x = 35$$
$$x = \frac{35}{0.025} = 1400 \text{ hits,}$$
which is in the domain of the second formula. Thus, 1400 hits per day will result in average sales of 30 books per day.

5.

t	1	2	3	4	5	6
$S(t)$	12.5	37.5	62.5	72.0	74.5	75.0

(a) Technology formulas:
(A): `300/(4+100*5^(-t))`
(B): `13.3*t+8.0`
(C): `-2.3*t^2+30.0*t-3.3`
(D): `7*3^(0.5*t)`

Here are the values for the three given models (rounded to 1 decimal place):

t	1	2	3	4	5	6
(A)	12.5	37.5	62.5	72.1	74.4	74.9
(B)	21.3	34.6	47.9	61.2	74.5	87.8
(C)	24.4	47.5	66.0	79.9	89.2	93.9
(D)	12.1	21.0	36.4	63.0	109.1	189.0

Model (A) gives an almost perfect fit, whereas the other models are not even close.

(b) Looking at the table, we see the following behavior as t increases:
(A) Leveling off; (B) Rising (C) Rising (they begin to fall after 7 months, however) (D) Rising

6.

c	$2000	$5000
h	1900	2050

Point: (2000, 1900)
Slope: $m = \dfrac{h_2 - h_1}{c_2 - c_1} = \dfrac{2050-1900}{5000-2000}$
$= \dfrac{150}{3000} = 0.05$
Intercept: $b = y_1 - mx_1$
$= 1900 - (0.05)(2000)$
$= 1900 - 100 = 1800$
Thus, the equation is
$$h = mc + b$$
$$h = 0.05c + 1800$$

(b) A budget of $6000 per month for banner ads corresponds to $c = 6000$
$h(6000) = 0.05(6000) + 1800$
$= 300 + 1800 = 2100$ hits per day

(c) We are given $h = 2500$ and want c.
$2500 = 0.05c + 1800$
$0.05c = 2500 - 1800 = 700$
$c = \dfrac{700}{0.05} = \$14000$ per month

7. $h = -0.000005c^2 + 0.085c + 1750$

(a) Currently, $c = \$6000$, so

Chapter 1 Review Test

$h = -0.000005(6000)^2 + 0.085(6000) + 1750$
$= -180 + 510 + 1750 = 2080$ hits per day
(b) Here is a portion of the graph of h:

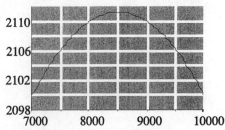

Technology formula:
`-0.000005*x^2+0.085*x+1750`

For $c = 8500$ or larger, we see that web site traffic is projected to decrease as advertising increases, and then drop toward zero. Thus, the model does not appear to give a reasonable prediction of traffic at expenditures larger than $8,500 per month.

8.
Cost function: $C = mx + b$
b = fixed cost = $900 per month
m = marginal cost = $4 per novel
Thus, the linear cost function is
$\quad C = 4x + 900$

Revenue function: $R = mx + b$
b = fixed revenue = 0
m = marginal revenue = $5.50 per novel
Thus, the linear revenue function is
$\quad R = 5.50x$

Profit function: $P = R - C$
$P = 5.50x - (4x + 900)$
$\quad = 5.50x - 4x - 900$
$\quad = 1.50x - 900$

(b) For break-even, $P = 0$
$\quad 1.50x - 900 = 0$

$1.50x = 900$
$x = \dfrac{900}{1.50} = 600$ novels per month

(c) $R = 5.00x$
$P = 5.00x - (4x + 900)$
$\quad = 5x - 4x - 900 = x - 900$
For break-even,
$\quad x - 900 = 0,$
$\quad x = 900$ novels per month

9.
(a) Demand: We are given two points:
$\quad (p, q) = (10, 350)$ and $(5.5, 620)$ Slope:

$$m = \frac{q_2 - q_1}{p_2 - p_1} = \frac{620 - 350}{5.5 - 10} = \frac{270}{-4.5} = -60$$

Intercept:
$\quad b = q_1 - mp_1 = 350 - (-60)(10)$
$\quad = 350 + 600 = 950$
Thus, the demand equation is
$\quad q = mp + b$
$\quad q = -60p + 950$
(b) When $p = \$15$, the demand is
$\quad q = -60(15) + 950 = -900 + 950$
$\quad = 50$ novels per month
(c) From Question 8, the cost function is
$\quad C = 4q + 900$
We use q for the monthly sales rather than x
$\quad = 4(-60p + 950) + 900$
We want everything expressed in terms of p, so we used the demand equation.
$\quad = -240p + 3800 + 900$
$\quad C = -240p + 4700$

To compute the profit in terms of price, we need the revenue as well:
$\quad R = pq = p(-60p + 950)$
$\quad = -60p^2 + 950p$

Profit: $P = R - C$
$P = -60p^2 + 950p - (-240p + 4700)$
$= -60p^2 + 1190p - 4700$

Now we compare profits for the three prices:
$P(5.50) = -60(5.5)^2 + 1190(5.5) - 4700$
$= \$30$
$P(10) = -60(10)^2 + 1190(10) - 4700$
$= \$1200$
$P(15) = -60(15)^2 + 1190(15) - 4700$
$= -\$350$ (loss)

Thus, charging $10 will result in the largest monthly profit of $1200.

(b) We are given $p = \$8$, so the demand is
$q = -53.3945(8) + 909.8165$
≈ 483 novels per month (rounded)

10.

p	5.50	10	12	15
q	620	350	300	100

(a) Calculation of the regression line:

	x	y	xy	x^2
	5.5	620	3410	30.25
	10	350	3500	100
	12	300	3600	144
	15	100	1500	225
Σ	42.5	1370	12010	499.25

$n = 4$ (number of data points)

Slope: $m = \dfrac{n(\Sigma xy) - (\Sigma x)(\Sigma y)}{n(\Sigma x^2) - (\Sigma x)^2}$

$= \dfrac{4(12010) - (42.5)(1370)}{4(499.25) - (42.5)^2}$

$= \dfrac{-10185}{190.75} \approx 53.3945$

Intercept: $b = \dfrac{\Sigma y - m(\Sigma x)}{n}$

$\approx \dfrac{1370 - (-53.394495)(42.5)}{4}$

$\approx \dfrac{3639.26606}{4} \approx 909.8165$

Thus, the regression equation is
$q = -53.3945p + 909.8165$
(q plays the role of y and p plays the role of x)

Chapter 2
2.1

1. $a = 1$, $b = 3$, $c = 2$; $-b/(2a) = -3/2$;
$f(-3/2) = -1/4$, so: vertex: $(-3/2, -1/4)$
y intercept $= c = 2$
$x^2 + 3x + 2 = (x + 1)(x + 2)$, so:
x intercepts: $-2, -1$
$a > 0$ so the parabola opens upward

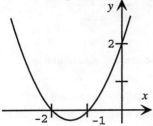

3. $a = -1$, $b = 4$, $c = -4$; $-b/(2a) = 2$;
$f(2) = 0$, so: vertex: $(2, 0)$
y intercept $= c = -4$
$-x^2 + 4x - 4 = -(x - 2)^2$, so:
x intercept: 2
$a < 0$ so the parabola opens downward

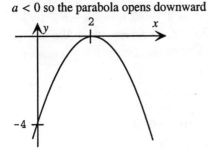

5. $a = -1$, $b = -40$, $c = 500$; $-b/(2a) = -20$;
$f(-20) = 900$, so: vertex: $(-20, 900)$
y intercept $= c = 500$
$-x^2 - 40x + 500 = -(x + 50)(x - 10)$, so:
x intercepts: $-50, 10$
$a < 0$ so the parabola opens downward

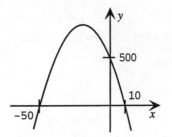

7. $a = 1$, $b = 1$, $c = -1$; $-b/(2a) = -1/2$;
$f(-1/2) = -5/4$, so: vertex: $(-1/2, -5/4)$
y intercept $= c = -1$
from the quadratic formula:
x intercepts: $-1/2 \pm \sqrt{5}/2$
$a > 0$ so the parabola opens upward

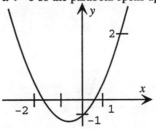

9. $a = 1$, $b = 0$, $c = 1$; $-b/(2a) = 0$;
$f(0) = 1$, so: vertex: $(0, 1)$
y intercept $= c = 1$
$b^2 - 4ac = -4 < 0$, so no x intercept
$a > 0$ so the parabola opens upward

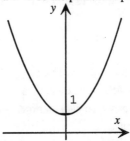

Section 2.1

11. $R = pq = p(-4p + 100) = -4p^2 + 100p$; maximum revenue when $p = -b/(2a) = \$12.50$

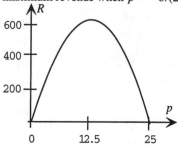

13. $R = pq = p(-2p + 400) = -2p^2 + 400p$; maximum revenue when $p = -b/(2a) = \$100$

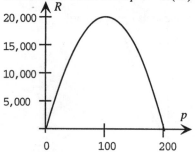

15. $y = -0.7955x^2 + 4.4591x - 1.6000$

17. $y = -1.1667x^2 - 6.1667x - 3$

19. (a) Positive because the data suggest a curve that is concave up.
(b) (A) because $18{,}000 > 0$ and the y intercept is 320,000.
(c) $-b/(2a) \approx 0.28$, which is in 1995. The parabola rises to the left of the vertex and thus predicts increasing sales as we go back in time, contradicting common sense.

21.

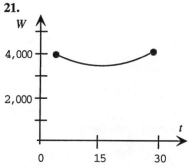

$-b/(2a) = 15$, which represents 1985. $W(15) = 3525$ pounds was the average weight that year.

23. $-b/(2a) = 5000$ pounds. The model is not trustworthy for vehicle weights larger than 5000 pounds because it predicts increasing fuel economy with increasing weight. Also, 5000 is close to the upper limit of the domain of the function.

25. The given data points are $(p, q) = (80, 100)$ and $(100, 90)$. The line passing through these points is $q = -p/2 + 140$. Revenue is $R = pq = -p^2/2 + 140p$. Maximum revenue occurs when $p = -b/(2a) = \$140$; the corresponding revenue is $R = \$9800$.

27. The given data points are $(p, q) = (40, 200{,}000)$ and $(60, 160{,}000)$. The line passing through these points is $q = -2000p + 280{,}000$. Revenue is $R = pq = -2000p^2 + 280{,}000p$. Maximum revenue occurs when $p = -b/(2a) = 70$ houses; the corresponding revenue is $R = \$9{,}800{,}000$.

29. (a) The data points are $(x, q) = (2, 280)$ and $(1.5, 560)$. Thus, $q = -560x + 1400$ and $R = xq = -560x^2 + 1400x$. **(b)** $P = R - C = -560x^2 + 1400x - 30$. The largest monthly profit occurs when $x = -b/(2a) = \$1.25$ and then $P = \$845$ per month.

31. As a function of q, $C = 0.5q + 20$. Substituting, we get $C = 0.5(-400x + 1200) + 20 = -200x + 620$. The profit is $P = R - C = xq - C = -400x^2 + 1400x - 620$. The profit is largest when $x = -b/(2a) = \$1.75$ per hit; the corresponding profit is $P = \$605$ per month.

33. (a) The data points are $(p, q) = (10, 300)$ and $(15, 250)$, so $q = -10p + 400$. (b) $R = pq = -10p^2 + 400p$. (c) $C = 3q + 3000 = 3(-10p + 400) + 3000 = -30p + 4200$ (d) $P = R - C = -10p^2 + 430p - 4200$. The maximum profit occurs when $p = -b/(2a) = \$21.50$.

35. Use technology to find $R(t) = 95t^2 + 685t + 1350$ million dollars. The revenue in 2000 is then $R(0) = \$1350$ million.

37. Use technology to find $I(t) = 7.32t^2 - 34.5t + 401$ thousand dollars. The average after-tax income in 1999 is then $I(9) \approx \$683,000$.

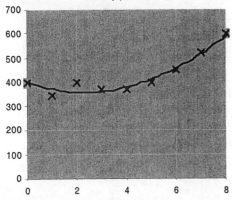

39. The x coordinate of the vertex represents the unit price that leads to the maximum revenue, the y coordinate of the vertex gives the maximum possible revenue, the x intercepts give the unit prices that result in zero revenue, and the y intercept gives the revenue resulting from zero unit price (which is obviously zero).

41. Graph the data to see whether the points suggest a curve rather than a straight line. If the curve suggested by the graph is concave up or concave down, then a quadratic model would be a likely candidate.

43. If $q = mp + b$ (with $m < 0$), then the revenue is given by $R = pq = mp^2 + bp$. This is the equation of a parabola with $a = m < 0$, and so is concave down. Thus, the vertex is the highest point on the parabola, showing that there is a single highest value for R, namely the y coordinate of the vertex.

45. Since $R = pq$, the demand must be given by $q = \dfrac{R}{p} = \dfrac{-50p^2 + 60p}{p} = -50p + 60$.

2.2

1. 4^x

x	-3	-2	-1	0	1	2	3
f(x)	$\frac{1}{64}$	$\frac{1}{16}$	$\frac{1}{4}$	1	4	16	64

3. 3^(-x)

x	-3	-2	-1	0	1	2	3
f(x)	27	9	3	1	$\frac{1}{3}$	$\frac{1}{9}$	$\frac{1}{27}$

5. 2*2^x or 2*(2^x)

x	-3	-2	-1	0	1	2	3
f(x)	$\frac{1}{4}$	$\frac{1}{2}$	1	2	4	8	16

7. -3*2^(-x)

x	-3	-2	-1	0	1	2	3
f(x)	-24	-12	-6	-3	$-\frac{3}{2}$	$-\frac{3}{4}$	$-\frac{3}{8}$

9. 2^x-1

x	-3	-2	-1	0	1	2	3
f(x)	$-\frac{7}{8}$	$-\frac{3}{4}$	$-\frac{1}{2}$	0	1	3	7

11. 2^(x-1)

x	-3	-2	-1	0	1	2	3
f(x)	$\frac{1}{16}$	$\frac{1}{8}$	$\frac{1}{4}$	$\frac{1}{2}$	1	2	4

13.

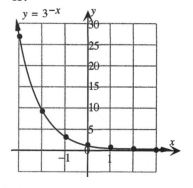

15.

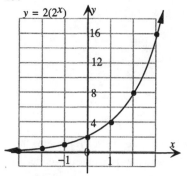

Section 2.2

17.

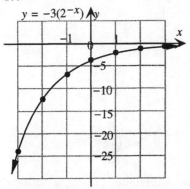

19. `e^(-2*x)` or `EXP(-2*x)`

x	−3	−2	−1	0	1	2	3
$f(x)$	403.4	54.60	7.389	1	0.1353	0.01832	0.002479

21. `1.01*2.02^(-4*x)`

x	−3	−2	−1	0	1	2	3
$f(x)$	4662	280.0	16.82	1.01	0.06066	0.003643	0.0002188

23. `50*(1+1/3.2)^(2*x)`

x	−3	−2	−1	0	1	2	3
$f(x)$	9.781	16.85	29.02	50	86.13	148.4	255.6

The following solutions also show some common errors you should avoid.

25. `2^(x-1)` *not* `2^x-1`

27. `2/(1-2^(-4*x))` *not* `2/1-2^-4*x` *and not* `2/1-2^(-4*x)`

29. `(3+x)^(3*x)/(x+1)` or `((3+x)^(3*x))/(x+1)` *not* `(3+x)^(3*x)/x+1` *and not* `(3+x^(3*x))/(x+1)`

31. `2*e^((1+x)/x)` or `2*EXP((1+x)/x)` *not* `2*e^1+x/x` *and not* `2*e^(1+x)/x` *and not* `2*EXP(1+x)/x`

Section 2.2

In the following solutions, f_1 is black and f_2 is gray.

33.

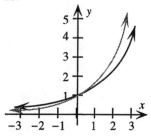

35. (Note that the x-axis shown here crosses at 200, not 0.)

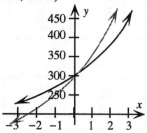

37.

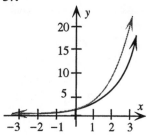

39. (Note that the x-axis shown here crosses at 900.)

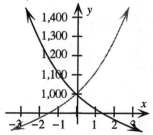

41. Each time x increases by 1, $f(x)$ is multiplied by 0.5. Also, $f(0) = 500$. So, $f(x) = 500(0.5)^x$.

43. Each time x increases by 1, $f(x)$ is multiplied by 3. Also, $f(0) = 10$. So, $f(x) = 10(3)^x$.

45. Each time x increases by 1, $f(x)$ is multiplied by $225/500 = 0.45$. Also, $f(0) = 500$. So, $f(x) = 500(0.45)^x$.

47. Write $f(x) = Ab^x$. We have $Ab^1 = -110$ and $Ab^2 = -121$. Dividing, $b = -121/(-110) = 1.1$. Substituting, $A(1.1) = -110$, so $A = -100$. Thus, $f(x) = -100(1.1)^x$.

49. Write $f(x) = Ab^x$. We have $Ab^1 = 3$ and $Ab^3 = 6$. Dividing, $b^2 = 6/3 = 2$, so $b = \sqrt{2} \approx 1.4142$. Substituting, $A\sqrt{2} = 3$, so $A = 3/\sqrt{2} \approx 2.1213$. Thus, $y = 2.1213\,(1.4142^x)$.

51. Write $f(x) = Ab^x$. We have $Ab^{-1} = 5$ and $Ab^4 = 20$. Dividing, $b^5 = 20/5 = 4$, so $b = \sqrt[5]{4} \approx 1.3195$. Substituting, $A/\sqrt[5]{4} = 5$, so $A = 5\sqrt[5]{4} \approx 6.5975$. Thus, $y = 6.5975(1.3195^x)$.

53. Write $f(x) = Ab^x$. We have $Ab^2 = 3$ and $Ab^6 = 2$. Dividing, $b^4 = 2/3$, so $b = \sqrt[4]{2/3} \approx 0.9036$.

Section 2.2

Substituting, $A\left(\sqrt[4]{2/3}\right)^2 = 3$, so $A = 3/\left(\sqrt[4]{2/3}\right)^2$
≈ 3.6742. Thus, $y = 3.6742(0.9036^x)$.

55. $y = 1.0442(1.7564)^x$

57. $y = 15.1735(1.4822)^x$

59. From the compound interest formula, the deposit would be worth $5000(1 + 0.0415)^5 =$ €6127.

61. We are looking for the value of P that makes $P(1 + 0.0415)^5 = 250{,}000$. Thus, $P = 250{,}000/(1 + 0.0415)^5 =$ €204,006.

63. $C(t) = 104(0.999879)^t$, so $C(10{,}000) \approx 31.0$ grams, $C(20{,}000) \approx 9.25$ grams, and $C(30{,}000) \approx 2.76$ grams.

65. We are looking for t such that $4.06 = 46(0.999879)^t$. Among the values suggested we find that $C(15{,}000) \approx 7.5$, $C(20{,}000) \approx 4.09$ and $C(25{,}000) \approx 2.23$. Thus, the answer is 20,000 years to the nearest 5000 years.

67. If y represents the size of the culture at time t, then $y = Ab^t$. We are told that the initial size is 1000, so $A = 1000$. We are told that the size doubles every 3 hours, so $b^3 = 2$, or $b = 2^{1/3}$. Thus, $y = 1000(2^{1/3})^t = 1000(2^{t/3})$. There will be $1000(2^{48/3}) = 65{,}536{,}000$ bacteria after 2 days.

69. Let y represent the amount of aspirin in the bloodstream at time t hours. Then $y = Ab^t$. The initial value is given to us as $A = 300$ mg. We are told that half the amount is removed every 2 hours, so $b^2 = 0.5$, giving $b = \sqrt{0.5} \approx 0.7071$. Thus, $y = 300(0.7071)^t$. After 5 hours the amount left is $300(0.7071)^5 \approx 53$ mg.

71. (a) The line through $(t, P) = (0, 360)$ and $(3, 480)$ is $P = 40t + 360$ million dollars. **(b)** We need to find the exponential curve $P = Ab^t$ through $(0, 360)$ and $(3, 480)$. We see that $A = 360$ and $b^3 = 480/360 = 4/3$, so $b = \sqrt[3]{4/3} \approx 1.1006$. Thus, $P = 360(1.1006)^t$. Neither model is applicable to the data: The data move erratically up and down rather than increasing steadily either linearly or exponentially.

73. (a) Let y be the number of frogs in year t, with $t = 0$ representing two years ago; we seek a model of the form $y = Ab^t$. We are given the initial value of $A = 50{,}000$ and are told that $50{,}000b^2 = 75{,}000$. This gives $b = (75{,}000/50{,}000)^{1/2} = 1.5^{1/2}$. Thus, $y = 50{,}000(1.5^{1/2})^t = 50{,}000(1.5^{t/2})$. **(b)** When $t = 3$, $y = 50{,}000(1.5)^{3/2} = 91{,}856$ tags.

75. (a) Let $P = Ab^t$ million people. We are given the data points $(0, 180)$ and $(40, 281)$, so $A = 180$ while $b^{40} = 281/180$, which gives $b = \sqrt[40]{281/180} \approx 1.01120$. Thus, $P = 180(1.01120)^t$ million people. **(b)** When $t = 60$, $P = 180(1.01120)^{60} \approx 351$ million people.

77. From the continuous compounding formula, $1000e^{0.04 \times 10} = \$1491.82$, of which $\$491.82$ is interest.

79. (a)

year	1950	2000	2050	2100
$C(t)$ parts per million	561	669	799	953

(b) Testing the decades between 2000 ($t = 250$) and 2050, we find that $C(260) \approx 694$ and $C(270) \approx 718$. Testing individual years between $t = 260$ and $t = 270$, we find that the level surpasses 700 parts per million for the first time when $t = 263$. Thus, to the nearest decade, the level passes 700 in 2010 ($t = 260$).

81. (a) $y \approx 5.907(0.939^t)$

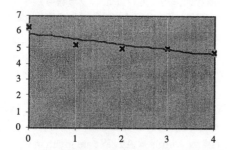

(b) $y(5) \approx 4.3$ million cameras (c) 5.7 million cameras, which is large compared with the other residuals, and shows the danger of extrapolating based on past information alone: The possible factors that may influence sales, such as introduction of new products, are not taken into account in the model.

83. (a) $y = 5.4433(1.0609)^t$

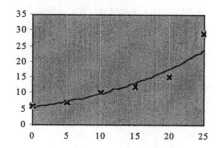

(b) Each year spending increases by a factor of 1.0609, which represents a 6.09% annual increase. (c) $y(18) \approx \$16$ billion

85. (B) An exponential function eventually becomes larger than any polynomial.

87. Exponential functions of the form $f(x) = Ab^x$ ($b > 0$) increase rapidly for large values of x. In real-life situations, such as population growth, this model is reliable only for relatively short periods of growth. Eventually, population growth tapers off because of pressures such as limited resources and overcrowding.

89. Quadratic models are better for revenue and profit functions where demand depends linearly on the price. Exponential models are better for compound interest and population growth.

91. Take the ratios y_2/y_1 and y_3/y_2. If they are the same, the points fit on an exponential curve.

93. This reasoning is suspect—the bank need not use its computer resources to update all the accounts every minute, but can instead use the continuous compounding formula to calculate the balance in any account at any time as needed.

2.3

1.

Exponential form	$10^4 = 10{,}000$	$4^2 = 16$	$3^3 = 27$	$5^1 = 5$	$7^0 = 1$	$4^{-2} = \dfrac{1}{16}$
Logarithmic form	$\log_{10} 10{,}000 = 4$	$\log_4 16 = 2$	$\log_3 27 = 3$	$\log_5 5 = 1$	$\log_7 1 = 0$	$\log_4 \dfrac{1}{16} = -2$

3.

Exponential form	$(0.5)^2 = 0.25$	$5^0 = 1$	$10^{-1} = 0.1$	$4^3 = 64$	$2^8 = 256$	$2^{-2} = \dfrac{1}{4}$
Logarithmic form	$\log_{0.5} 0.25 = 2$	$\log_5 1 = 0$	$\log_{10} 0.1 = -1$	$\log_4 64 = 3$	$\log_2 256 = 8$	$\log_2 \dfrac{1}{4} = -2$

5. $x = \log_3 5 = \log 5/\log 3 = 1.4650$

7. $-2x = \log_5 40 = \log 40/\log 5$, so $x = -(\log 40/\log 5)/2 = -1.1460$

9. $e^x = 2/4.16$, so $x = \ln(2/4.16) = -0.7324$

11. $1.06^{2x+1} = 11/5$, so $2x + 1 = \log_{1.06}(11/5)$, $x = (\log(11/5)/\log(1.06) - 1)/2 = 6.2657$

13. Let $y = \log_b x + C$. Substitute to get $2 = \log_b 1 + C$ and $4 = \log_b 9 + C$. The first equation gives $C = 2$ because $\log_b 1 = 0$ for every b. The second equation gives $4 = \log_b 9 + 2$, so $\log_b 9 = 2$, $b^2 = 9$, hence $b = 3$. Therefore, $y = \log_3 x + 2$.

15. Let $y = \log_b x + C$. Substitute to get $3 = \log_b 10 + C$ and $4 = \log_b 100 + C$. Subtract the first equation from the second to get $1 = \log_b 100 - \log_b 10 = \log_b(100/10) = \log_b 10$. So, $b^1 = 10$, i.e., $b = 10$. Substituting in the first equation gives $3 = \log 10 + C = 1 + C$, hence $C = 2$. Therefore, $y = \log x + 2$.

17. Let $y = \log_b x + C$. Substitute to get $25 = \log_b 10 + C$ and $41 = \log_b 20 + C$. Subtract the first equation from the second to get $16 = \log_b 20 - \log_b 10 = \log_b(20/10) = \log_b 2$. So, $b^{16} = 2$, i.e., $b = 2^{1/16} \approx 1.0443$. Substituting in the first equation gives $25 = \log_b 10 + C = \log(10)/\log(2^{1/16}) + C = 53.1508 + C$, hence $C = -28.1508$. Therefore, $y = \log_{1.0443} x - 28.1508$.

19. Substitute $A = 700$, $P = 500$, and $r = 0.10$ in the continuous compounding formula: $700 = 500e^{0.10t}$. Solve for t: $e^{0.10t} = 700/500 = 7/5$, $0.10t = \ln(7/5)$, $t = 10\ln(7/5) \approx 3.36$ years.

21. Substitute $A = 3P$ and $r = 0.10$ in the continuous compounding formula: $3P = Pe^{0.10t}$. Solve for t: $e^{0.10t} = (3P)/P = 3$, $0.10t = \ln 3$, $t = 10\ln 3 \approx 11$ years

23. If 99.95% has decayed, then 0.05% remains, so $C(t) = 0.0005A$. Therefore, $0.0005A = A(0.999879)^t$, $(0.999879)^t = 0.0005$, so $t = \log_{0.999879} 0.0005 = \log(0.0005)/\log(0.999879) \approx 62{,}814$ years old.

25. Substitute $A = 15{,}000$, $P = 10{,}000$, $r = 0.053$, and $m = 1$ into the compound interest formula: $15{,}000 = 10{,}000(1 + 0.053)^t$, $(1.053)^t = 1.5$, $t = \log_{1.053} 1.5 = \log(1.5)/\log(1.053) \approx 8$ years.

Section 2.3

27. Substitute $A = 20,000$, $P = 10,400$, $r = 0.052$, and $m = 12$ into the compound interest formula: $20,000 = 10,400(1 + 0.052/12)^{12t}$, $(1 + 0.052/12)^{12t} = 200/104$, $12t = \log(200/104)/\log(1 + 0.052/12) \approx 151$ months.

29. Substitute $A = 2P$, $r = 0.06$, and $m = 2$ into the compound interest formula: $2P = P(1 + 0.06/2)^{2t}$, $(1.03)^{2t} = 2$, $2t = \log 2/\log 1.03$, $t = (\log 2/\log 1.03)/2 \approx 12$ years.

31. Substitute $A = 2P$ and $r = 0.053$ in the continuous compounding formula: $2P = Pe^{0.053t}$, $e^{0.053t} = 2$, $0.053t = \ln 2$, $t = (\ln 2)/0.053 \approx 13.08$ years.

33. (a) If $y = Ab^t$, substitute $y = 3A$ and $t = 6$ to get $3A = Ab^6$, so $b = 3^{1/6} \approx 1.2009$. **(b)** Now substitute $y = 2A$ and solve for t: $2A = A(1.2009)^t$, so $t = \log 2/\log 1.2009 \approx 3.8$ months.

35. If $A(t)$ is the amount left after t million years, we first find a model of the form $A(t) = Pb^t$. If the half-life is 710 million years, we must have $b^{710} = 0.5$, so $b = (0.5)^{1/710}$. With $P = 10$ as given, our model is $A(t) = 10(0.5)^{t/710}$. Now we wish to find when $A(t) = 1$: $1 = 10(0.5)^{t/710}$, $t/710 = \log_{0.5}(0.1)$, $t = 710\log(0.1)/\log(0.5) \approx 2360$ million years.

37. We first find a model of the form $A(t) = Pb^t$. We are given the data points $(0, 1)$ and $(2, 0.7)$, so $P = 1$ and $b^2 = 0.7$, so $b = (0.7)^{1/2}$; the model is $A(t) = (0.7)^{t/2}$. We wish to find when $A(t) = 0.5$: $0.5 = (0.7)^{t/2}$, $t/2 = \log_{0.7}(0.5)$, $t = 2\log(0.5)/\log(0.7) \approx 3.89$ days.

39. Let y represent the amount of aspirin in the bloodstream at time t hours. Then $y = Ab^t$. The initial value is given to us as $A = 300$ mg. We are told that half the amount is removed every 2 hours, so $b^2 = 0.5$, giving $b = (0.5)^{1/2}$. Thus, $y = 300(0.5)^{t/2}$. We wish to find when $y = 100$: $100 = 300(0.5)^{t/2}$, $t/2 = \log_{0.5}(1/3)$, $t = 2\log(1/3)/\log(0.5) \approx 3.2$ hours.

41. (a) $E(t) = 0.4214 \ln t + 1.962$ **(b)** Yes; it predicts $3.24 billion. **(c) A:** The logarithm increases without bound.

43. $M(t) = 11.622 \ln t - 7.1358$. The model is unsuitable for large values of t because, for sufficiently large values of t, $M(t)$ will eventually become larger than 100%.

45. (a) Substitute: $8.2 = (2/3)(\log E - 4.4)$, $\log E = (3/2)\times 8.2 + 4.4 = 16.7$, $E = 10^{16.7} \approx 5.012\times 10^{16}$ joules of energy **(b)** Compute the energy released as in (a): $7.1 = (2/3)(\log E - 4.4)$, $\log E = (3/2)\times 7.1 + 4.4 = 15.05$, $E = 10^{15.05} \approx 1.122\times 10^{15}$ joules. Comparing, the energy released in 1989 was $100\times(1.122\times 10^{15})/(5.012\times 10^{16}) \approx 2.24\%$ of the energy released in 1906. **(c)** Solving for E in terms of R we get $E = 10^{1.5R + 4.4}$. This gives $E_2/E_1 = 10^{1.5R_2 + 4.4}/10^{1.5R_1 + 4.4} = 10^{1.5(R_2 - R_1)}$. **(d)** <u>1000</u>: If $R_2 - R_1 = 2$, then $E_2/E_1 = 10^3 = 1000$.

47. (a) By substitution: 75 dB, 69 dB, 61 dB **(b)** $D = 10 \log(320\times 10^7) - 10 \log r^2 \approx 95 - 20 \log r$ **(c)** Solve $0 = 95 - 20 \log r$: $\log r = 95/20$, $r = 10^{95/20} \approx 57,000$ feet (rounding up so that we're beyond the point where the decibel level is 0).

49. The logarithm of a negative number, were it defined, would be the power to which a base must be raised to give that negative number. But raising a base to a power never results in a negative

Section 2.3

number, so there can be no such number as the logarithm of a negative number.

51. $\log_4 y$

53. Any logarithmic curve $y = \log_b t + C$ will eventually surpass 100%, and hence not be suitable as a long-term predictor of market share.

55. Time is increasing logarithmically with population: Solving $P = Ab^t$ for t gives $t = \log_b(P/A) = \log_b P - \log_b A$, which is of the form $t = \log_b P + C$.

57. 8. Note that $b^{\log_b a} = a$ for any a and b, since $\log_b a$ is the power to which you raise b to get a.

59. x. To what power to you raise e to get e^x? x!

2.4

1. $N = 7$, $A = 6$, $b = 2$; `7/(1+6*2^-x)`

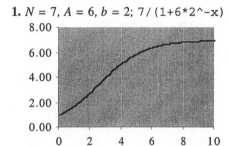

3. $N = 10$, $A = 4$, $b = 0.3$;
`10/(1+4*0.3^-x)`

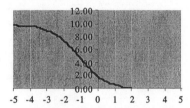

5. $N = 4$, $A = 7$, $b = 1.5$; `4/(1+7*1.5^-x)`

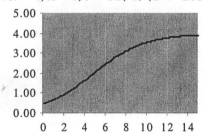

7. $N = 200$, the limiting value. $10 = f(0) = N/(1 + A) = 200/(1 + A)$, so $A = 19$. For small values of x, $f(x) \approx 10b^x$; to double with every increase of 1 in x we must therefore have $b = 2$. This gives $f(x) = \dfrac{200}{1 + 19 \cdot (2^{-x})}$.

9. $N = 6$, the limiting value. $3 = f(0) = N/(1 + A) = 6/(1 + A)$, so $A = 1$. $4 = f(1) = 6/(1 + b^{-1})$, $1 + b^{-1} = 6/4 = 3/2$, $b^{-1} = 1/2$, $b = 2$. So, $f(x) = \dfrac{6}{1 + 2^{-x}}$.

11. B: The limiting value is 9, eliminating (A). The initial value is $2 = 9/(1 + 3.5)$, not $9/(1 + 0.5)$, so the answer is (B), not (C).

13. B: The graph decreases, so $b < 1$, eliminating (C). The y intercept is $2 = 8/(1 + 3)$ as in (B).

15. C: The initial value is 2, as in (A) or (C). If b were 5, an increase in 1 in x would multiply the value of $f(x)$ by approximately 5 when x is small. However, increasing x from 0 to 10 does not quite double the value. Hence b must be smaller, as in (C).

17. $y = \dfrac{7.2}{1 + 2.4\,(1.05)^{-x}}$

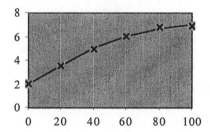

19. $y = \dfrac{97}{1 + 2.2(0.942)^{-x}}$

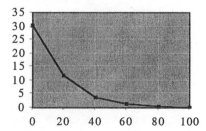

Section 2.4

21. $N = 10{,}000$, the susceptible population. $1000 = 10{,}000/(1 + A)$, so $A = 9$. In the initial stages, the rate of increase was 25% per day, so $b = 1.25$. Thus, $N(t) = \dfrac{10{,}000}{1 + 9(1.25)^{-t}}$. $N(7) \approx 3463$ cases.

23. $N = 3000$, the total available market. $100 = 3000/(1 + A)$, so $A = 29$. Sales are initially doubling every 5 days, so $b^5 = 2$, or $b = 2^{1/5}$. Thus, $N(t) = \dfrac{3000}{1 + 29(2^{1/5})^{-t}}$. Set $N(t) = 700$ and solve for t: $700 = 3000/[1 + 29(2^{1/5})^{-t}]$, $1 + 29(2^{1/5})^{-t} = 30/7$, $(2^{1/5})^{-t} = 23/(29 \times 7)$, $t = -\log[23/(29 \times 7)]/\log(2^{1/5}) \approx 16$ days.

25. (a) A: We can eliminate (C) and (D) because $N = 800$ is clearly too small. The value of $b = 1.96$ in (B) would imply that the values would initially almost double for every increase of 1 in t, but the growth is slower than that. **(b)** 1990 is in the initial period of the model, when the rate of growth is governed by b. Since $b = 1.16$, the rate of growth is 16% per year.

27. (a) For the extremely wealthy we look at large values of x, where $P(x)$ is close to $N = 91\%$. **(b)** $P(x) \approx [91/(1 + 5.35)](1.05)^x \approx 14.33(1.05)^x$. **(c)** Set $P(x) = 50$ and solve for x: $50 = 91/[1 + 5.35(1.05)^{-x}]$, $1 + 5.35(1.05)^{-x} = 91/50$, $x = -\log[41/(50 \times 5.35)]/\log(1.05) \approx \$38{,}000$.

29. (a) $N(t) = \dfrac{82.8}{1 + 21.8(7.14)^{-t}}$ The model predicts that books sales will level off at around 82.8 million books per year. **(b)** Not consistent; 15% of the market is represented by more than double the predicted value. This shows the difficulty in making long-term predictions from regression models obtained from a small amount of data. **(c)** Set $N(t) = 80$ and solve for t: $t = -\log[2.8/(80 \times 21.8)]/\log(7.14) \approx 3.3$, so book sales first exceed 80 million in 2001.

31. (a) $P(t) = \dfrac{5.7}{1 + 38.0\,(1.09)^{-t}}$. The graph suggests a good fit.

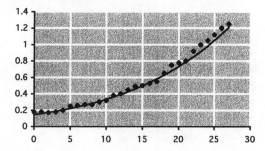

(b) Since $b = 1.09$, the growth rate is approximately 9% per year. **(c)** The long-term population is given by $N = 5.7$ million.

33. $N(t) = \dfrac{5}{1 + 1.081(1.056)^{-t}}$. Set $N(t) = 3.5$ and solve for t: $t = -\log[1.5/(3.5 \times 1.081)]/\log(1.056) \approx 17$, which represents the year 2010.

35. Just as diseases are communicated via the spread of a pathogen (such as a virus), new technology is communicated via the spread of information (such as advertising and publicity). Further, just as the spread of a disease is ultimately limited by the number of susceptible individuals, so the spread of a new technology is ultimately limited by the size of the potential market.

37. It can be used to predict where the sales of a new commodity might level off.

Chapter 2 Review Test

1. (a) $a = 1, b = 2, c = -3$, so $-b/(2a) = -1$; $f(-1) = -4$, so the vertex is $(-1, -4)$. y intercept: $c = -3$. $x^2 + 2x - 3 = (x + 3)(x - 1)$, so the x intercepts are -3 and 1. $a > 0$, so the parabola opens upward.

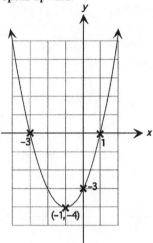

(b) $a = -1, b = -1, c = -1$, so $-b/(2a) = -1/2$; $f(-1/2) = -3/4$, so the vertex is $(-1/2, -3/4)$. y intercept: $c = -1$. $b^2 - 4ac = -3 < 0$, so there are no x intercepts. $a < 0$, so the parabola opens downward.

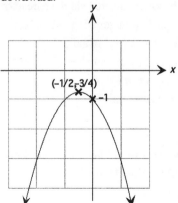

2. (a) $2 = Ab^1$ and $18 = Ab^3$. Dividing, $b^2 = 9$, so $b = 3$. Then $2 = 3A$, so $A = 2/3$: $f(x) = \frac{2}{3}3^x$

(b) $10 = Ab^1$ and $5 = Ab^3$. Dividing, $b^2 = 1/2$, so $b = 1/\sqrt{2}$. Then $10 = A/\sqrt{2}$, so $A = 10\sqrt{2}$: $f(x) = 10\sqrt{2}\left(\frac{1}{\sqrt{2}}\right)^x$

3. (a) $-2x = \log_3 4$, so $x = -\frac{1}{2}\log_3 4$ **(b)** $2x^2 - 1 = \log_2 2 = 1$, $x^2 = 1$, $x = \pm 1$ **(c)** $10^{3x} = 315/300 = 1.05$, $3x = \log 1.05$, $x = \frac{1}{3}\log 1.05$ **(d)** $(1 + i)^{mx} = A/P$, $mx = \log_{1+i}(A/P) = \ln(A/P)/\ln(1+i)$, $x = \dfrac{\ln(A/P)}{m \ln(1+i)}$

4. (a) $3000 = 2000(1 + 0.04/12)^{12t}$, $t = [\log(3/2)/\log(1 + 0.04/12)]/12 \approx 10.2$ years
(b) $3000 = 2000(1 + 0.0675/365)^{365t}$, $t = [\log(3/2)/\log(1 + 0.0675/365)]/365 \approx 6$ years
(c) $3000 = 2000e^{0.0375t}$, $t = \ln(3/2)/0.0375 \approx 10.8$ years

5. (a) $N = 900$ is given, $100 = 900/(1 + A)$ gives $A = 8$, initially increasing 50% per unit increase in x gives $b = 1.5$: $f(x) = \dfrac{900}{1 + 8(1.5)^{-x}}$ **(b)** $N = 25$ is given, $5 = 25/(1 + A)$ gives $A = 4$, $b = 1.1$ is given in the initial exponential form: $f(x) = \dfrac{25}{1 + 4(1.1)^{-x}}$ **(c)** $N = 20$ is given, $20/(1 + A) = 5$, so $A = 3$, decreasing at a rate of 20% per unit of x near 0 means $b = 1 - 0.2 = 0.8$: $f(x) = \dfrac{20}{1 + 3(0.8)^{-x}}$

6. (a) The largest volume will occur at the vertex: $-b/(2a) = 0.085/(2 \times 0.000005) \approx \$8500$ per month; substituting $c = 8500$ gives $h =$ an average of approximately 2100 hits per day.

(b) Solve $h = 0$ using the quadratic formula: $c \approx$ $29,049 per month (the other solution given by the quadratic formula is negative). (c) The fact that -0.000005, the coefficient of c^2, is negative.

7. (a) $R = pq = -60p^2 + 950p$. The maximum revenue occurs at the vertex: $p = -b/(2a) = 950/(2\times60) = \$7.92$ per novel. At that price the monthly revenue is $R = \$3760.42$. (b) $C = 900 + 4q = 900 + 4(-60p + 950) = -240p + 4700$, so $P = R - C = -60p^2 + 950p - (-240p + 4700) = -60p^2 + 1190p - 4700$. The maximum monthly profit occurs at the vertex: $p = -b/(2a) = 1190/(2\times60) = \$9.92$ per novel. At that price, the monthly profit is $P = \$1200.42$.

8. (a) Let S be the stock price and t be the time, in hours since the IPO. Model the stock price by $S = Ab^t$. Then $A = 10,000$ and $b^3 = 2$ (since the stock is doubling in price every 3 hours), so $b = 2^{1/3}$. After eight hours, $S = 10,000(2^{1/3})^8 = \$63,496.04$. (b) Solve $50,000 = 10,000(2^{t/3})$: $t = 3\log 5/\log 2 \approx 7.0$ hours. (c) After 10 hours the stock is worth $10,000(2^{10/3}) = \$100,793.68$. From this level it follows a new exponential curve with $A = 100,793.68$ and $b^4 = 2/3$ (it loses 1/3 of its value every 4 hours), hence $b = (2/3)^{1/4}$. Now solve $10,000 = 100,793.68(2/3)^{t/4}$ for $t = 4\log(10,000/100,793.68)/\log(2/3) \approx 22.8$. Adding the first 10 hours, the stock will be worth $10,000 again 32.8 hours after the IPO.

9. C: (A) is true because $L_1 = L_0 e^{-1/t} < L_0$. (B) is true because $L_{3t} = L_0 e^{-3} > 0$. (D) is true because increasing t decreases x/t, which increases $e^{-x/t}$ (makes it closer to 1), hence increases L_x for every x. (E) is true because $e^{-x/t}$ is never 0. On the other hand, (C) is false because $L_t = L_0 e^{-1} \approx 0.37L_0 < L_0/2$.

10. (a) Sales level off at $6053 + 4474 = 10,527$ per week. (b) Solve $10,000 = 6053 + 4474/(1 + e^{-0.55(t - 4.8)})$: $1 + e^{-0.55(t - 4.8)} = 4474/3947$, $-0.55(t - 4.8) = \ln(527/3947)$, $t = 4.8 - \ln(527/3947)/0.55 \approx 8.5$ weeks.

Chapter 2 Review Test

Chapter 3
3.1

1. $[f(3) - f(1)]/(3 - 1) = (-1 - 5)/2 = -3$

3. $[f(-1) - f(-1)]/[-1 - (-3)] = [-1.5 - (-2.1)]/2 = 0.3$

5. $[R(6) - R(2)]/(6 - 2) = (20.1 - 20.2)/4 = -\$25,000$ per month

7. $[q(5.5) - q(5)]/(5.5 - 5) = (300 - 400)/0.5 = -200$ items per dollar

9. $[S(6) - S(2)]/(6 - 2) = (15 - 20)/4 = -\$1.25$ per month

11. $[U(4) - U(0)]/(4 - 0) = (8 - 5)/4 = 0.75$ percentage point increase in unemployment per 1 percentage point increase in the deficit

13. $[f(3) - f(1)]/(3 - 1) = [6 - (-2)]/2 = 4$

15. $[f(0) - f(-2)]/[0 - (-2)] = (4 - 0)/2 = 2$

17. $[f(3) - f(2)]/(3 - 2) = [9/2 + 1/3 - (2 + 1/2)]/1 = 7/3$

19.

h	Ave. Rate of Change
1	2
0.1	0.2
0.01	0.02
0.001	0.002
0.0001	0.0002

21.

h	Ave. Rate of Change
1	−0.1667
0.1	−0.2381
0.01	−0.2488
0.001	−0.2499
0.0001	−0.24999

23.

h	Ave. Rate of Change
1	9
0.1	8.1
0.01	8.01
0.001	8.001
0.0001	8.0001

25. (a) $[P(10) - P(5)]/(10 - 5) = (132 - 117)/5 = 3$ million people per year. During the period 1995–2000, employment in the U.S. increased at an average rate of 3 million people per year. **(b)** $[P(11) - P(10)]/(11 - 10) = (132 - 132)/1 = 0$ people per year. During the period 2000–2001 the average rate of change of employment in the U.S. was zero people per year.

27. (a) Look for the largest increase in N over a period of two years; it occurs over the period 1998–2000. The number of companies that invested in venture capital each year was increasing most rapidly during the period 1998–2000, when it grew at an average rate of $[N(10) - N(8)]/(10 - 8) = (1700 - 400)/2 = 650$ companies per year. **(b)** Look for the least increase (or greatest decrease) in N over a period of two years; it occurs over the period 1999–2001. The number of companies that invested in venture

Section 3.1

capital each year was decreasing most rapidly during the period 1999–2001, when its rate of increase was $[N(11) - N(9)]/(11 - 9) = (900 - 1000)/2 = -50$, i.e., it decreased at an average rate of 50 companies per year.

29. (a) $(680 - 380)/4 = 75$ teams per year **(b)** It decreased: The slope of the graph from $t = 16$ to $t = 20$ is clearly less than the slope from $t = 14$ to $t = 18$.

31. (a) Check each interval ([7, 8], [7, 9], and so on). You will find the most negative average drop, in [7, 8] and [10, 11], is −40 thousand boats per year. So, during the periods 1997–1998 and 2000–2001, the number of recreational boats decreased at an average rate of 40 thousand boats per year. **(b)** The percentage rate of change is $[N(11) - N(9)]/N(9) = -50/590 \approx -0.0847$; the average rate of change is $[N(11) - N(9)]/(11 - 9) = -50/2 = -25$ thousand boats per year. Over the period 1999–2001, the number of boats sold decreased at an average rate of 25,000 boats per year, representing an 8.47% decrease over that period.

33. (a) $[N(1) - N(0)]/(1 - 0) = (600 - 350)/1 = 250$ million transactions per year; $[N(2) - N(1)]/(2 - 1) = (450 - 600)/1 = -150$ million transactions per year; $[N(2) - N(0)]/(2 - 0) = (450 - 350)/2 = 50$ million transactions per year. Over the period January 2000–January 2001, the (annual) number of on-line shopping transactions in the U.S. increased at an average rate of 250 million per year. From January 2001 to January 2002, this number decreased at an average rate of 150 million per year. From January 2000 to January 2002, this number increased at an average rate of 50 million per year. **(b)** The average rate of change of $N(t)$ over [0, 2] is the average of the rates of change over [0, 1] and [1, 2]: In this case 50 is the average of 250 and −150; in general the average of $[N(1) - N(0)]/(1 - 0)$ and $[N(2) - N(1)]/(2 - 1)$ equals $[N(2) - N(0)]/(2 - 0)$.

35. (a) (A): The slope of the graph is less than the slope of the regression line. **(b)** (C): The regression line passes through these two points so its slope is the same as the average rate of change. **(c)** (B): The graph is steeper than the regression line. **(d)** Approximately $(1.3 - 0.2)/(27 - 0) \approx 0.04$ million prisoners per year; this is roughly equal to the slope of the regression line.

37. From 1991 to 1995 the volatility decreased at an average rate of 0.2 points per year, so decreased a total of $4 \times 0.2 = 0.8$ points. Since its value in 1995 was 1.1, its value in 1991 must have been $1.1 + 0.8 = 1.9$. Similarly, from 1995 to 1999 the volatility increased a total of $4 \times 0.3 = 1.2$ to end at $1.1 + 1.2 = 2.3$. In between these points we've found almost anything could happen, but the graph might look something like the following:

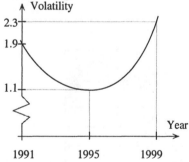

39. $[I(2) - I(0)]/(2 - 0) = (1600 - 1000)/2 = 300$. The index was increasing at an average rate of 300 points per day.

Section 3.1

41. $[n(10) - n(6)]/(10 - 6) = (159 - 127.8)/4 = 7.8$ thousand pools per year. The number of in-ground pools in the U.S. was growing at an average rate of about 7800 pools per year over the period 1996–2000.

43. $[I(8) - I(4)]/(8 - 4) = (593.64 - 380.4)/4 = 53.31$ thousand dollars per year. Between 1994 and 1998, the average after-tax income of the wealthiest 1% of U.S. taxpayers was increasing at an average rate of around $53,300 per year.

45. (a) $[f(6) - f(5)]/(6 - 5) = (27.6 - 18.75)/1 = 8.85$ manatee deaths per 100,000 boats; $[f(8) - f(7)]/(8 - 7) = (66.6 - 43.55)/1 = 23.05$ manatee deaths per 100,000 boats **(b)** More boats result in more manatee deaths per additional boat.

47. (a) $[R(2) - R(0)]/(2 - 0) = (760 - 150)/2 = \$305$ million per year. Over the period 1997–1999, annual advertising revenues increased at an average rate of $305 million per year. **(b)** (A): The average rate of change from 1998 to 1999 was larger than the average rate of change from 1997 to 1998. **(c)** $[R(3) - R(2)]/(2 - 0) = (1350 - 760)/1 = \$590$ million per year. The model projects annual advertising revenues to increase by $590 million per year in 2000.

49. The average rate of change of f over an interval $[a, b]$ can be determined numerically, using a table of values, graphically, by measuring the slope of the corresponding line segment through two points on the graph, or algebraically, using an algebraic formula for the function. Of these, the least precise is the graphical method, because it relies on reading coordinates of points on a graph.

51. Answers will vary. Here is one possibility:

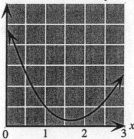

53. For every change of 1 in C, B changes by 3, so A changes by 2×3 = 6 units of quantity A per unit of quantity C

55. (A): The secant line given by $x = 1$ and $x = 1 + h$ is steeper for smaller values of h.

57. Yes. Here is an example, in which the average rate of growth for 2000–2003 is negative, but the average rates of growth for 2000–2001 and 2001–2002 are positive:

Year	2000	2001	2002	2003
Revenue (billion)	$10	$20	$30	$5

59. (A): This can be checked by algebra: $\{[f(2) - f(1)]/(2 - 1) + [f(3) - f(2)]/(3 - 2)\}/2 = [f(3) - f(1)]/(3 - 1)$.

Section 3.2

3.2

1. 6: The average rates of change are approaching 6 for both positive and negative values of h approaching 0.

3. −5.5: The average rates of change are approaching −5.5 for both positive and negative values of h approaching 0.

5.

h	1	0.1	0.01
Ave. rate	39	39.9	39.99

The average rates are approaching an instantaneous rate of 40 rupees per day.

7.

h	1	0.1	0.01
Ave. rate	140	66.2	60.602

The average rates are approaching an instantaneous rate of 60 rupees per day.

9.

h	10	1
C_{ave}	4.799	4.7999

$C'(1,000) = \$4.8$ per item

11.

h	10	1
C_{ave}	99.91	99.90

$C'(100) = \$99.9$ per item

In each of 13–17, the answer to (a) is the point at which the graph is rising with the steepest slope or falling with the shallowest slope; the answer to (b) is the point at which the graph is rising with the shallowest slope or falling with the steepest slope.

13. (a) R (b) P

15. (a) P (b) R

17. (a) Q (b) P

In each of 19 & 21, the answer is obtained by estimating the slope of the tangent line shown.

19. 1/2

21. 0

In each of 23 & 25, the answer may be obtained by estimating the slopes of the tangent lines to the given points and comparing these slopes to the given numbers.

23. (a) Q (b) R (c) P

25. (a) R (b) Q (c) P

27. (a) $(1, 0)$ (b) None; the graph never rises. (c) $(-2, 1)$

29. (a) $(-2, 0.3), (0, 0), (2, -0.3)$ (b) None; the graph never rises that steeply. (c) None; the graph never falls that steeply.

31. $(a, f(a)); f'(a)$.

33. (a) (A): The graph rises above the tangent line at $x = 2$. (b) (C): The secant line is roughly parallel to the tangent line at $x = 0$. (c) (B): The slopes of the tangent lines are decreasing. (d) (B): The slopes of the tangent lines decrease to 0 then increase again. (e) (C): The height of the graph is approximately 0.7 while the slope of the tangent line at $x = 4$ is approximately 1.

35. $[f(2 + 0.0001) - f(2 - 0.0001)]/0.0002 = -2$

37. $[f(-1 + 0.0001) - f(-1 - 0.0001)]/0.0002 \approx -1.5$

In each of 39–55 we use the "quick approximation" method of estimating the derivative using the balanced difference quotient. You could also use the ordinary difference quotient or a table of average rates of change with values of h approaching 0.

39. $[g(t + 0.0001) - g(t - 0.0001)]/0.0002 \approx -5$

41. $[y(2 + 0.0001) - y(2 - 0.0001)]/0.0002 = 16$

43. $[s(-2 + 0.0001) - s(-2 - 0.0001)]/0.0002 = 0$

45. $[R(20 + 0.0001) - R(20 - 0.0001)]/0.0002 \approx -0.0025$

47. (a) $[f(-1 + 0.0001) - f(-1 - 0.0001)]/0.0002 \approx 3$ **(b)** The equation of the line through $(-1, -1)$ with slope 3 is $y = 3x + 2$.

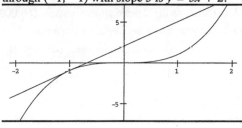

49. (a) $[f(2 + 0.0001) - f(2 - 0.0001)]/0.0002 \approx \frac{3}{4}$ **(b)** The equation of the line through $(2, 2.5)$ with slope $\frac{3}{4}$ is $y = \frac{3}{4}x + 1$.

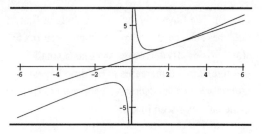

51. (a) $[f(4 + 0.0001) - f(4 - 0.0001)]/0.0002 \approx \frac{1}{4}$ **(b)** The equation of the line through $(4, 2)$ with slope $\frac{1}{4}$ is $y = \frac{1}{4}x + 1$.

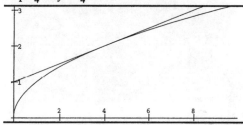

53. $[e^{0.0001} - e^{-0.0001}]/0.0002 \approx 1.000$

55. $[\ln(1 + 0.0001) - \ln(1 - 0.0001)]/0.0002 \approx 1.000$

57. (B): The derivative is the slope of the tangent line. It is not any particular average rate of change or difference quotient; these only *approximate* the derivative.

59. (C): The graph of f is a falling straight line, so must have the same negative slope (derivative) at every point; f' must be a negative constant.

61. (A): The function f decreases until $x = 0$, where it turns around and starts to increase. Its derivative must be negative until $x = 0$, where the derivative is 0; past that the derivative becomes positive. This is exactly what (A) illustrates.

Section 3.2

63. (F): The function f increases slowly at first, it becomes steeper around $x = 0$, then it returns to slowly rising. Its derivative starts as a small positive number, increases to become largest around $x = 0$, then decreases back toward 0. This is the behavior seen in (F).

65. The derivative:

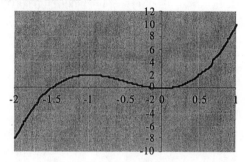

(a) None **(b)** $x = -1.5$, $x = 0$: These are the points where the derivative is 0 (crosses the x axis).

67. The derivative:

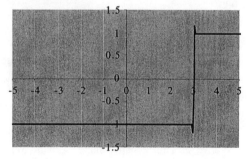

(a) $x = 3$: The sudden jump in value is a discontinuity and the derivative is not defined at $x = 3$. **(b)** None: The derivative is never 0.

69. The derivative:

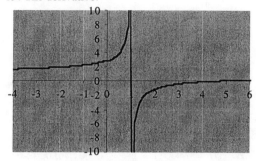

(a) $x = 1$: The sudden jump in value is a discontinuity and the derivative is not defined at $x = 1$. **(b)** $x = 4.2$: The derivative is 0 at approximately 4.2. The derivative is not 0 at $x = 1$; that is just a defect of the graphing technology.

71. $q(100) = 50{,}000$, $q'(100) = -500$ (use one of the quick approximations). A total of 50,000 pairs of sneakers can be sold at a price of $100, but the demand is decreasing at a rate of 500 pairs per $1 increase in the price.

73. (a) Sales in 2000 were approximately 160,000 pools per year and were increasing at a rate of 6000 per year. **(b)** Decreasing, since the slope is decreasing.

75. (a) (B): The graph is getting less steep. **(b)** (B): The graph goes from above the tangent line to below it, so the slope of the tangent line is greater than the average rate of change. **(c)** (A): From 0 to about 12 the graph is getting steeper, so the instantaneous rate of change is increasing; from that point on the graph is getting less steep, so the instantaneous rate of change is decreasing. **(d)** 1992: This is the point ($t = 12$) where the graph is steepest. **(e)** Reading values from the graph, we get the approximation $(1.2 - 0.8)/8 = 0.05$: In 1996, the total number of state prisoners

Section 3.2

was increasing at a rate of approximately 50,000 prisoners per year.

77. (a) $[s(4) - s(2)]/(4 - 2) = -96$ ft/sec (b) $[s(4 + 0.0001) - s(4 - 0.0001)]/0.0002 = -128$ ft/sec

79. (a) $[N(2) - N(1)]/(2 - 1) = -100$ million transactions per year. Over the period January 2001–January 2002, the (annual) number of on-line shopping transactions in the U.S. decreased at an average rate of 100 million per year.
(b) $[N(1 + 0.0001) - N(1 - 0.0001)]/0.0002 = 80$ million transactions per year. In January 2001, the (annual) number of on-line transactions was growing at a rate of about 80 million per year. (c) On-line sales slowed and then started to decrease sometime after January 2001.

81. (a) $[R(1 + 0.0001) - R(1 - 0.0001)]/0.0002 = \$305$ million per year. (b) (A): If we estimate the instantaneous rates of change at 1997, 1998, and 1999, we get $115 million per year, $305 million per year, and $495 million per year, respectively. We also know that the graph of R is a parabola opening upward, so it rises more and more steeply. (c) $[R(3 + 0.0001) - R(3 - 0.0001)]/0.0002 = \$685$ million/year. In December 2000, AOL's advertising revenue was projected to be increasing at a rate of $685 million per year.

83. $A(0) = 4.5$ million because A gives the number of subscribers; $A'(0) = 60{,}000$ because A' gives the rate at which the number of subscribers is changing.

85. (a) 60% of children can speak at the age of 10 months. At the age of 10 months, this percentage is increasing by 18.2 percentage points per month.
(b) As t increases, p approaches 100 percentage points (almost all children eventually learn to speak), and dp/dt approaches zero because the percentage stops increasing.

87. $S(5) \approx 109$, $\left.\dfrac{dS}{dt}\right|_{t=5} \approx [S(5 + 0.0001) - S(5 - 0.0001)]/0.0002 \approx 9.1$. After 5 weeks, sales are 109 pairs of sneakers per week, and sales are increasing at a rate of 9.1 pairs per week each week.

89. (a) $P(50) \approx 62$, $P'(50) \approx [P(50 + 0.0001) - P(50 - 0.0001)]/0.0002 \approx 0.96$; 62% of U.S. households with an income of $50,000 have a computer. This percentage is increasing at a rate of 0.96 percentage points per $1000 increase in household income. (b) As with any logistic function, the graph levels off for large values of x, so P' decreases toward zero.

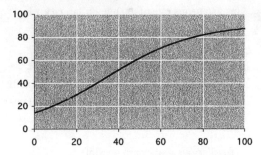

91.

Section 3.2

(a) (D): The graph of the derivative is rising. **(b)** 33 days after the egg was laid: That is where the graph of the derivative is highest. **(c)** 50 days after the egg was laid: That is where the graph of the derivative is lowest in the range $20 \le t \le 50$.

93. $L(0.95) \approx 31.2$ meters and $L'(0.95) \approx [L(0.95 + 0.0001) - L(0.95 - 0.0001)]/0.0002 \approx -304.2$ meters/warp. Thus, at a speed of warp 0.95, the spaceship has an observed length of 31.2 meters and its length is decreasing at a rate of 304.2 meters per unit warp, or 3.042 meters per increase in speed of 0.01 warp.

95. Company B. Although the company is currently losing money, the derivative is positive, showing that the profit is increasing. Company A, on the other hand, has profits that are declining.

97. (C) is the only graph in which the instantaneous rate of change on January 1 is greater than the one-month average rate of change.

99. The tangent to the graph is horizontal at that point, and so the graph is almost horizontal near that point.

101. Various graphs are possible.

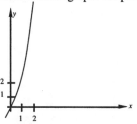

103. The difference quotient is not defined when $h = 0$ because there is no such number as $0/0$.

105. If $f(x) = mx + b$, then its average rate of change over any interval $[x, x+h]$ is
$$\frac{m(x+h) + b - (mx + b)}{h} = m.$$
Since this does not depend on h, the instantaneous rate is also equal to m.

107. Increasing, since the average rate of change appears to be rising as we get closer to 5 from the left (see the bottom row).

109. (B): His average speed was 60 miles per hour. If his instantaneous speed was always 55 mph or less, he could not have averaged more than 55 mph.

111. Answers will vary. Graph:

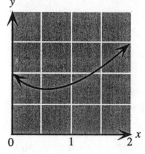

113. The derivative is positive (sales are still increasing) and decreasing toward zero (sales are leveling off).

Section 3.2

115. Answers may vary.

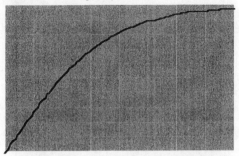

117.

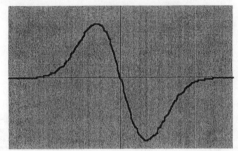

3.3

1. $f'(2) = \lim_{h \to 0} \dfrac{f(2+h) - f(2)}{h} = \lim_{h \to 0} \dfrac{(2+h)^2 + 1 - (2^2 + 1)}{h} = \lim_{h \to 0} \dfrac{4 + 4h + h^2 + 1 - 5}{h} =$
$\lim_{h \to 0} \dfrac{4h + h^2}{h} = \lim_{h \to 0} (4 + h) = 4$

$f'(x) = \lim_{h \to 0} \dfrac{f(x+h) - f(x)}{h} = \lim_{h \to 0} \dfrac{(x+h)^2 + 1 - (x^2 + 1)}{h} = \lim_{h \to 0} \dfrac{x^2 + 2xh + h^2 + 1 - x^2 - 1}{h} =$
$\lim_{h \to 0} \dfrac{2xh + h^2}{h} = \lim_{h \to 0} (2x + h) = 2x$

3. $f'(-1) = \lim_{h \to 0} \dfrac{f(-1+h) - f(-1)}{h} = \lim_{h \to 0} \dfrac{3(-1 + h) - 4 - (-3 - 4)}{h} = \lim_{h \to 0} \dfrac{-3 + 3h - 4 + 7}{h} =$
$\lim_{h \to 0} \dfrac{3h}{h} = \lim_{h \to 0} 3 = 3$

$f'(x) = \lim_{h \to 0} \dfrac{f(x+h) - f(x)}{h} = \lim_{h \to 0} \dfrac{3(x + h) - 4 - (3x - 4)}{h} = \lim_{h \to 0} \dfrac{3x + 3h - 4 - 3x + 4}{h} =$
$\lim_{h \to 0} \dfrac{3h}{h} = \lim_{h \to 0} 3 = 3$

5. $f'(1) = \lim_{h \to 0} \dfrac{f(1+h) - f(1)}{h} = \lim_{h \to 0} \dfrac{3(1+h)^2 + (1+h) - (3 + 1)}{h} = \lim_{h \to 0} \dfrac{3 + 6h + 3h^2 + 1 + h - 4}{h}$
$= \lim_{h \to 0} \dfrac{7h + 3h^2}{h} = \lim_{h \to 0} (7 + 3h) = 7$

$f'(x) = \lim_{h \to 0} \dfrac{f(x+h) - f(x)}{h} = \lim_{h \to 0} \dfrac{3(x+h)^2 + (x+h) - (3x^2 + x)}{h}$
$\lim_{h \to 0} \dfrac{3x^2 + 6xh + 3h^2 + x + h - 3x^2 - x}{h} = \lim_{h \to 0} \dfrac{6xh + 3h^2 + h}{h} = \lim_{h \to 0} (6x + 3h + 1) = 6x + 1$

7. $f'(-1) = \lim_{h \to 0} \dfrac{f(-1+h) - f(-1)}{h} = \lim_{h \to 0} \dfrac{2(-1+h) - (-1+h)^2 - (-2 - 1)}{h} =$
$\lim_{h \to 0} \dfrac{-2 + 2h - 1 + 2h - h^2 + 3}{h} = \lim_{h \to 0} \dfrac{4h - h^2}{h} = \lim_{h \to 0} (4 - h) = 4$

$f'(x) = \lim_{h \to 0} \dfrac{f(x+h) - f(x)}{h} = \lim_{h \to 0} \dfrac{2(x+h) - (x+h)^2 - (2x - x^2)}{h}$
$\lim_{h \to 0} \dfrac{2x + 2h - x^2 - 2xh - h^2 - 2x + x^2}{h} = \lim_{h \to 0} \dfrac{2h - 2xh - h^2}{h} = \lim_{h \to 0} (2 - 2x - h) = 2 - 2x$

Section 3.3

9. $f'(2) = \lim\limits_{h \to 0} \dfrac{f(2+h) - f(2)}{h} = \lim\limits_{h \to 0} \dfrac{(2+h)^3 + 2(2+h) - (8 + 4)}{h} =$

$\lim\limits_{h \to 0} \dfrac{8 + 12h + 6h^2 + h^3 + 4 + 2h - 12}{h} = \lim\limits_{h \to 0} \dfrac{14h + 6h^2 + h^3}{h} = \lim\limits_{h \to 0} (14 + 6h + h^2) = 14$

$f'(x) = \lim\limits_{h \to 0} \dfrac{f(x+h) - f(x)}{h} = \lim\limits_{h \to 0} \dfrac{(x+h)^3 + 2(x+h) - (x^3 + 2x)}{h} =$

$\lim\limits_{h \to 0} \dfrac{x^3 + 3x^2h + 3xh^2 + h^3 + 2x + 2h - x^3 - 2x}{h} = \lim\limits_{h \to 0} \dfrac{3x^2h + 3xh^2 + h^3 + 2h}{h} =$

$\lim\limits_{h \to 0} (3x^2 + 3xh + h^2 + 2) = 3x^2 + 2$

11. $f'(43) = \lim\limits_{h \to 0} \dfrac{f(43+h) - f(43)}{h} = \lim\limits_{h \to 0} \dfrac{m(43+h) + b - (43m + b)}{h} =$

$\lim\limits_{h \to 0} \dfrac{43m + mh + b - 43m - b}{h} = \lim\limits_{h \to 0} \dfrac{mh}{h} = \lim\limits_{h \to 0} m = m$

$f'(x) = \lim\limits_{h \to 0} \dfrac{f(x+h) - f(x)}{h} = \lim\limits_{h \to 0} \dfrac{m(x+h) + b - (mx + b)}{h} = \lim\limits_{h \to 0} \dfrac{mx + mh + b - mx - b}{h} =$

$\lim\limits_{h \to 0} \dfrac{mh}{h} = \lim\limits_{h \to 0} m = m$

13. $f'(1) = \lim\limits_{h \to 0} \dfrac{f(1+h) - f(1)}{h} = \lim\limits_{h \to 0} \dfrac{-1/(1+h) - (-1)}{h} = \lim\limits_{h \to 0} \dfrac{-1 + (1 + h)}{h(1 + h)} = \lim\limits_{h \to 0} \dfrac{h}{h(1 + h)} =$

$\lim\limits_{h \to 0} \dfrac{1}{1 + h} = 1$

$f'(x) = \lim\limits_{h \to 0} \dfrac{f(x+h) - f(x)}{h} = \lim\limits_{h \to 0} \dfrac{-1/(x+h) - (-1/x)}{h} = \lim\limits_{h \to 0} \dfrac{-x + (x + h)}{hx(x + h)} = \lim\limits_{h \to 0} \dfrac{h}{hx(x + h)} =$

$\lim\limits_{h \to 0} \dfrac{1}{x(x + h)} = \dfrac{1}{x^2}$

15. $R'(2) = \lim\limits_{h \to 0} \dfrac{R(2+h) - R(2)}{h} = \lim\limits_{h \to 0} \dfrac{-0.3(2+h)^2 - (-0.3 \times 2^2)}{h} = \lim\limits_{h \to 0} \dfrac{-1.2 - 1.2h - 0.3h^2 + 1.2}{h}$

$= \lim\limits_{h \to 0} \dfrac{-1.2h - 0.3h^2}{h} = \lim\limits_{h \to 0} (-1.2 - 0.3h) = -1.2$

17. $U'(3) = \lim\limits_{h \to 0} \dfrac{U(3+h) - U(3)}{h} = \lim\limits_{h \to 0} \dfrac{5.1(3+h)^2 + 5.1 - (5.1 \times 9 + 5.1)}{h} =$

$\lim\limits_{h \to 0} \dfrac{45.9 + 30.6h + 5.1h^2 + 5.1 - 51}{h} = \lim\limits_{h \to 0} \dfrac{30.6h + 5.1h^2}{h} = \lim\limits_{h \to 0} (30.6 + 5.1h) = 30.6$

Section 3.3

19. $U'(1) = \lim_{h\to 0} \dfrac{U(1+h) - U(1)}{h} = \lim_{h\to 0} \dfrac{-1.3(1+h)^2 - 4.5(1+h) - (-1.3 - 4.5)}{h} =$

$\lim_{h\to 0} \dfrac{-1.3 - 2.6h - 1.3h^2 - 4.5 - 4.5h + 5.8}{h} = \lim_{h\to 0} \dfrac{-7.1h - 1.3h^2}{h} = \lim_{h\to 0} (-7.1 - 1.3h) = -7.1$

21. $L'(1.2) = \lim_{h\to 0} \dfrac{L(1.2+h) - L(1.2)}{h} = \lim_{h\to 0} \dfrac{4.25(1.2+h) - 5.01 - (4.25 \times 1.2 - 5.01)}{h} =$

$\lim_{h\to 0} \dfrac{5.1 + 4.25h - 5.01 - 5.1 + 5.01}{h} = \lim_{h\to 0} \dfrac{4.25h}{h} = \lim_{h\to 0} 4.25 = 4.25$

23. $q'(2) = \lim_{h\to 0} \dfrac{q(2+h) - q(2)}{h} = \lim_{h\to 0} \dfrac{2.4/(2+h) - 2.4/2}{h} = \lim_{h\to 0} \dfrac{4.8 - 2.4(2+h)}{2h(2+h)} =$

$\lim_{h\to 0} \dfrac{-2.4h}{2h(2+h)} = \lim_{h\to 0} \dfrac{-1.2}{h(2+h)} = -0.6$

25. Find the slope by finding the derivative:

$m = f'(2) = \lim_{h\to 0} \dfrac{f(2+h) - f(2)}{h} = \lim_{h\to 0} \dfrac{(2+h)^2 - 3 - (2^2 - 3)}{h} = \lim_{h\to 0} \dfrac{4 + 4h + h^2 - 3 - 1}{h} =$

$\lim_{h\to 0} \dfrac{4h + h^2}{h} = \lim_{h\to 0} (4 + h) = 4$. The tangent line has slope 4 and goes through $(2, f(2)) = (2, 1)$, so has equation $y = 4x - 7$.

27. Find the slope by finding the derivative:

$m = f'(3) = \lim_{h\to 0} \dfrac{f(3+h) - f(3)}{h} = \lim_{h\to 0} \dfrac{-2(3+h) - 4 - (-2 \times 3 - 4)}{h} = \lim_{h\to 0} \dfrac{-6 - 2h - 4 + 10}{h} =$

$\lim_{h\to 0} \dfrac{-2h}{h} = \lim_{h\to 0} (-2) = -2$. The tangent line has slope -2 and goes through $(3, f(3)) = (3, -10)$, so has equation $y = -2x - 4$.

29. Find the slope by finding the derivative:

$m = f'(-1) = \lim_{h\to 0} \dfrac{f(-1+h) - f(-1)}{h} = \lim_{h\to 0} \dfrac{(-1+h)^2 - (-1+h) - [(-1)^2 - (-1)]}{h} =$

$\lim_{h\to 0} \dfrac{1 - 2h + h^2 + 1 - h - 2}{h} = \lim_{h\to 0} \dfrac{-3h + h^2}{h} = \lim_{h\to 0} (-3 + h) = -3$. The tangent line has slope -3 and goes through $(-1, f(-1)) = (-1, 2)$, so has equation $y = -3x - 1$.

Section 3.3

31. (a) $f'(1) = 1/3$:

h	$\dfrac{f(a+h) - f(a)}{h}$
−1	1
−0.1	0.34510615
−0.01	0.33445066
−0.001	0.33344451
−0.0001	0.33334445
1	0.25992105
0.1	0.32280115
0.01	0.33222835
0.001	0.33322228
0.0001	0.33332222

(b) f is not differentiable at 0:

h	$\dfrac{f(a+h) - f(a)}{h}$
−1	1
−0.1	4.64158883
−0.01	21.5443469
−0.001	100
−0.0001	464.158883
1	1
0.1	4.64158883
0.01	21.5443469
0.001	100
0.0001	464.158883

33. (a) Not differentiable at 1:

h	$\dfrac{f(a+h) - f(a)}{h}$
−1	0
−0.1	−4.4814047
−0.01	−21.472292
−0.001	−99.966656
−0.0001	−464.14341
1	−1.259921
0.1	−4.7914199
0.01	−21.615923
0.001	−100.03332
0.0001	−464.17435

(b) Not differentiable at 0:

h	$\dfrac{f(a+h) - f(a)}{h}$
−1	1.25992105
−0.1	4.79141986
−0.01	21.6159233
−0.001	100.033322
−0.0001	464.174355
1	0
0.1	4.48140475
0.01	21.4722917
0.001	99.9666555
0.0001	464.143411

Section 3.3

35. (a) Not differentiable at -1:

h	$\dfrac{f(a+h) - f(a)}{h}$
-1	2
-0.1	7.30957344
-0.01	40.8107171
-0.001	252.188643
-0.0001	1585.89319
1	2
0.1	7.30957344
0.01	40.8107171
0.001	252.188643
0.0001	1585.89319

(b) $f'(1) \approx 1.1149$

h	$\dfrac{f(a+h) - f(a)}{h}$
-1	1.14869835
-0.1	1.11723866
-0.01	1.11510027
-0.001	1.11489282
-0.0001	1.11487213
1	1.09703258
0.1	1.11263904
0.01	1.11464078
0.001	1.11484687
0.0001	1.11486754

37. (a) $\lim\limits_{h\to 0} \dfrac{f(1+h) - f(1)}{h} = \lim\limits_{h\to 0} \dfrac{|1+h| + (1+h) - (|1| + 1)}{h} = \lim\limits_{h\to 0} \dfrac{1 + h + 1 + h - 2}{h}$ (because $|1+h| = 1 + h$ for h near 0) $= \lim\limits_{h\to 0} \dfrac{2h}{h} = 2$, so $f'(1) = 2$ **(b)** $\lim\limits_{h\to 0} \dfrac{f(0+h) - f(0)}{h} = \lim\limits_{h\to 0} \dfrac{|h| + h - 0}{h} = \lim\limits_{h\to 0} \left(\dfrac{|h|}{h} + 1\right)$ does not exist (see Example 5), so f is not differentiable at 0.

39. (a) $\lim\limits_{h\to 0} \dfrac{f(-3+h) - f(-3)}{h} = \lim\limits_{h\to 0} \dfrac{|-3 + h - 3| - |-3 + 3|}{h} = \lim\limits_{h\to 0} \dfrac{|h|}{h}$ does not exist, so f is not differentiable at -3. **(b)** $\lim\limits_{h\to 0} \dfrac{f(3+h) - f(3)}{h} = \lim\limits_{h\to 0} \dfrac{|3 + h + 3| - |3 + 3|}{h} = \lim\limits_{h\to 0} \dfrac{|6 + h| - 6}{h} = \lim\limits_{h\to 0} \dfrac{6 + h - 6}{h}$ (because $|6 + h| = 6 + h$ for h near 0) $= \lim\limits_{h\to 0} \dfrac{h}{h} = 1$, so $f'(3) = 1$.

41. (a) $\lim\limits_{h\to 0} \dfrac{f(1+h) - f(1)}{h} = \lim\limits_{h\to 0} \dfrac{|1 + h| + |1 + h - 1| - (|1| + |1 - 1|)}{h} = \lim\limits_{h\to 0} \dfrac{1 + h + |h| - 1}{h}$ (because $|1 + h| = 1 + h$ for h near 0) $= \lim\limits_{h\to 0} \dfrac{h + |h|}{h} = \lim\limits_{h\to 0} \left(1 + \dfrac{|h|}{h}\right)$ does not exist, so f is not differentiable at 1. **(b)** $\lim\limits_{h\to 0} \dfrac{f(0+h) - f(0)}{h} = \lim\limits_{h\to 0} \dfrac{|0 + h| + |0 + h - 1| - (|0| + |0 - 1|)}{h} = \lim\limits_{h\to 0} \dfrac{|h| - h + 1 - 1}{h}$ (because $|h - 1| = -h + 1$ for h near 0) $= \lim\limits_{h\to 0} \dfrac{|h| - h}{h} = \lim\limits_{h\to 0} \left(\dfrac{|h|}{h} - 1\right)$ does not exist, so f is not differentiable at 0.

Section 3.3

43. $n'(10) = \lim\limits_{h \to 0} \dfrac{n(10+h) - n(10)}{h} =$

$\lim\limits_{h \to 0} \dfrac{-0.45(10+h)^2 + 15(10+h) + 54 - (-0.45 \times 10^2 + 15 \times 10 + 54)}{h} = \lim\limits_{h \to 0} \dfrac{6h - 0.45h^2}{h} = 6$; the number of in-ground swimming pools built in the U.S. was increasing by 6000 pools per year in 2000.

45. $s'(t) = \lim\limits_{h \to 0} \dfrac{s(t+h) - s(t)}{h} = \lim\limits_{h \to 0} \dfrac{400 - 16(t+h)^2 - (400 - 16t^2)}{h} = \lim\limits_{h \to 0} \dfrac{-32th - 16h^2}{h} = -32t$ ft/sec; $s'(4) = -128$ ft/sec

47. $N'(2) = \lim\limits_{h \to 0} \dfrac{N(2+h) - N(2)}{h} =$

$\lim\limits_{h \to 0} \dfrac{18{,}000(2+h)^2 - 10{,}000(2+h) + 320{,}000 - (18{,}000 \times 2^2 - 10{,}000 \times 2 + 320{,}000)}{h} =$

$\lim\limits_{h \to 0} \dfrac{62{,}000h + 18{,}000h^2}{h} = 62{,}000$ SUVs per year each year

49. $f'(8) = \lim\limits_{h \to 0} \dfrac{f(8+h) - f(8)}{h} = \lim\limits_{h \to 0} \dfrac{3.55(8+h)^2 - 30.2(8+h) + 81 - (3.55 \times 8^2 - 30.2 \times 8 + 81)}{h} =$

$\lim\limits_{h \to 0} \dfrac{26.6h + 3.55h^2}{h} = 26.6$ manatee deaths per 100,000 additional boats. At a level of 800,000 boats, the number of manatee deaths is increasing at a rate of 26.6 manatees per 100,000 additional boats.

51. The algebraic method, since it gives the exact value of the derivative. The other two approaches give only approximate values (except in some special cases).

53. Since the algebraic computation of $f'(a)$ is exact, and not an approximation, it makes no difference whether one uses the balanced difference quotient or the ordinary difference quotient in the algebraic computation.

55. The computation results in a limit that cannot be evaluated.

3.4

1. $5x^4$

3. $-4x^{-3}$

5. $-0.25x^{-0.75}$

7. $8x^3+9x^2$

9. $-1-1/x^2$

11. $\dfrac{dy}{dx} = 10(0) = 0$ (constant multiple and power rule)

13. $\dfrac{dy}{dx} = \dfrac{d}{dx}(x^2) + \dfrac{d}{dx}(x)$ (sum rule) $= 2x + 1$ (power rule)

15. $\dfrac{dy}{dx} = \dfrac{d}{dx}(4x^3) + \dfrac{d}{dx}(2x) - \dfrac{d}{dx}(1)$ (sum and difference) $= 4\dfrac{d}{dx}(x^3) + 2\dfrac{d}{dx}(x) - \dfrac{d}{dx}(1)$ (constant multiples) $= 12x^2 + 2$ (power rule)

17. $\dfrac{dy}{dx} = \dfrac{d}{dx}(x^{104}) - \dfrac{d}{dx}(99x^2) + \dfrac{d}{dx}(x)$ (sums and differences) $= 104x^{103} - 99\dfrac{d}{dx}(x^2) + 1$ (constant multiples and power rule) $= 104x^{103} - 198x + 1$ (power rule)

19. $f'(x) = 2x - 3$

21. $f'(x) = 1 + 0.5x^{-0.5}$

23. $g'(x) = -2x^{-3} + 3x^{-2}$

25. $g'(x) = \dfrac{d}{dx}(x^{-1} - x^{-2}) = -x^{-2} + 2x^{-3} = -\dfrac{1}{x^2} + \dfrac{2}{x^3}$

27. $h'(x) = \dfrac{d}{dx}(2x^{-0.4}) = -0.8x^{-1.4} = -\dfrac{0.8}{x^{1.4}}$

29. $h'(x) = \dfrac{d}{dx}(x^{-2} + 2x^{-3}) = -2x^{-3} - 6x^{-4} = -\dfrac{2}{x^3} - \dfrac{6}{x^4}$

31. $r'(x) = \dfrac{d}{dx}\left(\dfrac{2}{3}x^{-1} - \dfrac{1}{2}x^{-0.1}\right) = -\dfrac{2}{3}x^{-2} + \dfrac{0.1}{2}x^{-1.1} = -\dfrac{2}{3x^2} + \dfrac{0.1}{2x^{1.1}}$

33. $r'(x) = \dfrac{d}{dx}\left(\dfrac{2}{3}x - \dfrac{1}{2}x^{0.1} + \dfrac{4}{3}x^{-1.1} - 2\right) = \dfrac{2}{3} - \dfrac{0.1}{2}x^{-0.9} - \dfrac{4.4}{3}x^{-2.1} = \dfrac{2}{3} - \dfrac{0.1}{2x^{0.9}} - \dfrac{4.4}{3x^{2.1}}$

35. $t'(x) = \dfrac{d}{dx}(|x| + x^{-1}) = |x|/x - x^{-2} = |x|/x - 1/x^2$

37. $s'(x) = \dfrac{d}{dx}(x^{1/2} + x^{-1/2}) = \dfrac{1}{2}x^{-1/2} - \dfrac{1}{2}x^{-3/2} = \dfrac{1}{2\sqrt{x}} - \dfrac{1}{2x\sqrt{x}}$

39. $s'(x) = \dfrac{d}{dx}(x^3 - 1) = 3x^2$

41. $t'(x) = \dfrac{d}{dx}(x - 2x^2) = 1 - 4x$

43. $2.6x^{0.3} + 1.2x^{-2.2}$

45. $1.2(1-|x|/x)$

47. $3at^2 - 4a$ (Remember to treat a as a constant, i.e., a number.)

Section 3.4

49. $5.15x^{9.3} - 99x^{-2}$

51. $\dfrac{ds}{dt} = \dfrac{d}{dt}(2.3 + 2.1t^{-1.1} - \dfrac{1}{2}t^{0.6}) =$
$-2.31t^{-2.1} + 0.3t^{-0.4} = -\dfrac{2.31}{t^{2.1}} - \dfrac{0.3}{t^{0.4}}$

53. $4\pi r^2$

In 55–59, we need to find the derivative at the indicated value of x or t.

55. $f'(x) = 3x^2$, so $f'(-1) = 3$

57. $f'(x) = -2$, so $f'(2) = -2$

59. $g'(t) = \dfrac{d}{dt}t^{-5} = -5t^{-6} = -\dfrac{5}{t^6}$, so $g'(1) = -5$

61. $f'(x) = 3x^2$, so $f'(-1) = 3$. The line with slope 3 passing through $(-1, f(-1)) = (-1, -1)$ is $y = 3x + 2$.

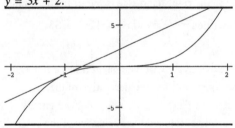

63. $f(x) = x + x^{-1}$, so $f'(x) = 1 - x^{-2} = 1 - \dfrac{1}{x^2}$;
$f'(2) = 1 - 1/4 = 3/4$. The line with slope 3/4 passing through $(2, f(2)) = (2, 5/2)$ is $y = \dfrac{3}{4}x + 1$

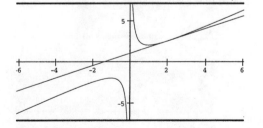

65. $f(x) = x^{1/2}$, so $f'(x) = \dfrac{1}{2}x^{-1/2} = \dfrac{1}{2\sqrt{x}}$; $f'(4) = 1/4$. The line with slope 1/4 passing through $(4, f(4)) = (4, 2)$ is $y = \dfrac{1}{4}x + 1$.

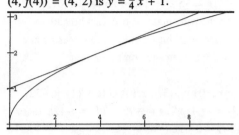

In 67–71 we need to find all values of x (if any) where the derivative is 0.

67. $y' = 4x + 3 = 0$ when $x = -3/4$

69. $y' = 2$ is never 0, so there are no such values of x

71. $y' = 1 - x^{-2} = 1 - 1/x^2 = 0$ when $x^2 = 1$, so $x = 1$ or -1

73. $\dfrac{d}{dx}x^4 = \lim_{h \to 0}\dfrac{(x+h)^4 - x^4}{h} =$
$\lim_{h \to 0}\dfrac{x^4 + 4x^3h + 6x^2h^2 + 4xh^3 + h^4 - x^4}{h} =$
$\lim_{h \to 0}\dfrac{4x^3h + 6x^2h^2 + 4xh^3 + h^4}{h} =$
$\lim_{h \to 0}(4x^3 + 6x^2h + 4xh^2 + h^3) = 4x^3$

75. (a) $s'(t) = 3.04t + 9.45$

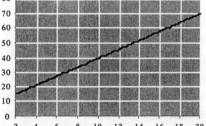

Section 3.4

(b) From the graph, the first year in which $s'(t)$ was larger than 50 was 1994–1995 ($t = 14$).

77. $P'(t) = -5.2t + 13$. In January 2002, $P'(2) = 2.6$, so the percentage of people who had purchased anything on-line was increasing at a rate of 2.6 percentage points per year.

79. (a) $s'(t) = -32t$; $s'(0) = 0$, $s'(1) = -32$, $s'(2) = -64$, $s'(3) = -96$, $s'(4) = -128$ ft/sec (b) $s(t) = 0$ when $400 - 16t^2 = 0$, so $t^2 = 400/16 = 25$, so at $t = 5$ seconds; the stone is traveling at the velocity $s'(5) = -160$, so downward at 160 ft/sec.

81. (a) $n'(t) = -0.90t + 15$; $n'(10) = 6$, so annual sales of in-ground pools were increasing at a rate of 6000 pools per year (b) (B): $n'(t)$ is positive but decreasing in the range in question.

83. (a) $f'(x) = 7.1x - 30.2$ manatees per 100,000 boats. (b) $f'(x)$ is increasing; the number of manatees killed per additional 100,000 boats increases as the number of boats increases. (c) $f'(8) = 26.6$ manatees per 100,000 additional boats. At a level of 800,000 boats, the number of manatee deaths is increasing at a rate of 26.6 manatees per 100,000 additional boats.

85. (a) $U(t) - E(t)$ measures the excess of SUV sales in the U.S. over sales in Europe; $U'(t) - E'(t) = [U(t) - E(t)]'$ measures the rate at which that difference was changing. (b) $U'(t) - E'(t) = 440t - 420 - (36t - 10) = 404t - 410$; this derivative is rising with increasing t. The gap between SUV sales in the U.S. and Europe was increasing during the last five years of the 1900s. (c) This is not a valid interpretation: The graphs use different y axis scales. Sales in Europe were increasing at an average rate of 62,000 vehicles per year, while sales in the U.S. were increasing at an average rate of 460,000 vehicles per year. However, SUV sales in Europe were rising at a greater *percentage* rate than in the U.S, so the graph of European sales appears to rising at a proportionally higher rate than the graph of U.S. sales.

87. After graphing the curve $y = 3x^2$, draw the line passing through $(-1, 3)$ with slope -6.

89. The slope of the tangent line of g is twice the slope of the tangent line of f because $g(x) = 2f(x)$, so $g'(x) = 2f'(x)$.

91. $g'(x) = -f'(x)$

93. The left-hand side is not equal to the right-hand side. The *derivative* of the left-hand side is equal to the right-hand side, so your friend should have written $\frac{d}{dx}(3x^4 + 11x^5) = 12x^3 + 55x^4$.

95. The derivative of a constant times a function is the constant times the derivative of the function, so $f'(x) = (2)(2x) = 4x$. Your enemy mistakenly computed the *derivative* of the constant times the derivative of the function. (The derivative of a product of two functions is not the product of the derivative of the two functions. The rule for taking the derivative of a product is discussed in the next chapter.).

Section 3.4

97. Answers may vary; here is one possibility. At one point its derivative is not defined but it has a tangent line: The tangent line at that point is vertical so has undefined slope.

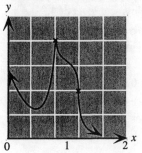

Section 3.5

3.5

1. $C'(x) = 5 - 0.0002x$; $C'(1000) = \$4.80$ per item

3. $C'(x) = 100 - 1000/x^2$; $C'(100) = \$99.90$ per item

5. $C'(x) = 4$; $R'(x) = 8 - x/500$; $P'(x) = 4 - x/500$; $P'(x) = 0$ when $x = 2000$. Thus, at a production level of 2000, the profit is stationary (neither increasing nor decreasing) with respect to the production level. This may indicate a maximum profit at a production level of 2000.

7. (a) (B): The slope of the graph decreases and then increases. **(b)** (C): This is where the slope of the graph is least. **(c)** (C): At $x = 50$, the height of the graph is about 3000, so the cost is $3000. The tangent line at that point passes roughly through (100, 4000), so has a slope of roughly 20; hence the cost is increasing at a rate of about $20 per item.

9. (a) $C'(x) = 1200 - 0.004x$. The cost is going up at a rate of $C'(4) = \$1{,}199{,}984$ per television commercial. The exact cost of airing the fifth television commercial is $C(5) - C(4) = \$1{,}199{,}982$. **(b)** $\bar{C}(x) = 150/x + 1200 - 0.002x$; $\bar{C}(4) = \$1{,}237{,}492$ per television commercial. The average cost of airing the first four television commercials is $1,237,492.

11. The profit on the sale of 1000 DVDs is $3000 and is decreasing at a rate of $3 per additional DVD sold.

13. $dP/dx = 5 - 1/(2\sqrt{x})$, so at $x = 50$, $P \approx \$257.07$ and $dP/dx \approx 5.07$. Your current profit is $257.07 per month and this would increase at a rate of $5.07 per additional magazine in sales.

15. $P'(n) = 400 - n$, so $P'(50) = \$350$. This means that, at an employment level of 50 workers, the firm's daily profit will increase at a rate of $350 per additional worker it hires.

17. (a) (B): $C'(x) = -0.002x + 0.3$ decreases as x increases. **(b)** (B): $\bar{C}(x) = -0.001x + 0.3 + 500/x$ decreases as x increases. **(c)** (C): $C'(100) = 0.1$, $\bar{C}(100) = 5.2$.

19. (a) When $q = 400$, $p = \$2.50$ per pound. **(b)** $R(q) = pq = 20{,}000/q^{0.5}$ **(c)** $R(400) = \$1000$. This is the monthly revenue that will result from setting the price at $2.50 per pound. $R'(q) = -10{,}000/q^{1.5}$, so $R'(400) = -\$1.25$ per pound of tuna. Thus, at a demand level of 400 pounds per month, the revenue is decreasing at a rate of $1.25 per pound. **(d)** The fishery should raise the price to reduce the demand and hence increase revenue.

21. (a) $C(x) = 500{,}000 + 685{,}000x - 10{,}000\sqrt{x}$; $C'(x) = 685{,}000 - \dfrac{5000}{\sqrt{x}}$; $\bar{C}(x) = \dfrac{500{,}000}{x} + 685{,}000 - \dfrac{10{,}000}{\sqrt{x}}$ **(b)** $C'(3) = \$682{,}000$ per spot, $\bar{C}(3) = \$846{,}000$ per spot. Since the marginal cost is less than the average cost, the cost of the fourth ad is lower than the average cost of the first three, so the average cost will decrease as x increases.

23. (a) $C'(q) = 200q$ so $C'(10) = \$2000$ per one-pound reduction in emissions. **(b)** $S'(q) = 500$. Thus $S'(q) = C'(q)$ when $500 = 200q$, or $q = 2.5$ pounds per day reduction. **(c)** $N(q) = C(q) - S(q) = 100q^2 - 500q + 4000$. This is a parabola with lowest point (vertex) given by $q = 2.5$. The net

Section 3.5

cost at this production level is $N(2.5) = \$3375$ per day. The value of q is the same as that for part (b). The net cost to the firm is minimized at the reduction level for which the cost of controlling emissions begins to increase faster than the subsidy. This is why we get the answer by setting these two rates of increase equal to each other.

25. $M'(10) \approx 0.0002557$ mpg/mph. This means that, at a speed of 10 mph, the fuel economy is increasing at a rate of 0.0002557 miles per gallon per 1-mph increase in speed. $M'(60) = 0$ mpg/mph. This means that, at a speed of 60 mph, the fuel economy is neither increasing nor decreasing with increasing speed. $M'(70) \approx -0.00001799$. This means that, at 70 mph, the fuel economy is decreasing at a rate of 0.00001799 miles per gallon per 1-mph increase in speed. Thus 60 mph is the most fuel-efficient speed for the car.

27. (c): If the marginal cost were lower in one plant than another, moving some production from the higher cost plant to the lower would result in a lower cost for the same production level.

29. (d): The marginal product per dollar of salary is $2/1.5 \approx 1.33$ times as high for a junior professor as compared to a senior professor. Therefore, discharging senior professors and hiring more junior professors will result in a higher quantity of output for the same amount of money.

31. (B): (In most cases) This is why we use the marginal cost as an estimate of the actual cost of the item.

33. Cost is often measured as a function of the number of items x. Thus, $C(x)$ is the cost of producing (or purchasing, as the case may be) x items.
(a) The average cost function $\bar{C}(x)$ is given by $\bar{C}(x) = C(x)/x$. The marginal cost function is the derivative, $C'(x)$, of the cost function.
(b) The average cost $\bar{C}(r)$ is the slope of the line through the origin and the point on the graph where $x = r$. The marginal cost of the rth unit is the slope of the tangent to the graph of the cost function at the point where $x = r$.
(c) The average cost function $\bar{C}(x)$ gives the average cost of producing the first x items. The marginal cost function $C'(x)$ is the rate at which cost is changing with respect to the number of items x, or the incremental cost per item, and approximates the cost of producing the $(x+1)$st item.

35. The marginal cost: If the average cost is rising, then the cost of the next piano must be larger than the average cost of the pianos already built.

37. Not necessarily. For example, it may be the case that the marginal cost of the 101st item is larger than the average cost of the first 100 items (even though the marginal cost is decreasing). Thus, adding this additional item will *raise* the average cost.

39. The circumstances described suggest that the average cost function is at a relatively low point at the current production level, and so it would be appropriate to advise the company to maintain current production levels; raising or lowering the production level will result in increasing average costs.

Section 3.6

3.6

1. 0:

x	$f(x)$
−0.1	0.01111111
−0.01	0.00010101
−0.001	1.001×10^{-6}
−0.0001	1.0001×10^{-6}
0	
0.0001	9.999×10^{-9}
0.001	9.99×10^{-7}
0.01	9.901×10^{-5}
0.1	0.00909091

3. 4:

x	$f(x)$
1.9	3.9
1.99	3.99
1.999	3.999
1.9999	3.9999
2	
2.0001	4.0001
2.001	4.001
2.01	4.01
2.1	4.1

5. Does not exist:

x	$f(x)$
−1.1	−22.1
−1.01	−202.01
−1.001	−2002.001
−1.0001	−20002
−1	
−0.9999	19998.0001
−0.999	1998.001
−0.99	198.01
−0.9	18.1

7. 1.5:

x	$f(x)$
10	2.66
100	1.58969231
1000	1.50877143
10,000	1.50087521
100,000	1.5000875

9. 0.5:

x	$f(x)$
−10	53.1578947
−1000	1
−100,000	0.505
−10,000,000	0.50005
−1,000,000,000	0.5000005

11. Diverges to $+\infty$:

x	$f(x)$
10	76.9423077
100	258.966135
1000	2059.17653
10000	20059.1977
100000	200059.2

13. 0:

x	$f(x)$
10	0.79988005
100	0.02600019
1000	0.00206
10000	0.0002006
100000	2.0006×10^{-5}

Section 3.6

15. 1:

x	f(x)
1.9	0.90483742
1.99	0.99004983
1.999	0.9990005
1.9999	0.9999
2	
2.0001	1.00010001
2.001	1.0010005
2.01	1.01005017
2.1	1.10517092

17. 0:

x	f(x)
10	0.000453999
100	3.72008×10^{-42}
1000	0
10000	0
100000	0

(The last three values of e^{-x}, while not mathematically 0, are too small to be represented in Excel, which just gives the values as 0.)

19. (a) −2 (b) −1

21. (a) 2 (b) 1 (c) 0 (d) +∞

23. (a) 0 (b) 2: As x approaches 0 from the right, $f(x)$ approaches the solid dot at height 2. (c) −1: As x approaches 0 from the left, $f(x)$ approaches the open dot at height −1. The fact that $f(0) = 2$ is irrelevant. (d) Does not exist: Parts (b) and (c) show that the one-sided limits, though they both exist, do not agree. (e) 2: The solid dot indicates the actual value of $f(0)$. (f) +∞

25. (a) 1 (b) 1: Similar to Exercise 23. (c) 2 (d) Does not exist (e) 1 (f) 2

27. (a) 1 (b) +∞ (c) +∞ (d) +∞ (e) Not defined (f) −1

29. (a) −1: Approaching from the left or the right, the value of $f(x)$ approaches the height of the open dot, −1. (b) +∞ (c) −∞ (d) Does not exist (e) 2 (f) 1: The value of the function is given by the closed dot on the graph.

31. $\lim_{t \to 1-} C(t) = 0.06$, $\lim_{t \to 1+} C(t) = 0.08$, so that $\lim_{t \to 1} C(t)$ does not exist.

33. $\lim_{t \to +\infty} I(t) = +\infty$, $\lim_{t \to +\infty} (I(t)/D(t)) \approx 0.024$. In the long term, sales of imported bottled water will rise without bound, but will level off at around 2.4% of sales of domestic bottled water. In the real world, sales cannot rise without bound. Thus, the given models should not be extrapolated far into the future.

35. (a) $\lim_{t \to +\infty} n(t) \approx 80$. On-line book sales can be expected to level off at 80 million per year in the long term. (b) $\lim_{t \to +\infty} n'(t) \approx 0$. The annual number of books sold on-line can be expected to stop changing in the long term.

37. 470. This suggests that students whose parents earn an exceptionally large income score an average of 470 on the verbal SAT test.

39. To approximate $\lim_{x \to a} f(x)$ numerically, choose values of x closer and closer to and on either side of $x = a$, and evaluate $f(x)$ for each of them. The limit (if it exists) is then the number that these values of $f(x)$ approach. A disadvantage of this method is that it may never give the exact value of the limit, but only an approximation.

Section 3.6

(However, we can make the approximation as accurate as we like.)

41. It is possible for $\lim_{x \to a} f(x)$ to exist even though $f(a)$ is not defined. An example is $\lim_{x \to 1} \frac{x^2-3x+2}{x-1}$.

43. Any situation in which there is a sudden change can be modeled by a function in which $\lim_{t \to a^+} f(t)$ is not the same as $\lim_{t \to a^-} f(t)$ One example is the value of a stock market index before and after a crash: $\lim_{t \to a^-} f(t)$ is the value immediately before the crash at time $t = a$, while $\lim_{t \to a^+} f(t)$ is the value immediately after the crash. Another example might be the price of a commodity that is suddenly increased from one level to another.

45. An example is $f(x) = (x-1)(x-2)$.

3.7

1. Continuous on its domain

3. Continuous on its domain

5. Discontinuous at $x = 0$: $\lim_{x \to 0^+} f(x) \neq \lim_{x \to 0^-} f(x)$ so $\lim_{x \to 0} f(x)$ does not exist.

7. Discontinuous at $x = -1$: $\lim_{x \to -1^+} f(x) \neq \lim_{x \to -1^-} f(x)$ so $\lim_{x \to -1} f(x)$ does not exist.

9. Continuous on its domain: Note that $f(0)$ is not defined, so 0 is not in the domain of f.

11. Discontinuous at $x = -1$ and 0: $\lim_{x \to -1} f(x) = -1 \neq f(-1)$ and $\lim_{x \to 0} f(x)$ does not exist. Note that $f(0)$ is defined [$f(0) = 2$] so 0 is in the domain of f.

13. (A), (B), (D), (E): Note that 1 is not in the domain in (B) and (D) and that the domain in (E) is $(-\infty, -1] \cup (1, +\infty)$; the "horizontal break" in the graph in (E) does not make the function discontinuous. (C) is discontinuous at 1 because the limit there does not equal the function's value.

In Exercises 15–21, either a graph of the function or a table of values can be used to compute the indicated limit.

15. $\lim_{x \to 1} \dfrac{x^2 - 2x + 1}{x - 1} = 0$, so setting $f(1) = 0$ makes it continuous at 1.

17. $\lim_{x \to 0} \dfrac{x}{3x^2 - x} = -1$, so setting $f(0) = -1$ makes it continuous at 0.

19. $\lim_{x \to 0} \dfrac{3}{3x^2 - x}$ is undefined, so no value of $f(0)$ will make it continuous at 0.

21. $\lim_{x \to 0} \dfrac{1 - e^x}{x} = -1$, so setting $f(0) = -1$ will make it continuous at 0.

Note: The vertical lines near discontinuities in some of the graphs below is typical behavior of graphing technology.

23. Continuous on its domain:

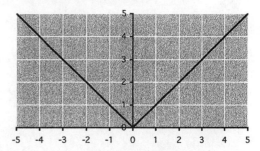

25. Continuous on its domain: Note that 1 and -1 are not in the domain of g.

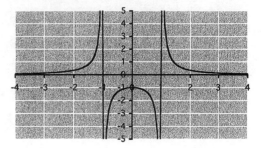

27. Discontinuity at $x = 0$:

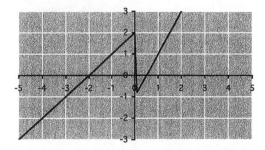

29. Discontinuity at $x = 0$: The limit of $h(x)$ as $x \to 0$ does not exist, but 0 is now in the domain of h.

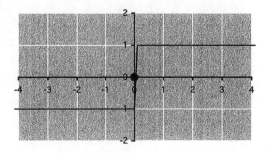

31. Continuous on its domain:

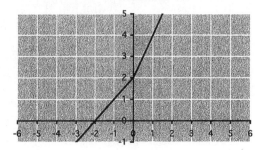

33. Not unless the domain of the function consists of all real numbers. (It is impossible for a function to be continuous at points not in its domain.) For example, $f(x) = 1/x$ is continuous on its domain—the set of nonzero real numbers—but not at $x = 0$.

35. True. If the graph of a function has a break in its graph at any point a, then it cannot be continuous at the point a.

37. Answers may vary.

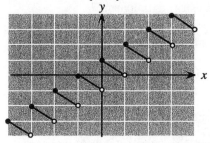

39. Answers may vary. The price of OHaganBooks.com stock suddenly drops by $10 as news spreads of a government investigation. Let $f(x)$ = the price of OHaganBooks.com stock.

Section 3.8

3.8

1. $\lim\limits_{x \to 0} (x + 1) = 0 + 1 = 1$

3. $\lim\limits_{x \to 2} \dfrac{2 + x}{x} = \dfrac{2 + 2}{2} = 2$

5. $\lim\limits_{x \to -1} \dfrac{x + 1}{x} = \dfrac{-1 + 1}{-1} = 0$

7. $\lim\limits_{x \to 8} (x - \sqrt[3]{x}) = 8 - \sqrt[3]{8} = 6$

9. $\lim\limits_{h \to 1} (h^2 + 2h + 1) = 1^2 + 2(1) + 1 = 4$

11. $\lim\limits_{h \to 3} 2 = 2$

13. $\lim\limits_{h \to 0} \dfrac{h^2}{h + h^2} = \lim\limits_{h \to 0} \dfrac{h}{1 + h} = \dfrac{0}{1 + 0} = 0$

15. $\lim\limits_{x \to 1} \dfrac{x^2 - 2x + 1}{x^2 - x} = \lim\limits_{x \to 1} \dfrac{(x-1)^2}{x(x-1)} = \lim\limits_{x \to 1} \dfrac{x - 1}{x} = \dfrac{1 - 1}{1} = 0$

17. $\lim\limits_{x \to 2} \dfrac{x^3 - 8}{x - 2} = \lim\limits_{x \to 2} \dfrac{(x - 2)(x^2 + 2x + 4)}{x - 2} = \lim\limits_{x \to 2} (x^2 + 2x + 4) = 2^2 + 2(2) + 4 = 12$

19. $\lim\limits_{x \to 0+} \dfrac{1}{x^2}$ diverges to $+\infty$ because it has the form "$k/0$" and $1/x^2$ is positive.

21. $\lim\limits_{x \to -1} \dfrac{x^2 + 1}{x + 1}$ does not exist: The limit from the left is $-\infty$ whereas the limit from the right is $+\infty$.

23. $\lim\limits_{x \to +\infty} \dfrac{3x^2 + 10x - 1}{2x^2 - 5x} = \lim\limits_{x \to +\infty} \dfrac{3x^2}{2x^2} = \lim\limits_{x \to +\infty} \dfrac{3}{2} = 3/2$

25. $\lim\limits_{x \to +\infty} \dfrac{x^5 - 1000x^4}{2x^5 + 10{,}000} = \lim\limits_{x \to +\infty} \dfrac{x^5}{2x^5} = \lim\limits_{x \to +\infty} \dfrac{1}{2} = 1/2$

27. $\lim\limits_{x \to +\infty} \dfrac{10x^2 + 300x + 1}{5x + 2} = \lim\limits_{x \to +\infty} \dfrac{10x^2}{5x} = \lim\limits_{x \to +\infty} (2x)$ diverges to $+\infty$

29. $\lim\limits_{x \to +\infty} \dfrac{10x^2 + 300x + 1}{5x^3 + 2} = \lim\limits_{x \to +\infty} \dfrac{10x^2}{5x^3} = \lim\limits_{x \to +\infty} \dfrac{2}{x} = 0$

Exercises 31–37 can be solved almost identically to Exercises 23–30. For variety, we give their solutions using L'Hospital's rule.

31. $\lim\limits_{x \to -\infty} \dfrac{3x^2 + 10x - 1}{2x^2 - 5x} = \lim\limits_{x \to -\infty} \dfrac{6x + 10}{4x - 5} = \lim\limits_{x \to -\infty} \dfrac{6}{4} = 3/2$

33. $\lim\limits_{x \to -\infty} \dfrac{x^5 - 1000x^4}{2x^5 + 10{,}000} = \lim\limits_{x \to -\infty} \dfrac{5x^4 - 4000x^3}{10x^4} = \lim\limits_{x \to -\infty} \dfrac{5 - 4000/x}{10} = \dfrac{5}{10} = 1/2$

35. $\lim\limits_{x \to -\infty} \dfrac{10x^2 + 300x + 1}{5x + 2} = \lim\limits_{x \to -\infty} \dfrac{20x + 300}{5}$ diverges to $-\infty$

Section 3.8

37. $\lim\limits_{x \to -\infty} \dfrac{10x^2 + 300x + 1}{5x^3 + 2} =$
$\lim\limits_{x \to -\infty} \dfrac{20x + 300}{15x^2} = \lim\limits_{x \to -\infty} \dfrac{20}{30x} = 0$

39. The only possible discontinuity is at $x = 0$. There, $\lim\limits_{x \to 0^-} f(x) = 0 + 2 = 2$ whereas $\lim\limits_{x \to 0^+} f(x) = 2(0) - 1 = -1$. Since these disagree, there is a discontinuity at $x = 0$.

41. The only possible discontinuities are at $x = 0$ and $x = 2$. We have $\lim\limits_{x \to 0^-} g(x) = 0 + 2 = 2$ and $\lim\limits_{x \to 0^+} g(x) = 2(0) + 2 = 2 = g(0)$, hence g is continuous at 0. We have $\lim\limits_{x \to 2^-} g(x) = 2(2) + 2 = 8$ and $\lim\limits_{x \to 2^+} g(x) = 2^2 + 2 = 8 = g(2)$. Hence, g is continuous everywhere.

43. The only possible discontinuity is at $x = 0$. We have $\lim\limits_{x \to 0^-} h(x) = 0 + 2 = 2 \neq h(0) = 0$, so there is a discontinuity at $x = 0$. (Note that we don't have to bother computing the limit from the right, which also equals 2.)

45. The only possible discontinuities are at $x = 0$ and $x = 2$. We have $\lim\limits_{x \to 0^-} f(x) = \lim\limits_{x \to 0^-} (1/x) = -\infty$, so there is a discontinuity at $x = 0$. On the other hand, $\lim\limits_{x \to 2^-} f(x) = 2 = f(2)$ and $\lim\limits_{x \to 2^+} f(x) = 2^{2-1} = 2$ also, so f is continuous at 2.

47. $\lim\limits_{t \to +\infty} \dfrac{P(t)}{C(t)} = \lim\limits_{t \to +\infty} \dfrac{1.745t + 29.84}{1.097t + 10.65} =$
$\lim\limits_{t \to +\infty} \dfrac{1.745}{1.097} \approx 1.59$ by L'Hospital's rule. If the trend continues indefinitely, the annual spending on police will be 1.59 times the annual spending on courts in the long run.

49. $\lim\limits_{t \to +\infty} I(t) = \lim\limits_{t \to +\infty} (0.0004t^2 + 0.013t + 0.06) = +\infty$; $\lim\limits_{t \to +\infty} \dfrac{I(t)}{D(t)} = \lim\limits_{t \to +\infty} \dfrac{0.0004t^2 + 0.013t + 0.06}{0.0166t^2 - 0.050t + 0.27} = \lim\limits_{t \to +\infty} \dfrac{0.0004t^2}{0.0166t^2} = \dfrac{0.0004}{0.0166} \approx 0.024$. In the long term, sales of imported bottled water will rise without bound, but will level off at around 2.4% of sales of domestic bottled water. In the real world, sales cannot rise without bound. Thus, the given models should not be extrapolated far into the future.

51. $\lim\limits_{t \to +\infty} p(t) = \lim\limits_{t \to +\infty} 100\left(1 - \dfrac{12{,}200}{t^{4.48}}\right) = 100(1 - 0) = 100$. $p'(t) = 100[12{,}200(4.48)t^{-5.48}] = 5{,}465{,}600/t^{5.48}$, so $\lim\limits_{t \to +\infty} p'(t) = \lim\limits_{t \to +\infty} 5{,}465{,}600/t^{5.48} = 0$. The percentage of children who learn to speak approaches 100% as their age increases, with the number of additional children learning to speak approaching zero.

53. (a) Yes: $\lim\limits_{t \to 8^-} C(t) = \lim\limits_{t \to 8^+} C(t) = 1.24$.
(b) No: $C'(8)$ does not exist. $C'(t) = 0.08$ for $0 < t < 8$ and $C'(t) = 0.355$ for $8 < t < 11$. Hence, $\lim\limits_{t \to 8^-} C'(t) = 0.08$ whereas $\lim\limits_{t \to 8^+} C'(t) = 0.355$. Until 1998, the cost of a Super Bowl ad was increasing at a rate of $80,000 per year. Immediately thereafter, it was increasing at a rate of $355,000 per year.

55. To evaluate $\lim\limits_{x \to a} f(x)$ algebraically, first check whether $f(x)$ is a closed-form function. Then check whether $x = a$ is in its domain. If so,

78

the limit is just $f(a)$; that is, it is obtained by substituting $x = a$. If not, then try to first simplify $f(x)$ in such a way as to transform it into a new function such that $x = a$ is in its domain, and then substitute. A disadvantage of this method is that it is sometimes extremely difficult to evaluate limits algebraically, and rather sophisticated methods are often needed.

57. She is wrong. Closed-form functions are continuous only at points in their domains, and $x = 2$ is not in the domain of the closed-form function $f(x) = 1/(x-2)^2$.

59. The statement may not be true. For example, if $f(x) = \begin{cases} x+2 & \text{if } x < 0 \\ 2x-1 & \text{if } x \geq 0, \end{cases}$
then $f(0)$ is defined and equals -1, and yet $\lim_{x \to 0} f(x)$ does not exist. The statement can be corrected by requiring that f be a closed-form function: "If f is a closed form function, and $f(a)$ is defined, then $\lim_{x \to a} f(x)$ exists and equals $f(a)$."

61. Answers may vary. For example, $f(x) = \begin{cases} 0 & \text{if } x \text{ is any number other than 1 or 2} \\ 1 & \text{if } x = 1 \text{ or } 2 \end{cases}$

Chapter 3 Review Test
1. (a)

h	Ave. Rate of Change
1	−0.5
0.01	−0.9901
0.001	−0.9990

Slope ≈ −1

(b)

h	Ave. Rate of Change
1	23
0.01	6.8404
0.001	6.7793

Slope ≈ 6.8

(c)

h	Ave. Rate of Change
1	6.3891
0.01	2.0201
0.001	2.0020

Slope ≈ 2

(d)

h	Ave. Rate of Change
1	0.6931
0.01	0.9950
0.001	0.9995

Slope ≈ 1

2. (a) (i) P (ii) Q (iii) R (iv) S **(b)** (i) None (ii) R (iii) Q (iv) P **(c)** (i) Q (ii) None (iii) None (iv) None **(d)** (i) None (ii) R (iii) P and S (iv) Q

3. (a) (B): The graph starts on the tangent line and falls below it. **(b)** (B): The graph starts above the tangent line and ends below it. **(c)** (B): The graph is getting less steep. **(d)** (A): The graph gets steeper until $x = 0$ and then gets less steep. **(e)** (C): The value of $f(2)$ is the height of the graph at $x = 2$, which is about 2.5; the rate of change is the slope of the tangent line at that point, which is approximately 1.5.

4. (a) $f'(x) = \lim_{h \to 0} \dfrac{f(x+h) - f(x)}{h} =$

$\lim_{h \to 0} \dfrac{(x+h)^2 + (x+h) - (x^2 + x)}{h} =$

$\lim_{h \to 0} \dfrac{x^2 + 2xh + h^2 + x + h - x^2 - x}{h} =$

$\lim_{h \to 0} \dfrac{2xh + h^2 + h}{h} = \lim_{h \to 0} (2x + h + 1) =$

$2x + 1$ **(b)** $f'(x) = \lim_{h \to 0} \dfrac{f(x+h) - f(x)}{h} =$

$\lim_{h \to 0} \dfrac{1/(x+h) + 1 - (1/x + 1)}{h} =$

$\lim_{h \to 0} \dfrac{1/(x+h) - 1/x}{h} = \lim_{h \to 0} \dfrac{x - (x+h)}{hx(x+h)} =$

$\lim_{h \to 0} \dfrac{-h}{hx(x+h)} = \lim_{h \to 0} \dfrac{-1}{x(x+h)} = -\dfrac{1}{x^2}$

5. (a) $f'(x) = 50x^4 + 2x^3 - 1$ **(b)** $f'(x) = (10x^{-5} + x^{-4}/2 - x^{-1} + 2)' = -50x^{-6} - 4x^{-5}/2 + x^{-2} = -50/x^6 - 2/x^5 + 1/x^2$ **(c)** $f'(x) = (3x^3 + 3x^{1/3})' = 9x^2 + x^{-2/3}$ **(d)** $f'(x) = (2x^{-2.1} - x^{0.1}/2)' = -4.2x^{-3.1} - 0.1x^{-0.9}/2 = -4.2/x^{3.1} - 0.1/(2x^{0.9})$

6. (a) $\dfrac{d}{dx}\left(x + \dfrac{1}{x^2}\right) = \dfrac{d}{dx}(x + x^{-2}) = 1 - 2x^{-3} = 1 - \dfrac{2}{x^3}$ **(b)** $\dfrac{d}{dx}\left(2x - \dfrac{1}{x}\right) = \dfrac{d}{dx}(2x - x^{-1}) = 2 + x^{-2} = 2 + \dfrac{1}{x^2}$ **(c)** $\dfrac{d}{dx}\left(\dfrac{4}{3x} - \dfrac{2}{x^{0.1}} + \dfrac{x^{1.1}}{3.2} - 4\right) =$

$\dfrac{d}{dx}\left(\dfrac{4}{3}x^{-1} - 2x^{-0.1} + \dfrac{1}{3.2}x^{1.1} - 4\right) =$

$-\dfrac{4}{3}x^{-2} + 0.2x^{-1.1} + \dfrac{1.1}{3.2}x^{0.1} =$

$-\dfrac{4}{3x^2} + \dfrac{0.2}{x^{1.1}} + \dfrac{1.1x^{0.1}}{3.2}$ (d) $\dfrac{d}{dx}\left(\dfrac{4}{x} + \dfrac{x}{4} - |x|\right) = \dfrac{d}{dx}\left(4x^{-1} + \dfrac{1}{4}x - |x|\right) = -4x^{-2} + 1/4 - |x|/x = -4/x^2 + 1/4 - |x|/x$

7. *The derivatives are the ones found in Exercise 5. The technology formulas are indicated below.*
(a) 50*x^4+2*x^3-1

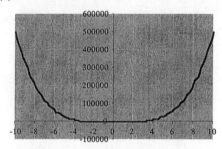

(b) -50/x^6-2/x^5+1/x^2

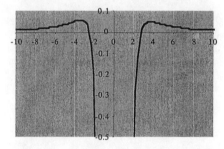

(c) 9*x^2+1/(x^2)^(1/3)

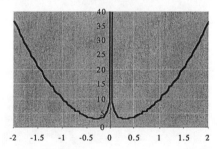

(d) -4.2/x^3.1-0.1/(2*x^0.9)

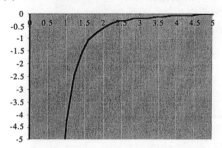

8. (a) $(9000 - 6500)/5 = 500$ books per week
(b) [3, 4] (600 books per week), [4, 5] (700 books per week) (c) [3, 5], when the average rate of increase was 650 books per week

9. (a) $w'(t) = -11.1t^2 + 149.2t + 135.5$, so $w'(1) \approx 274$ books per week (b) $w'(7) = 636$ books per week (c) It would not be realistic to use the function w through week 20: It begins to decrease after $t = 14$. Graph:

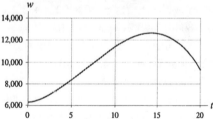

10. (a) Since we don't yet know how to find $s'(6)$ algebraically, we estimate it:

h	Ave. Rate of Change
1	496.78652
0.1	547.825452
0.01	552.276972
0.001	552.713694
0.0001	552.75728

So, $s'(6) \approx 553$ books per week.

(b)

h	Ave. Rate of Change
1	11.8911425
0.1	15.0069534
0.01	15.3764323
0.001	15.4140372
0.0001	15.4178043

$s'(14) \approx 15$ books per week. **(c)** Sales level off at 10,527 books per week, with a zero rate of change. Graph:

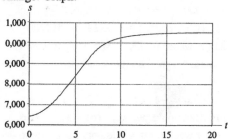

11. (a) $C'(x) = -0.00004x + 3.2$, so $C'(8000) = \$2.88$ per book **(b)** $\bar{C}(x) = -0.00002x + 3.2 + 5400/x$, $\bar{C}(8000) = \$3.715$ per book **(c)** $\bar{C}'(x) = -0.00002 - 5400/x^2$, so $\bar{C}'(8000) \approx -\$0.000104$ per book, per additional book sold. **(d)** At a sales level of 8000 books per week, the cost is increasing at a rate of $2.88 per book (so that the 8001st book costs approximately $2.88 to sell), and it costs an average of $3.715 per book to sell the first 8000 books. Moreover, the average cost is decreasing at a rate of $0.000104 per book, per additional book sold.

Chapter 4
4.1

The solutions to Exercises 1–11 show the calculation of the derivative using the product or quotient rule as appropriate.

1. Product rule: $f'(x) = (0)x + 3(1) = 3$

3. Product rule: $g'(x) = (1)x^2 + x(2x) = 3x^2$

5. Product rule: $h'(x) = (1)(x + 3) + x(1) = 2x + 3$

7. Product rule: $r'(x) = (0)x^{2.1} + 100(2.1x^{1.1}) = 210x^{1.1}$

9. Quotient rule: $s'(x) = \dfrac{(0)x - 2(1)}{x^2} = -\dfrac{2}{x^2}$

11. Quotient rule: $u'(x) = \dfrac{(2x)3 - x^2(0)}{3^2} = \dfrac{2x}{3}$

13. $\dfrac{dy}{dx} = 3(4x^2 - 1) + 3x(8x) = 36x^2 - 3$

15. $\dfrac{dy}{dx} = 3x^2(1 - x^2) + x^3(-2x) = 3x^2 - 5x^4$

17. $\dfrac{dy}{dx} = 2(2x + 3) + (2x + 3)(2) = 8x + 12$

19. $\dfrac{dy}{dx} = \sqrt{x} + \dfrac{x}{2\sqrt{x}} = \sqrt{x} + \dfrac{\sqrt{x}}{2} = \dfrac{3\sqrt{x}}{2}$

21. $\dfrac{dy}{dx} = (x^2 - 1) + (x + 1)(2x) = (x + 1)(3x - 1)$

23. $\dfrac{dy}{dx} = (x^{-0.5} + 4)(x - x^{-1}) + (2x^{0.5} + 4x - 5) \cdot (1 + x^{-2})$

25. $\dfrac{dy}{dx} = (4x - 4)(2x^2 - 4x + 1) + (2x^2 - 4x + 1)(4x - 4) = 8(2x^2 - 4x + 1)(x - 1)$

27. $\dfrac{dy}{dx} = \left(\dfrac{1}{3.2} - \dfrac{3.2}{x^2}\right)(x^2+1) + \left(\dfrac{x}{3.2} + \dfrac{3.2}{x}\right)(2x)$

29. $\dfrac{dy}{dx} = 2x(2x + 3)(7x + 2) + 2x^2(7x + 2) + 7x^2(2x + 3)$

31. $\dfrac{dy}{dx} = 5.3(1 - x^{2.1})(x^{-2.3} - 3.4) - 2.1x^{1.1}(5.3x - 1)(x^{-2.3} - 3.4) - 2.3x^{-3.3}(5.3x - 1)(1 - x^{2.1})$

33. $\dfrac{dy}{dx} = \dfrac{1}{2\sqrt{x}}\left(\sqrt{x} + \dfrac{1}{x^2}\right) + (\sqrt{x} + 1)\left(\dfrac{1}{2\sqrt{x}} - \dfrac{2}{x^3}\right)$

35. $\dfrac{dy}{dx} = \dfrac{2(3x - 1) - 3(2x + 4)}{(3x - 1)^2} = \dfrac{-14}{(3x - 1)^2}$

37. $\dfrac{dy}{dx} = \dfrac{(4x + 4)(3x - 1) - 3(2x^2 + 4x + 1)}{(3x - 1)^2} = \dfrac{6x^2 - 4x - 7}{(3x - 1)^2}$

39. $\dfrac{dy}{dx} = \dfrac{(2x - 4)(x^2 + x + 1) - (x^2 - 4x + 1)(2x + 1)}{(x^2 + x + 1)^2} = \dfrac{5x^2 - 5}{(x^2 + x + 1)^2}$

41. $\dfrac{dy}{dx} = \dfrac{(0.23x^{-0.77} - 5.7)(1 - x^{-2.9}) - 2.9x^{-3.9}(x^{0.23} - 5.7x)}{(1 - x^{-2.9})^2}$

Section 4.1

43. $\dfrac{dy}{dx} = \dfrac{\frac{1}{2}x^{-1/2}(x^{1/2} - 1) - \frac{1}{2}x^{-1/2}(x^{1/2} + 1)}{(x^{1/2} - 1)^2} = \dfrac{-1}{\sqrt{x}\,(\sqrt{x} - 1)^2}$

45. $\dfrac{dy}{dx} = \dfrac{d}{dx}\left[\dfrac{\frac{1}{x^2}(x + 1)}{x(1 + x)}\right] = \dfrac{d}{dx}\left(\dfrac{1}{x^3}\right) = -\dfrac{3}{x^4}$ (sometimes it pays to simplify first)

47. $\dfrac{dy}{dx} = \dfrac{[(x + 1) + (x + 3)](3x - 1) - 3(x + 3)(x + 1)}{(3x - 1)^2} = \dfrac{3x^2 - 2x - 13}{(3x - 1)^2}$

49. $\dfrac{dy}{dx} = \dfrac{[(x + 1)(x + 2) + (x + 3)(x + 2) + (x + 3)(x + 1)](3x - 1) - 3(x + 3)(x + 1)(x + 2)}{(3x - 1)^2} =$
$\dfrac{6x^3 + 15x^2 - 12x - 29}{(3x - 1)^2}$

51. $\dfrac{d}{dx}[(x^2 + x)(x^2 - x)] = (2x + 1)(x^2 - x) + (x^2 + x)(2x - 1) = 4x^3 - 2x$

53. $\dfrac{d}{dx}[(x^3 + 2x)(x^2 - x)]\Big|_{x=2} = [(3x^2 + 2)(x^2 - x) + (x^3 + 2x)(2x - 1)]\Big|_{x=2} =$
$(5x^4 - 4x^3 + 6x^2 - 4x)\Big|_{x=2} = 64$

55. $\dfrac{d}{dt}[(t^2 - t^{0.5})(t^{0.5} + t^{-0.5})]\Big|_{t=1} = [(2t - 0.5t^{-0.5})(t^{0.5} + t^{-0.5}) + (t^2 - t^{0.5})(0.5t^{-0.5} - 0.5t^{-1.5})]\Big|_{t=1} =$
$(2.5t^{1.5} + 1.5t^{0.5} - 1)\Big|_{t=1} = 3$

57. $f'(x) = 2x(x^3 + x) + (x^2 + 1)(3x^2 + 1) = 5x^4 + 6x^2 + 1$, so $f'(1) = 12$ is the slope. The tangent line passes through $(1, f(1)) = (1, 4)$, so its equation is $y = 12x - 8$.

59. $f'(x) = \dfrac{(x + 2) - (x + 1)}{(x + 2)^2} = \dfrac{1}{(x + 2)^2}$, so $f'(0) = 1/4$. The tangent line passes through $(0, f(0)) = (0, 1/2)$, so its equation is $y = x/4 + 1/2$.

61. $f'(x) = \dfrac{2x(x) - (x^2 + 1)}{x^2} = \dfrac{x^2 - 1}{x^2}$, so $f'(-1) = 0$. The tangent line passes through $(-1, f(-1)) = (-1, -2)$, so its equation is $y = -2$.

63. $S'(x) = 20 - 2x$, so $S'(5) = 10$: Sales are increasing at a rate of 1000 units per month). $p'(x) = -2x$, so $p'(5) = -10$: The price is dropping at a rate of $10 per sound system per month. $R'(x) = 100S'(x)p(x) + 100S(x)p'(x) = 100(20 - 2x)(1000 - x^2) + 100(20x - x^2)(-2x)$, so $R'(5) = 900{,}000$: Revenue is increasing at a rate of $900,000 per month)

65. Let $S(t)$ be the number of T-shirts sold per day. If $t = 0$ is now, we are told that $S(0) = 20$ and $S'(0) = -3$. Let $p(t)$ be the price of T-shirts. We are told that $p(0) = 7$ and $p'(0) = 1$. The

84

Section 4.1

revenue is then $R(t) = S(t)p(t)$, so $R'(0) = S'(0)p(0) + S(0)p'(0) = -3(7) + 20(1) = -1$. So, revenue is decreasing at a rate of $1 per day.

67. The cost per passenger is $Q(t) = C(t)/P(t) = (10{,}000 + t^2)/(1000 + t^2)$. So, $Q'(t) = \dfrac{2t(1000 + t^2) - 2t(10{,}000 + t^2)}{(1000 + t^2)^2} = \dfrac{-18{,}000t}{(1000 + t^2)^2}$ and $Q'(6) \approx -0.10$. The cost per passenger is decreasing at a rate of $0.10 per month.

69. $M'(x) = \dfrac{3000(3600x^{-2} - 1)}{(x + 3600x^{-1})^2}$, so $M'(10) \approx 0.7670$ mpg/mph. This means that, at a speed of 10 mph, the fuel economy is increasing at a rate of 0.7670 miles per gallon per one mph increase in speed. $M'(60) = 0$ mpg/mph. This means that, at a speed of 60 mph, the fuel economy is neither increasing nor decreasing with increasing speed. $M'(70) \approx -0.0540$. This means that, at 70 mph, the fuel economy is decreasing at a rate of 0.0540 miles per gallon per one mph increase in speed. 60 mph is the most fuel-efficient speed for the car. (In the next chapter we shall discuss how to locate largest values in general.)

71. Let $R(t) = P(t)Q(t)$ be the revenue. Saudi Arabia's daily revenue was $R(1) = \$158.4$ million. $R'(t) = P'(t)Q(t) + P(t)Q'(t) = -20t(-70.8t^2 + 78) + (-10t^2 + 32)(-141.6t)$, so $R'(1) = -3259.2$, hence Saudi Arabia's daily revenue was decreasing at a rate of $3259.20 million per year.

73. Let $t = 0$ represent 1995. The annual cost is the line through $(0, 80{,}000)$ and $(12, 120{,}000)$, which is $C(t) = 10{,}000t/3 + 80{,}000$. The number of personnel is the line through $(0, 1.5)$ and $(8, 1.4)$, which is $P(t) = -0.0125t + 1.5$. The total personnel cost is then $C(t)P(t)$ and its rate of change is $C'(t)P(t) + C(t)P'(t) = (10{,}000/3)\cdot(-0.0125t + 1.5) + (10{,}000t/3 + 80{,}000)\cdot(-0.0125)$. In 2002 ($t = 7$) the rate of change was $(10{,}000/3)(-0.0125(7) + 1.5) + (10{,}000(7)/3 + 80{,}000)(-0.0125) \approx 3420$. Hence, the total personnel cost was increasing at a rate of $3420 million per year.

75. $R'(p) = -\dfrac{5.625}{(1 + 0.125p)^2}$ so $R'(4) = -2.5$ thousand organisms per hour, per 1000 organisms. This means that the reproduction rate of organisms in a culture containing 4000 organisms is declining at a rate of 2500 organisms per hour, per 1000 additional organisms.

77. Let $P(t)$ be the number of eggs; then $P(t) = 30 - t$. The total oxygen consumption is $P(t)C(t)$ and its rate of change is $P'(t)C(t) + P(t)C'(t) = -(-0.0163t^4 + 1.096t^3 - 10.704t^2 + 3.576t) + (30 - t)(-0.0652t^3 + 3.288t^2 - 21.408t + 3.576)$. At $t = 25$ this is approximately -1634. Thus, oxygen consumption is decreasing at a rate of 1634 milliliters per day. This must be due to the fact that the number of eggs is decreasing, because $C'(25)$ is positive.

79. (a) $W(t) - S(t)$ represents sales of non-sparkling water, while $S(t)/W(t)$ represents the fraction of bottled water that is sparkling. Its derivative is the rate of change of this fraction.
(b) $\dfrac{d}{dt}\left(\dfrac{S(t)}{W(t)}\right) = [(-0.0026t + 0.030)(0.0010t^3 - 0.028t^2 + 0.28t - 0.70) - (-0.0013t^2 + 0.030t - 0.0069)(0.0030t^2 - 0.056t + 0.28)]/(0.0010t^3 - 0.028t^2 + 0.28t - 0.70)^2$. At $t = 10$ this is approximately -0.023 per year. The percentage of total sales represented by sparkling water was declining by 2.3 points per year in 1990.

Section 4.1

81. The analysis is suspect, as it seems to be asserting that the annual increase in revenue, which we can think of as *dR/dt*, is the product of the annual increases, *dp/dt* in price, and *dq/dt* in sales. However, since $R = pq$, the product rule implies that *dR/dt* is not the product of *dp/dt* and *dq/dt*, but is instead $\dfrac{dR}{dt} = \dfrac{dp}{dt} \cdot q + p \cdot \dfrac{dq}{dt}$.

83. Answers will vary. $q = -p + 1000$ is one example: $R(p) = pq = -p^2 + 1000p$, so $R'(p) = -2p + 1000$ and $R'(100) = 800 > 0$.

85. Mine; it is increasing twice as fast as yours. The rate of change of revenue is given by $R'(t) = p'(t)q(t)$ because $q'(t) = 0$ for both of us. Thus, in this case, $R'(t)$ does not depend on the selling price $p(t)$.

87. (A): If the marginal product was greater than the average product it would force the average product to increase. (See the formula for the rate of change of the average given in the solution to Exercise 82.)

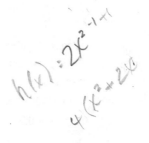

4.2

1. $2(2x + 1)(2) = 4(2x + 1)$

3. $-(x - 1)^{-2}(1) = -(x - 1)^{-2}$

5. $-2(2 - x)^{-3}(-1) = 2(2 - x)^{-3}$

7. $0.5(2x + 1)^{-0.5}(2) = (2x + 1)^{-0.5}$

9. $-(4x - 1)^{-2}(4) = -4(4x - 1)^{-2}$

11. $-(3x - 1)^{-2}(3) = -3/(3x - 1)^2$

13. $4(x^2 + 2x)^3 \frac{d}{dx}(x^2 + 2x) = 4(x^2 + 2x)^3(2x + 2)$

15. $-(2x^2 - 2)^{-2} \frac{d}{dx}(2x^2 - 2) = -4x(2x^2 - 2)^{-2}$

17. $-5(x^2 - 3x - 1)^{-6} \frac{d}{dx}(x^2 - 3x - 1) =$
$-5(2x - 3)(x^2 - 3x - 1)^{-6}$

19. $-3(x^2 + 1)^{-4} \frac{d}{dx}(x^2 + 1) = -6x/(x^2 + 1)^4$

21. $1.5(0.1x^2 - 4.2x + 9.5)^{0.5} \frac{d}{dx}(0.1x^2 - 4.2x + 9.5) = 1.5(0.2x - 4.2)(0.1x^2 - 4.2x + 9.5)^{0.5}$

23. $4(s^2 - s^{0.5})^3 \frac{d}{ds}(s^2 - s^{0.5}) = 4(2s - 0.5s^{-0.5})(s^2 - s^{0.5})^3$

25. $\frac{d}{dx}\sqrt{1 - x^2} = \frac{d}{dx}(1 - x^2)^{1/2} =$
$\frac{1}{2}(1 - x^2)^{-1/2}(-2x) = -\frac{x}{\sqrt{1 - x^2}}$

27. $\left(-\frac{1}{2}\right)2[(x + 1)(x^2 - 1)]^{-3/2} \frac{d}{dx}[(x + 1)(x^2 - 1)] = -[(x + 1)(x^2 - 1)]^{-3/2}[(x^2 - 1) + (x + 1)(2x)] = -[(x + 1)(x^2 - 1)]^{-3/2}(3x - 1)(x + 1)$

29. $2(3.1x - 2)^1(3.1) - (-2)(3.1x - 2)^{-3}(3.1) = 6.2(3.1x - 2) + 6.2/(3.1x - 2)^3$

31. $2[(6.4x - 1)^2 + (5.4x - 2)^3] \frac{d}{dx}[(6.4x - 1)^2 + (5.4x - 2)^3] = 2[(6.4x - 1)^2 + (5.4x - 2)^3][12.8(6.4x - 1) + 16.2(5.4x - 2)^2]$

33. $\frac{d}{dx}[(x^2 - 3x)^{-2}](1 - x^2)^{0.5} + (x^2 - 3x)^{-2} \frac{d}{dx}[(1 - x^2)^{0.5}] = -2(x^2 - 3x)^{-3}(2x - 3)(1 - x^2)^{0.5} - x(x^2 - 3x)^{-2}(1 - x^2)^{-0.5}$

35. $2\left(\frac{2x + 4}{3x - 1}\right) \frac{d}{dx}\left(\frac{2x + 4}{3x - 1}\right) = \left(\frac{4(x + 2)}{3x - 1}\right) \cdot \frac{2(3x - 1) - (2x + 4)(3)}{(3x - 1)^2} = \frac{-56(x + 2)}{(3x - 1)^3}$

37. $3\left(\frac{z}{1 + z^2}\right)^2 \frac{d}{dz}\left(\frac{z}{1 + z^2}\right) = \left(\frac{3z^2}{(1 + z^2)^2}\right) \cdot \frac{(1 + z^2) - z(2z)}{(1 + z^2)^2} = \frac{3z^2(1 - z^2)}{(1 + z^2)^4}$

39. $3[(1 + 2x)^4 - (1 - x)^2]^2 \frac{d}{dx}[(1 + 2x)^4 - (1 - x)^2] = 3[(1 + 2x)^4 - (1 - x)^2]^2[8(1 + 2x)^3 + 2(1 - x)]$

Section 4.2

41. $4.3[2 + (x + 1)^{-0.1}]^{3.3} \frac{d}{dx}[2 + (x + 1)^{-0.1}] =$
$-0.43(x + 1)^{-1.1}[2 + (x + 1)^{-0.1}]^{3.3}$

43. $-(\sqrt{2x+1} - x^2)^{-2} \frac{d}{dx}[(2x + 1)^{1/2} - x^2] =$
$-(\sqrt{2x+1} - x^2)^{-2}\left(\frac{1}{2}(2x + 1)^{-1/2}(2) - 2x\right) =$
$-\dfrac{\dfrac{1}{\sqrt{2x+1}} - 2x}{(\sqrt{2x+1} - x^2)^2}$

45. $3\{1 + [1 + (1 + 2x)^3]^3\}^2$
$\frac{d}{dx}\{1 + [1 + (1 + 2x)^3]^3\} = 9\{1 + [1 + (1 + 2x)^3]^3\}^2[1 + (1 + 2x)^3]^2 \frac{d}{dx}[1 + (1 + 2x)^3] =$
$27\{1 + [1 + (1 + 2x)^3]^3\}^2[1 + (1 + 2x)^3]^2(1 + 2x)^2(2) = 54(1 + 2x)^2[1 + (1 + 2x)^3]^2\{1 + [1 + (1 + 2x)^3]^3\}^2$

47. $\frac{dy}{dt} = 100x^{99}\frac{dx}{dt} - 99x^{-2}\frac{dx}{dt} = (100x^{99} - 99x^{-2})\frac{dx}{dt}$

49. $\frac{ds}{dt} = \frac{d}{dt}(r^{-3} + r^{0.5}) = -3r^{-4}\frac{dr}{dt} + 0.5r^{-0.5}\frac{dr}{dt}$
$= (-3r^{-4} + 0.5r^{-0.5})\frac{dr}{dt}$

51. $\frac{dV}{dt} = 4\pi r^2 \frac{dr}{dt}$

53. $\frac{dy}{dt} = 3x^2 \frac{dx}{dt} - x^{-2}\frac{dx}{dt}$, so $\left.\frac{dy}{dt}\right|_{t=1} =$
$3(2)^2(-1) - 2^{-2}(-1) = -47/4$

55. $c = 100(0.42 + 0.02t)^2 - 160(0.42 + 0.02t) + 110$; $\frac{dc}{dt} = 200(0.02)(0.42 + 0.02t) -$
$160(0.02)$; $\left.\frac{dc}{dt}\right|_{t=8} = 200(0.02)(0.42 + 0.02(8)) - 160(0.02) = -\$0.88$ per trade per month

57. We are given $dy/dt = -3$ and we need to find dx/dt. From the linear relation we have $\frac{dy}{dt} =$
$\frac{dy}{dx}\frac{dx}{dt} = 1.5\frac{dx}{dt}$, so $-3 = 1.5\frac{dx}{dt}$. Hence, $\frac{dx}{dt} = \frac{-3}{1.5} = -2$ murders per 100,000 residents/yr each year.

59. (a) $R(p) = pq = -4p^2/3 + 80p$; $\frac{dR}{dp} = -8p/3 + 80$; when $q = 60$, $60 = -4p/3 + 80$, so $p = (60 - 80)(-3/4) = 15$, hence $\left.\frac{dR}{dp}\right|_{q=60} =$
$-8(15)/3 + 80 = \$40$ per \$1 increase in price
(b) Solving for p, we get $p = (3/4)(80 - q)$, so $\frac{dp}{dq} = -3/4$, hence $\left.\frac{dp}{dq}\right|_{q=60} = -3/4 = -\$0.75$ per ruby. **(c)** $\left.\frac{dR}{dq}\right|_{q=60} = \frac{dR}{dp}\frac{dp}{dq} = 40(-0.75) = -\$30$ per ruby. Thus, at a demand level of 60 rubies per week, the weekly revenue is decreasing at a rate of \$30 per additional ruby demanded.

61. $\frac{dP}{dq} = 5000 - 0.5q$ and $\frac{dq}{dn} = 30 + 0.02n$.
When $n = 10$, $q = 30(10) + 0.01(10)^2 = 301$.
Hence, $\left.\frac{dP}{dn}\right|_{n=10} = \frac{dP}{dq}\frac{dq}{dn} = (5000 - 0.5(301))(30 + 0.02(10)) = 146{,}454.9$. At an employment level of 10 engineers, Paramount will increase its profit at a rate of \$146,454.90 per additional engineer hired.

63. $\frac{dM}{dt} = 2.48$ while $\frac{dB}{dt} = 15{,}700$. The chain rule gives $\frac{dM}{dt} = \frac{dM}{dB}\frac{dB}{dt}$, hence $\frac{dM}{dB} = \frac{dM/dt}{dB/dt} =$
$2.48/15{,}700 = 0.000158$ manatees per boat, or

15.8 manatees per 100,000 boats. Approximately 15.8 more manatees are killed each year for each additional 100,000 registered boats.

65. $\dfrac{dA}{dt} = 2\pi r \dfrac{dr}{dt} = 2\pi(3)(2) = 12\pi$ mi²/h

67. $\dfrac{dV}{dt} = 4\pi r^2 \dfrac{dr}{dt} = 4\pi(10)^2(0.5) = 200\pi$ ft²/week; $C = 1000V$, so $\dfrac{dC}{dt} = 1000\dfrac{dV}{dt} =$ $200{,}000\pi$/week $\approx \$628{,}000$/week

69. (a) $q'(4) \approx (q(4 + 0.0001) - q(4 - 0.0001))/0.0002 \approx 333$ units per month (b) $R = 800q$, so $dR/dq = \$800$/unit: The marginal revenue is the selling price. (c) $\dfrac{dR}{dt} = \dfrac{dR}{dq}\dfrac{dq}{dt} \approx 800(333) \approx \$267{,}000$ per month.

71. Keeping r and p fixed, $\dfrac{dM}{dt} = 1.2y^{-0.4}r^{-0.3}p\,\dfrac{dy}{dt}$, so $\dfrac{dM/dt}{M} = \dfrac{1.2y^{-0.4}r^{-0.3}p\,\dfrac{dy}{dt}}{2y^{0.6}r^{-0.3}p} = 0.6\dfrac{dy/dt}{y}$. We are given $\dfrac{dy/dt}{y} = 5\%$ per year, so $\dfrac{dM/dt}{M} = 0.6(5) = 3\%$ per year

73. Keeping r fixed, $\dfrac{dM}{dt} = 1.2y^{-0.4}r^{-0.3}p\,\dfrac{dy}{dt} + 2y^{0.6}r^{-0.3}\dfrac{dp}{dt}$, so $\dfrac{dM/dt}{M} = 0.6\dfrac{dy/dt}{y} + \dfrac{dp/dt}{p} = 0.6(5) + 5 = 8\%$ per year.

75. the glob squared, times the derivative of the glob.

77. The derivative of a quantity cubed is three times the (original) quantity squared, times the derivative of the quantity. Thus, the correct answer is $3(3x^3 - x)^2(9x^2 - 1)$.

79. Following the calculation thought experiment, pretend that you are evaluating the function at a specific value of x. If the last operation you would perform is addition or subtraction, look at each summand separately. If the last operation is multiplication, use the product rule first; if it is division, use the quotient rule first; if it is any other operation (such as raising a quantity to a power or taking a radical of a quantity) then use the chain rule first.

81. An example is
$$f(x) = \sqrt{x + \sqrt{x + \sqrt{x + \sqrt{x + \sqrt{x + 1}}}}}\,.$$

4.3

1. $1/(x-1)$

3. $1/(x \ln 2)$

5. $2x/(x^2 + 3)$

7. e^{x+3}

9. $-e^{-x}$

11. $4^x \ln 4$

13. $2^{x^2-1} \, 2x \ln 2$

15. $(1) \ln x + x \dfrac{d}{dx} \ln x = \ln x + x\left(\dfrac{1}{x}\right) = 1 + \ln x$

17. $2x \ln x + (x^2 + 1)\left(\dfrac{1}{x}\right) = 2x \ln x + \dfrac{x^2 + 1}{x}$

19. $5(x^2 + 1)^4(2x)\ln x + (x^2 + 1)^5\left(\dfrac{1}{x}\right) =$
    $10x(x^2 + 1)^4 \ln x + \dfrac{(x^2 + 1)^5}{x}$

21. $3/(3x - 1)$

23. $4x/(2x^2 + 1)$

25. $(2x - 0.63x^{-0.7})/(x^2 - 2.1x^{0.3})$

27. $\dfrac{d}{dx}[\ln(-2x + 1) + \ln(x + 1)] =$
    $\dfrac{-2}{-2x + 1} + \dfrac{1}{x + 1}$

29. $\dfrac{d}{dx}[\ln(3x + 1) - \ln(4x - 2)] =$
    $\dfrac{3}{3x + 1} - \dfrac{4}{4x - 2}$

31. $\dfrac{d}{dx}[\ln(x + 1) + \ln(x - 3) - \ln(-2x - 9)] =$
    $\dfrac{1}{x + 1} + \dfrac{1}{x - 3} - \dfrac{-2}{-2x - 9} = \dfrac{1}{x + 1} + \dfrac{1}{x - 3} - \dfrac{2}{2x + 9}$

33. $\dfrac{d}{dx}[1.3\ln(4x - 2)] = \dfrac{5.2}{4x - 2}$

35. $\dfrac{d}{dx}[\ln(x + 1)^2 - \ln(3x - 4)^3 - \ln(x - 9)] =$
    $\dfrac{d}{dx}[2\ln(x + 1) - 3\ln(3x - 4) - \ln(x - 9)] =$
    $\dfrac{2}{x + 1} - \dfrac{9}{3x - 4} - \dfrac{1}{x - 9}$

37. $\dfrac{1}{(x + 1) \ln 2}$

39. $\dfrac{1 - 1/t^2}{(t + 1/t) \ln 3}$

41. $2(\ln|x|)\left(\dfrac{1}{x}\right) = \dfrac{2\ln|x|}{x}$

43. $\dfrac{d}{dx}\{2\ln x - [\ln(x - 1)]^2\} = \dfrac{2}{x} -$
    $2[\ln(x - 1)]\dfrac{1}{x - 1} = \dfrac{2}{x} - \dfrac{2\ln(x - 1)}{x - 1}$

45. $e^x + xe^x = e^x(1 + x)$

47. $\dfrac{1}{x + 1} + 9x^2 e^x + 3x^3 e^x = \dfrac{1}{x + 1} + 3e^x(x^3 + 3x^2)$

49. $e^x \ln|x| + e^x\left(\dfrac{1}{x}\right) = e^x(\ln|x| + 1/x)$

51. $2e^{2x+1}$

53. $(2x - 1)e^{x^2-x+1}$

Section 4.3

55. $2xe^{2x-1} + x^2(2e^{2x-1}) = 2xe^{2x-1}(1 + x)$

57. $2e^{2x-1}\dfrac{d}{dx}e^{2x-1} = 2e^{2x-1}(2e^{2x-1}) = 4(e^{2x-1})^2$ OR
$\dfrac{d}{dx}[(e^{2x-1})^2] = \dfrac{d}{dx}(e^{4x-2}) = 4e^{4x-2} = 4(e^{2x-1})^2$

59. $\dfrac{(e^x - e^{-x})(e^x - e^{-x}) - (e^x + e^{-x})(e^x + e^{-x})}{(e^x - e^{-x})^2} =$
$\dfrac{e^{2x} - 2 + e^{-2x} - (e^{2x} + 2 + e^{-2x})}{(e^x - e^{-x})^2} = \dfrac{-4}{(e^x - e^{-x})^2}$

61. $\dfrac{d}{dx}[e^{(3x-1)+(x-2)+x}] = \dfrac{d}{dx}e^{5x-3} = 5e^{5x-3}$

63. $\dfrac{d}{dx}(x \ln x)^{-1} = -(x \ln x)^{-2}[\ln x + x(1/x)] =$
$-\dfrac{\ln x + 1}{(x \ln x)^2}$

65. Note that $\ln(e^x) = x$, so $f(x) = x^2 - 2\ln(e^x) = x^2 - 2x$, so $f'(x) = 2x - 2 = 2(x - 1)$

67. $\dfrac{1}{\ln x}\dfrac{d}{dx}\ln x = \dfrac{1}{x \ln x}$

69. $\dfrac{d}{dx}\ln(\ln x)^{1/2} = \dfrac{d}{dx}\left(\dfrac{1}{2}\ln(\ln x)\right) = \dfrac{1}{2}\left(\dfrac{1}{\ln x}\right)\left(\dfrac{1}{x}\right) =$
$\dfrac{1}{2x \ln x}$

71. $\dfrac{dy}{dx} = e^x\log_2 x + \dfrac{e^x}{x \ln 2} = \dfrac{e}{\ln 2}$ when $x = 1$. So, the equation of the line is $y = (e/\ln 2)(x - 1) \approx 3.92(x - 1)$.

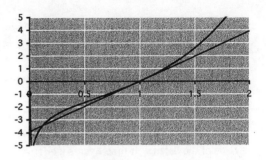

73. $\dfrac{dy}{dx} = \dfrac{d}{dx}\left(\dfrac{1}{2}\ln(2x + 1)\right) = \dfrac{1}{2x + 1} = 1$ when $x = 0$. The tangent line has slope 0 and passes through $(0, 0)$, so its equation is $y = x$.

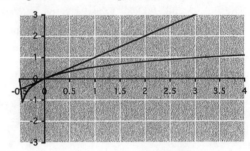

75. $\dfrac{dy}{dx} = 2xe^{x^2} = 2e$ when $x = 1$. The line at right angles to the graph has slope $-1/(2e)$ and passes through $(1, e)$, so its equation is
$y = -[1/(2e)](x - 1) + e \approx -0.1839x + 2.9022$.

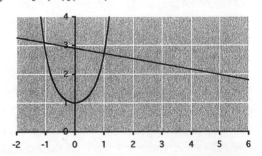

91

Section 4.3

77. From the continuous compounding formula, the value of the balance at time t years is $A(t) = 10{,}000e^{0.04t}$. Its derivative is $A'(t) = 400e^{0.04t}$, so, after 3 years, the balance is growing at the rate of $A'(3) = \$451.00$ per year.

79. From the compound interest formula, the value of the balance at time t years is $A(t) = 10{,}000(1 + 0.04/2)^{2t} = 10{,}000(1.02)^{2t}$. Its derivative is $A'(t) = [20{,}000 \ln(1.02)](1.02)^{2t}$, so, after 3 years, the balance is growing at the rate of $A'(3) = \$446.02$ per year.

81. The population t years after 1995 was given by $P(t) = 4{,}000{,}000(2^{t/10})$. Its derivative is $P'(t) = [400{,}000 \ln 2]\, 2^{t/10}$, so, at the start of 1995, the population was growing at the rate of $P'(0) \approx 277{,}000$ people/year.

83. The amount of Plutonium-239 left after t years is $P(t) = 10(0.5)^{t/24{,}400}$. Its derivative is $P'(t) = [(10/24{,}400)\ln 0.5](0.5)^{t/24{,}400}$, so, after 100 years, the rate of change is $P'(100) \approx -0.000283$ g/year. That is, the Plutonium is decaying at the rate of 0.000283 g/year.

85. $P'(t) = -92(0.0277)e^{-0.0277t}$, so $P'(22) \approx -1.4$. This indicates that, in ancient Rome, the percentage of people surviving was decreasing at a rate of 1.4 percentage points per year at age 22. That is, approximately 1.4% of the original birth cohort died that year.

87. (a) (A): The data would best be modeled by a function that approaches a value in the high 400s exponentially. The function in (A) is the only one that does that: the one in (B) decays to 0, the one in (C) increases without bound, and the one in (D) decreases without bound. **(b)** $S'(x) = 136(0.0000264)e^{-0.0000264x}$, so $S'(45{,}000) \approx 0.001$. At an income level of \$45,000, the average verbal SAT increases by approximately $1000(0.001) = 1$ point for each \$1000 increase in income. **(c)** $S'(x)$ decreases with increasing x, so that as parental income increases, the effect on SAT scores decreases.

89. $R(t) = 350e^{-0.1t}(39t + 68)$ million dollars, so $R(2) \approx \$42{,}000$ million $= \$42$ billion. $R'(t) = -35e^{-0.1t}(39t + 68) + 350(39)e^{-0.1t} = (11{,}270 - 1365t)e^{-0.1t}$, so $R'(2) \approx \$7000$ million per year $= \$7$ billion per year.

91. Write $P(t) = 150{,}000{,}000(1 + 14{,}999e^{-0.3466t})^{-1}$. Then $P'(t) = -150{,}000{,}000(1 + 14{,}999e^{-0.3466t})^{-2}[14{,}999(-0.3466)e^{-0.3466t}] \approx 779{,}800{,}000{,}000 e^{-0.3466t}/(1 + 14{,}999e^{-0.3466t})^2$. So, $P'(20) \approx 3{,}110{,}000$ cases/week; $P'(30) \approx 11{,}200{,}000$ cases/week; $P'(40) \approx 722{,}000$ cases/week.

93. (a) $W'(t) = -1500[1 + 0.77(1.16)^{-t}]^{-2} \cdot [0.77(-1)\ln(1.16)(1.16)^{-t}] \approx \dfrac{171.425(1.16)^{-t}}{(1 + 0.77(1.16)^{-t})^2}$, $W'(6) \approx 41$ to two significant digits. The constants in the model are specified to two and three significant digits, so we cannot expect the answer to be accurate to more than two digits. In other words, all digits from the third on are probably meaningless. The answer tells us that in 1996, the number of authorized wiretaps was increasing at a rate of approximately 41 wiretaps per year. **(b)** (A): As seen in the following graph, the derivative is positive (W is increasing…) but getting smaller (at a decreasing rate).

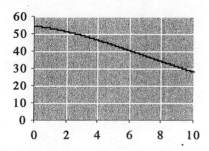

95. (a)

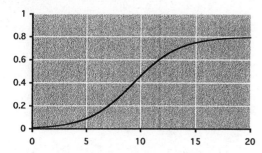

$p'(10) \approx 0.09$, so the percentage of firms using numeric control is increasing at a rate of 9 percentage points per year after 10 years.
(b) $\lim_{t \to +\infty} p(t) = 0.80$. Thus, in the long run, 80% of all firms will be using numeric control.
(c) $p'(t) = 0.3816 e^{4.46 - 0.477t}/(1 + e^{4.46 - 0.477t})^2$.
$p'(10) = 0.0931$. Graph:

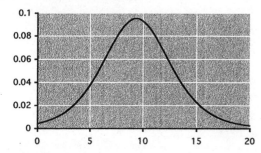

(d) $\lim_{t \to +\infty} p'(t) = 0$. Thus, in the long run, the percentage of firms using numeric control will stop increasing.

97. e raised to the glob, times the derivative of the glob.

99. 2 raised to the glob, times the derivative of the glob, times the natural logarithm of 2.

101. The power rule does not apply when the exponent is not constant. The derivative of 3 raised to a quantity is 3 raised to the quantity, times the derivative of the quantity, times ln 3. Thus, the correct answer is $3^{2x} 2 \ln 3$

103. No. If $N(t)$ is exponential, so is its derivative.

105. If $f(x) = e^{kx}$, then the fractional rate of change is $\dfrac{f'(x)}{f(x)} = \dfrac{ke^{kx}}{e^{kx}} = k$, the fractional growth rate.

107. If $A(t)$ is growing exponentially, then $A(t) = A_0 e^{kt}$ for constants A_0 and k. Its percentage rate of change is then $\dfrac{A'(t)}{A(t)} = \dfrac{kA_0 e^{kt}}{A_0 e^{kt}} = k$, a constant.

Section 4.4

4.4

1. $2 + 3\frac{dy}{dx} = 0, \frac{dy}{dx} = -2/3; y = (7 - 2x)/3,$

$\frac{dy}{dx} = -2/3$

3. $2x - 2\frac{dy}{dx} = 0, \frac{dy}{dx} = x; y = (x^2 - 6)/2, \frac{dy}{dx} = x$

5. $2 + 3\frac{dy}{dx} = y + x\frac{dy}{dx}, (3 - x)\frac{dy}{dx} = y - 2,$

$\frac{dy}{dx} = (y - 2)/(3 - x); y = -2x/(3 - x),$

$\frac{dy}{dx} = \frac{-2(3 - x) - 2x}{(3 - x)^2} = \frac{-2}{3 - x} - \frac{2x}{(3 - x)^2} =$

$\frac{-2}{3 - x} + \frac{y}{3 - x} = \frac{y - 2}{3 - x}$

7. $e^x y + e^x \frac{dy}{dx} = 0, \frac{dy}{dx} = -y; y = 1/e^x = e^{-x},$

$\frac{dy}{dx} = -e^{-x} = -y$

9. $\frac{dy}{dx} \ln x + \frac{y}{x} + \frac{dy}{dx} = 0, (1 + \ln x)\frac{dy}{dx} = -\frac{y}{x},$

$\frac{dy}{dx} = -\frac{y}{x(1 + \ln x)}; y = 2/(1 + \ln x), \frac{dy}{dx} =$

$-2(1 + \ln x)^{-2}\left(\frac{1}{x}\right) = -\frac{2}{1 + \ln x} \cdot \frac{1}{x(1 + \ln x)} =$

$-\frac{y}{x(1 + \ln x)}$

11. $2x + 2y\frac{dy}{dx} = 0, \frac{dy}{dx} = -x/y$

13. $2xy + x^2\frac{dy}{dx} - 2y\frac{dy}{dx} = 0, (x^2 - 2y)\frac{dy}{dx} = -2xy, \frac{dy}{dx} = -2xy/(x^2 - 2y)$

15. $3y + 3x\frac{dy}{dx} - \frac{1}{3}\frac{dy}{dx} = -\frac{2}{x^2}, (9x^3 - x^2)\frac{dy}{dx} = -(6 + 9x^2 y), \frac{dy}{dx} = -(6 + 9x^2 y)/(9x^3 - x^2)$

17. $2x\frac{dx}{dy} - 6y = 0, \frac{dx}{dy} = 3y/x$

19. $2p\frac{dp}{dq} - q\frac{dp}{dq} - p = 10pq^2\frac{dp}{dq} + 10p^2 q,$

$(2p - q - 10pq^2)\frac{dp}{dq} = p + 10p^2 q, \frac{dp}{dq} = (p + 10p^2 q)/(2p - q - 10pq^2)$

21. $e^y + xe^y\frac{dy}{dx} - e^x\frac{dy}{dx} - ye^x = 0,$

$(xe^y - e^x)\frac{dy}{dx} = ye^x - e^y,$

$\frac{dy}{dx} = (ye^x - e^y)/(xe^y - e^x)$

23. $e^s\left(t\frac{ds}{dt} + s\right) = 2s\frac{ds}{dt}, (2s - te^{st})\frac{ds}{dt} = se^{st},$

$\frac{ds}{dt} = se^{st}/(2s - te^{st})$

25. $\frac{e^x y^2 - 2ye^x(dy/dx)}{y^4} = e^y\frac{dy}{dx}, (2e^x + y^3 e^y)\frac{dy}{dx} =$

$ye^x, \frac{dy}{dx} = ye^x/(2e^x + y^3 e^y)$

27. $\frac{2y - 1}{y^2 - y}\frac{dy}{dx} + 1 = \frac{dy}{dx},$

$[(2y - 1) - (y^2 - y)]\frac{dy}{dx} = -(y^2 - y),$

$\frac{dy}{dx} = (y - y^2)/(-1 + 3y - y^2)$

29. $\frac{y + x\frac{dy}{dx} + 2y\frac{dy}{dx}}{xy + y^2} = e^y\frac{dy}{dx},$

$[x + 2y - e^y(xy + y^2)]\frac{dy}{dx} = -y,$

$\frac{dy}{dx} = -y/(x + 2y - xye^y - y^2 e^y)$

31. $\ln y = \ln[(x^3 + x)\sqrt{x^3 + 2}] = \ln(x^3 + x) +$

Section 4.4

$\frac{1}{2}\ln(x^3+2)$, $\frac{1}{y}\frac{dy}{dx} = \frac{3x^2+1}{x^3+x} + \frac{1}{2}\frac{3x^2}{x^3+2}$, $\frac{dy}{dx} =$
$(x^3+x)\sqrt{x^3+2}\left(\frac{3x^2+1}{x^3+x} + \frac{1}{2}\frac{3x^2}{x^3+2}\right)$

33. $\ln y = \ln(x^x) = x \ln x$, $\frac{1}{y}\frac{dy}{dx} = \ln x + x\left(\frac{1}{x}\right) =$
$1 + \ln x$, $\frac{dy}{dx} = x^x(1 + \ln x)$

35. $8x + 4y\frac{dy}{dx} = 0$, $\frac{dy}{dx} = -2x/y$; at $(1, -2)$,
$\frac{dy}{dx} = -2(1)/(-2) = 1$

37. $4x - 2y\frac{dy}{dx} = y + x\frac{dy}{dx}$, $\frac{dy}{dx} = \frac{4x-y}{x+2y}$; at
$(-1, 2)$, $\frac{dy}{dx} = \frac{4(-1)-2}{-1+2(2)} = -2$

39. $0.9x^{-0.7}y^{0.7} + 2.1x^{0.3}y^{-0.3}\frac{dy}{dx} = 0$, $\frac{dy}{dx} = -\frac{3y}{7x}$;
when $x = 20$, $3(20)^{0.3}y^{0.7} = 10$, $y = [10/(3\times 20^{0.3})]^{1/0.7} \approx 1.54663$, so $\frac{dy}{dx} \approx -\frac{3(1.54663)}{7(20)} \approx -0.03314$

41. $0.4x^{-0.6}y^{0.6} + 0.6x^{0.4}y^{-0.4}\frac{dy}{dx} - 0.4x = 0$, $\frac{dy}{dx} =$
$\frac{0.4x - 0.4x^{-0.6}y^{0.6}}{0.6x^{0.4}y^{-0.4}} = \frac{2}{3}\cdot\frac{x^{1.6}y^{0.4} - y}{x}$; when $x = 20$,
$20^{0.4}y^{0.6} - 0.2(20)^2 = 100$, $y = \left(\frac{100 + 0.2(20)^2}{20^{0.4}}\right)^{1/0.6} \approx 778.815$, so $\frac{dy}{dx} \approx$
$\frac{2}{3}\cdot\frac{20^{1.6}(778.815)^{0.4} - 778.815}{20} \approx 31.73$

43. $\left(y + x\frac{dy}{dx}\right)e^{xy} - 1 = 4$, $\frac{dy}{dx} = \frac{5 - ye^{xy}}{xe^{xy}}$;
when $x = 3$, $e^{3y} - 3 = 12$, $e^{3y} = 15$, $3y = \ln 15$,
$y = \frac{1}{3}\ln 15 \approx 0.902683$, so $\frac{dy}{dx} \approx$

$\frac{5 - (0.902683)(15)}{3(15)} \approx -0.1898$ (using the fact that $e^{3y} = 15$)

45. $\frac{1 + dy/dx}{x+y} - 1 = 6x$, $\frac{dy}{dx} = (6x+1)(x+y) - 1$; when $x = 0$, $\ln y = 0$, $y = e^0 = 1$, so $\frac{dy}{dx} = (0+1)(0+1) - 1 = 0$

47. (a) Set $p = 5$ and solve for q: $5q - 2000 = q$, $q = 500$ T-shirts (b) $q + p\frac{dq}{dp} = \frac{dq}{dp}$, $\frac{dq}{dp} = \frac{q}{1-p}$, $\frac{dq}{dp}\bigg|_{p=5} = \frac{500}{1-5} = -125$ T-shirts per dollar. Thus, when the price is set at \$5, the demand is dropping by 125 T-shirts per \$1 increase in price.

49. Set $C = 200{,}000$ and differentiate the equation with respect to e: $0 = 100k\frac{dk}{de} + 120e$, $\frac{dk}{de} = -\frac{120e}{100k} = -\frac{6e}{5k}$. When $e = 15$, $200{,}000 = 15{,}000 + 50k^2 + 60(15)^2 = 50k^2 + 28{,}500$, $50k^2 = 171{,}500$, $k \approx 58.57$, so $\frac{dk}{de}\bigg|_{e=15} = -\frac{6(15)}{5(58.57)} \approx -0.307$ carpenters per electrician. This means that, for a \$200,000 house whose construction employs 15 electricians, adding one more electrician would cost as much as approximately 0.307 additional carpenters. In other words, one electrician is worth approximately 0.307 carpenters.

51. (a) Set $x = 3.0$ and $g = 80$ and solve for t: $80 = 12t - 0.2t^2 - 90$, $0.2t^2 - 12t + 170 = 0$, $t \approx 22.93$ hours by the quadratic formula. (The other root is rejected because it is larger than 30.)
(b) Set $g = 80$ and differentiate the equation with respect to x: $0 = 4x\frac{dt}{dx} + 4t - 0.4t\frac{dt}{dx} - 20x$,

$\dfrac{dt}{dx} = \dfrac{4t - 20x}{0.4t - 4x} = \dfrac{t - 5x}{0.1t - x}$, so $\dfrac{dt}{dx}\bigg|_{x=3.0} \approx$

$\dfrac{22.93 - 5(3.0)}{0.1(22.93) - 3.0} \approx -11.2$ hours per grade point.

This means that, for a 3.0 student who scores 80 on the examination, 1 grade point is worth approximately 11.2 hours.

53. With P constant at 20,000, $0 = 0.6x^{-0.4}y^{0.4} + 0.4x^{0.6}y^{-0.6}\dfrac{dy}{dx}$, $\dfrac{dy}{dx} = -\dfrac{3}{2}\dfrac{y}{x}$. When $x = 100$, $20,000 = 100^{0.6}y^{0.4}$, $y = (20,000/100^{0.6})^{1/0.4} \approx 56,568,500$, so $\dfrac{dy}{dx}\bigg|_{x=100} = -\dfrac{3}{2}\dfrac{56,568,500}{100} \approx$

$-\$848,528$ per worker. The annual expenditure to maintain production at 20,000 units per day is decreasing at a rate of $848,528 per additional worker employed. In other words, each additional worker is worth approximately $848,528 to the company in annual expense.

55. $0 = 1.2y^{-0.4}r^{-0.3}p - 0.6y^{0.6}r^{-1.3}p\dfrac{dr}{dy}$, $\dfrac{dr}{dy} =$

$2\dfrac{r}{y}$, so $\dfrac{dr}{dt} = \dfrac{dr}{dy}\dfrac{dy}{dt} = 2\dfrac{r}{y}\dfrac{dy}{dt}$ by the chain rule.

57. Let $y = f(x)g(x)$. Then $\ln y = \ln f(x) + \ln g(x)$, and $\dfrac{1}{y}\dfrac{dy}{dx} = \dfrac{f'(x)}{f(x)} + \dfrac{g'(x)}{g(x)}$, so $\dfrac{dy}{dx} =$

$y\left(\dfrac{f'(x)}{f(x)} + \dfrac{g'(x)}{g(x)}\right) = f(x)g(x)\left(\dfrac{f'(x)}{f(x)} + \dfrac{g'(x)}{g(x)}\right) =$

$f'(x)g(x) + f(x)g'(x)$

59. Writing $y = f(x)$ specifies y as an explicit function of x. This can be regarded as an equation giving y as an *implicit* function of x. The procedure of finding dy/dx by implicit differentiation is then the same as finding the derivative of y as an explicit function of x: we take d/dx of both sides.

61. True. The answer follows by differentiating both sides of the equation $y = f(x)$ with respect to y: $1 = f'(x) \cdot \dfrac{dx}{dy}$, giving $\dfrac{dx}{dy} = \dfrac{1}{f'(x)} = \dfrac{1}{dy/dx}$.

Chapter 4 Review Test

1. (a) $f'(x) = e^x(x^2 - 1) + e^x(2x) = e^x(x^2 + 2x - 1)$

(b) $f'(x) = \dfrac{2x(x^2 - 1) - (x^2 + 1)(2x)}{(x^2 - 1)^2} = \dfrac{-4x}{(x^2 - 1)^2}$

(c) $f'(x) = 10(x^2 - 1)^9(2x) = 20x(x^2 - 1)^9$

(d) $f'(x) = -10(x^2 - 1)^{-11}(2x) = \dfrac{-20x}{(x^2 - 1)^{11}}$

(e) $f'(x) = e^x(x^2 + 1)^{10} + e^x \cdot 10(x^2 + 1)^9(2x) = e^x(x^2 + 1)^9(x^2 + 20x + 1)$

(f) $f'(x) = 3\left[\dfrac{x-1}{3x+1}\right]^2 \dfrac{(3x+1) - 3(x-1)}{(3x+1)^2} = \dfrac{12(x-1)^2}{(3x+1)^4}$

(g) $f'(x) = 2xe^{x^2 - 1}$

(h) $f'(x) = 2xe^{x^2-1} + (x^2 + 1)(2x)e^{x^2-1} = 2x(x^2 + 2)e^{x^2-1}$

(i) $f'(x) = \dfrac{2x}{x^2 - 1}$

(j) $f'(x) = \dfrac{\dfrac{2x}{x^2-1}(x^2 - 1) - 2x \ln(x^2 - 1)}{(x^2 - 1)^2} = \dfrac{2x - 2x \ln(x^2 - 1)}{(x^2 - 1)^2}$

2. (a) $2x - 2y\dfrac{dy}{dx} = 1, \dfrac{dy}{dx} = \dfrac{2x-1}{2y}$

(b) $2y + 2x\dfrac{dy}{dx} + 2y\dfrac{dy}{dx} = \dfrac{dy}{dx}, \dfrac{dy}{dx} = -\dfrac{2y}{2(x+y) - 1}$

(c) $\left(y + x\dfrac{dy}{dx}\right)e^{xy} + y + x\dfrac{dy}{dx} = 0, \dfrac{dy}{dx} = \dfrac{-y(e^{xy} + 1)}{x(e^{xy} + 1)} = -\dfrac{y}{x}$

(d) $\left(\dfrac{1}{y/x}\right)\dfrac{(dy/dx)x - y}{x^2} = \dfrac{dy}{dx}, x\dfrac{dy}{dx} - y = xy\dfrac{dy}{dx}, \dfrac{dy}{dx} = \dfrac{y}{x(1-y)}$

3. (a) $y' = 1 - 2e^{2x-1}$; set $y' = 0$: $1 - 2e^{2x-1} = 0$, $2x - 1 = \ln(1/2) = -\ln 2$, $x = (1 - \ln 2)/2$

(b) $y' = 2xe^{x^2} = 0$ when $x = 0$

(c) $y' = \dfrac{(x+1) - x}{(x+1)^2} = \dfrac{1}{(x+1)^2}$ is never 0, so there are no points at which the tangent line is horizontal.

(d) $y' = \dfrac{3}{2}x^{1/2} - \dfrac{1}{2}x^{-1/2} = 0$, $3x^{1/2} = x^{-1/2}$, $3x = 1$, $x = 1/3$

4. (a) Let $q(t)$ be the number of books sold per week and let $p(t)$ be the price per book. We are given that $q(0) = 1000$ and $q'(0) = 200$ (taking $t = 0$ as now), and that $p(0) = 20$ and $p'(0) = -1$. Since $R(t) = p(t)q(t)$, we have $R'(0) = p'(0)q(0) + p(0)q'(0) = (-1)(1000) + 20(200) = \$3000$ per week (rising).

(b) $(-1)(1000) + 20q'(0) = 5000$, $q'(0) = 300$ books per week

(c) $R = pq$ gives $R' = p'q + pq'$. Thus, $R'/R = R'/(pq) = (p'q + pq')/pq = p'/p + q'/q$

5. (a) Let P be the stock price and E the earnings. We are given $P = 100$, $P' = 50$, $E = 1$, and $E' = 0.10$. $\dfrac{d}{dt}\left(\dfrac{P}{E}\right) = \dfrac{P'E - PE'}{E^2} = \dfrac{50(1) - 100(0.10)}{1^2} = 40$, so the P/E ratio was rising at a rate of 40 units per year.

(b) $\dfrac{P' - 100(0.10)}{1^2} = 100$, $P' = \$110$ per year

(c) $Q = P/E$ gives $Q' = (P'E - PE')/E^2$. Thus, $Q'/Q = Q'/(P/E) = (P'E - PE')/PE = P'/P - E'/E$.

6. (a) $s'(t) = \dfrac{2460.7 \, e^{-0.55(t-4.8)}}{(1 + e^{-0.55(t-4.8)})^2}$ **(b)** $s'(6) \approx 553$ books per week **(c)** $s'(15) \approx 15$ books per week.

7. If $H(t)$ is the daily number of hits t weeks into the year, then $H(t) = 1000(1.05)^t$. So, $H'(t) = 1000 \ln(1.05) (1.05)^t$ and $H'(52) = 1000 \cdot \ln(1.05) (1.05)^{52} \approx 616.8$ hits per day per week.

8. (a) $100q + 100p\dfrac{dq}{dp} + 2q\dfrac{dq}{dp} = 0, \dfrac{dp}{dq} =$
$\dfrac{-100q}{100p + 2q}$; when $p = 40$ and $q = 1000, \dfrac{dp}{dq} =$
$\dfrac{-100(1000)}{100(40) + 2(1000)} \approx -16.67$ copies per $1. The demand for the gift edition of *Lord of the Rings* is dropping at a rate of about 16.67 copies per $1 increase in the price.

(b) Since $R = pq, \dfrac{dR}{dp} = q + p\dfrac{dq}{dp} \approx 1000 + 40(-16.67) \approx \$333$ per dollar increase in price. The derivative is positive, so the price should be raised.

Chapter 5
5.1

1. Absolute min.: $(-3, -1)$, relative max: $(-1, 1)$, relative min: $(1, 0)$, absolute max: $(3, 2)$

3. Absolute min: $(3, -1)$ and $(-3, -1)$, absolute max: $(1, 2)$

5. Absolute min: $(-3, 0)$ and $(1, 0)$, absolute max: $(-1, 2)$ and $(3, 2)$

7. Relative min: $(-1, 1)$

9. Absolute min: $(-3, -1)$, relative max: $(-2, 2)$, relative min: $(1, 0)$, absolute max: $(3, 3)$

11. Relative max: $(-3, 0)$, absolute min: $(-2, -1)$, stationary non-extreme point: $(1, 1)$.

13. Stationary minimum at $x = -1$

15. Stationary minima at $x = -2$ and $x = 2$, stationary maximum at $x = 0$

17. Singular minimum at $x = 0$, stationary non-extreme point at $x = 1$

19. Stationary minimum at $x = -2$, singular non-extreme point at $x = -1$, singular non-extreme point at $x = 1$, stationary maximum at $x = 2$

21. $f'(x) = 2x - 4$; $f'(x) = 0$ when $2x - 4 = 0$, $x = 2$; $f'(x)$ is defined for all x in $[0, 3]$. Thus, we have the stationary point at $x = 2$ and the endpoints $x = 0$ and $x = 3$:

x	0	2	3
$f(x)$	1	-3	-2

The graph must decrease from $x = 0$ until $x = 2$, then increase again until $x = 3$.

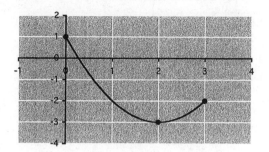

This gives an absolute max at $(0, 1)$, an absolute min at $(2, -3)$, and a relative max at $(3, -2)$.

23. $g'(x) = 3x^2 - 12$; $g'(x) = 0$ when $3x^2 - 12 = 0$, $x = \pm 2$; $g'(x)$ is defined for all x in $[-4, 4]$. Thus, we have stationary points at $x = \pm 2$ and the endpoints $x = \pm 4$:

x	-4	-2	2	4
$g(x)$	-16	16	-16	16

The graph increases from $x = -4$ until $x = -2$, then decreases until $x = 2$, then increases until $x = 4$.

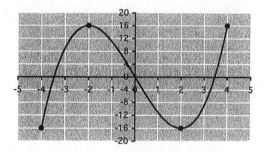

This gives an absolute min at $(-4, -16)$, an absolute max at $(-2, 16)$, an absolute min at $(2, -16)$, and an absolute max at $(4, 16)$.

25. $f'(t) = 3t^2 + 1$; $f'(t) = 0$ when $3t^2 + 1 = 0$, which has no solution; $f'(t)$ is defined for all t in $[-2, 2]$. Thus, there are no critical points in the domain, just the endpoints ± 2:

t	-2	2
$f(t)$	-10	10

The graph increases from $x = -2$ until $x = 2$.

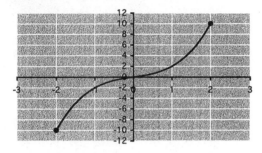

This gives an absolute min at $(-2, -10)$ and an absolute max at $(2, 10)$.

27. $h'(t) = 6t^2 + 6t$; $h'(t) = 0$ when $6t^2 + 6t = 0$, $t = 0$ or $t = -1$; $h'(t)$ is defined for all t in $[-2, \infty)$. Thus, we have stationary points at $t = 0$ and $t = -1$ and the endpoint at $t = -2$. In addition to these we test one point to the right of $t = 0$:

t	-2	-1	0	1
$h(t)$	-4	1	0	5

The graph increases from $x = -2$ until $x = -1$, then decreases until $x = 0$, then increases from that point on. (Remember that $x = 1$ is just a test point to see whether the graph is increasing or decreasing to the right of $x = 0$.)

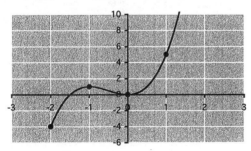

This gives an absolute min at $(-2, -4)$, a relative max at $(-1, 1)$, and relative min at $(0, 0)$.

29. $f'(x) = 4x^3 - 12x^2$; $f'(x) = 0$ when $4x^3 - 12x^2 = 0$, $4x^2(x - 3) = 0$, $x = 0$ or $x = 3$; $f'(x)$ is defined for all x in $[-1, \infty)$. Thus, we have stationary points at $x = 0$ and $x = 3$ and the endpoint at $x = -1$. In addition to these we test one point to the right of $x = 3$:

x	-1	0	3	4
$f(x)$	5	0	-27	0

The graph decreases from $x = -1$ until $x = 0$, continues to decrease until $x = 3$, then increases from that point on.

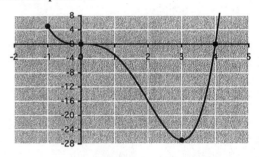

This gives a relative max at $(-1, 5)$ and an absolute min at $(3, -27)$.

31. $g'(t) = t^3 - 2t^2 + t$; $g'(t) = 0$ when $t^3 - 2t^2 + t = 0$, $t(t - 1)^2 = 0$, $t = 0$ or $t = 1$; $g'(t)$ is defined for all t. Thus, we have stationary points at $t = 0$ and $t = 1$. Since we have no endpoints, we test a point to the left of $t = 0$ and a point to the right of $t = 1$:

t	-1	0	1	2
$g(t)$	17/12	0	1/12	2/3

The graph decreases until $t = 0$, increases until $t = 1$, then continues to increase to the right of $t = 1$.

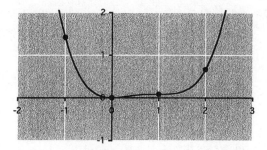

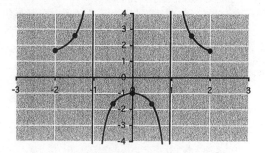

This gives an absolute min at $(0, 0)$ and no other extrema.

This gives a relative min at $(-2, 5/3)$, a relative max at $(0, -1)$, and a relative min at $(2, 5/3)$.

33. $f'(t) = \dfrac{2t(t^2 - 1) - 2t(t^2 + 1)}{(t^2 - 1)^2} = \dfrac{-4t}{(t^2 - 1)^2}$;
$f'(t) = 0$ when $t = 0$; $f'(t)$ is not defined when $t = \pm 1$, but these points are not in the domain of f. Thus, we have a stationary point at $t = 0$ and the endpoints at $t = \pm 2$. We also test points on either side of $t = \pm 1$ to see how f behaves near these points where it goes to $\pm\infty$ (that is, it has vertical asymptotes there, because the denominator goes to 0):

t	-2	$-3/2$	$-1/2$	0
$f(t)$	$5/3$	$13/5$	$-5/3$	-1
t	$1/2$	$3/2$	2	
$f(t)$	$-5/3$	$13/5$	$5/3$	

The graph increases from $t = -2$, approaching the vertical asymptote at $t = -1$; on the other side of the asymptote it increases until $t = 0$ then decreases as it approaches another vertical asymptote at $t = 1$; on the other side of the second asymptote it decreases until $t = 2$.

35. $f'(x) = \frac{3}{2}x^{1/2} - \frac{1}{2}x^{-1/2}$; $f'(x) = 0$ when $\frac{3}{2}x^{1/2} - \frac{1}{2}x^{-1/2} = 0$, $3x - 1 = 0$, $x = 1/3$; $f'(x)$ is not defined at $x = 0$. Thus, we have a stationary point at $x = 1/3$ and a singular point at $x = 0$, which is also an endpoint. In addition, we test a point to the right of $x = 1/3$:

x	0	$1/3$	1
$f(x)$	0	$-2\sqrt{3}/9$	0

The graph decreases from $t = 0$ until $t = 1/3$, then increases from that point on.

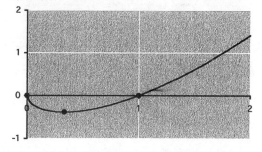

This gives a relative max at $(0, 0)$ and an absolute min at $(1/3, -2\sqrt{3}/9)$.

37. Since the domain of g is not specified, we take it to be the set of all x for which g is defined, which is $[0, \infty)$. $g'(x) = 2x - 2x^{-1/2}$; $g'(x) = 0$ when $2x - 2x^{-1/2} = 0$, $x = x^{-1/2}$, $x^{3/2} = 1$, $x = 1$;

101

$g'(x)$ is not defined when $x = 0$, which is also an endpoint. Thus, we have a singular endpoint at $x = 0$ and a stationary point at $x = 1$. We test one point to the right of $x = 1$:

x	0	1	2
$g(x)$	0	-3	-1.7

The graph decreases from $x = 0$ to $x = 1$, then increases from that point on.

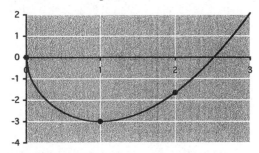

Thus, there is a relative max at $(0, 0)$ and an absolute min at $(1, -3)$.

39. The domain of g is $(-\infty, \infty)$. $g'(x) = \dfrac{3x^2(x^2 + 3) - x^3(2x)}{(x^2 + 3)^2} = \dfrac{x^4 + 9x^2}{(x^2 + 3)^2}$; $g'(x) = 0$ when $x^4 + 9x^2 = 0$, $x^2(x^2 + 9) = 0$, $x = 0$; $g'(x)$ is always defined. Thus, we have only one stationary point, at $x = 0$. In addition, we test a point on either side:

x	-1	0	1
$g(x)$	$-1/4$	0	$1/4$

The graph is always increasing.

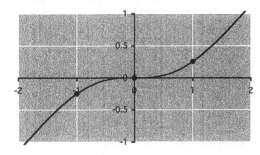

Thus, there are no relative extrema.

41. $f'(x) = 1 - 1/x$; $f'(x) = 0$ when $1 - 1/x = 0$, $x = 1$; $f'(x)$ is defined for all x in the domain of f. We test the one stationary point at $x = 1$ and a point on either side:

x	$1/2$	1	2
$g(x)$	1.19	1	1.31

The graph decreases until $x = 1$ and then increases from that point on.

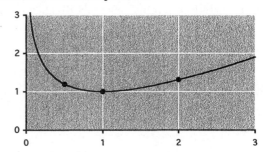

Thus, there is an absolute min at $(1, 1)$.

43. $g'(t) = e^t - 1$; $g'(t) = 0$ when $e^t - 1 = 0$, $e^t = 1$, $t = \ln 1 = 0$; $g'(t)$ is always defined. We need to test only the stationary point and the endpoints:

t	-1	0	1
$g(t)$	1.37	1	1.72

The graph decreases from $t = -1$ to $t = 0$ and then increases to $t = 1$.

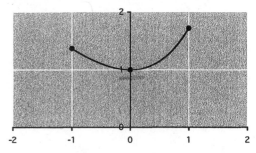

Thus, there is a relative max at $(-1, e^{-1} + 1)$, an absolute min at $(0, 1)$, and an absolute max at $(1, e - 1)$.

Section 5.1

45. The domain of f is all $x \neq 4$. $f'(x) = \dfrac{4x(x+4) - (2x^2 - 24)}{(x+4)^2} = \dfrac{2x^2 + 16x + 24}{(x+4)^2}$; $f'(x) = 0$ when $2x^2 + 16x + 24 = 0$; $2(x+2) \cdot (x+6) = 0$; $x = -2$ or -6; $f'(x)$ is defined for all x in the domain of f. Thus, we have stationary points at $x = -2$ and $x = -6$. We test points on either side of the stationary points and between them and the asymptote at $x = -4$:

x	-7	-6	-5
$f(x)$	-24.7	-24	-26

x	-3	-2	-1
$f(x)$	-6	-8	-7.3

The graph increases to $x = -6$ then decreases approaching the asymptote at $x = -4$. On the other side of the asymptote the graph decreases to $x = -2$ and then increases again.

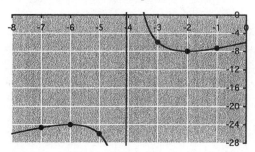

Thus, there is a relative max at $(-6, -24)$ and a relative min at $(-2, -8)$.

47. $f'(x) = e^{1-x^2} + x(-2x)e^{1-x^2} = (1 - 2x^2)e^{1-x^2}$; $f'(x) = 0$ when $1 - 2x^2 = 0$, $x = \pm 1/\sqrt{2}$; $f'(x)$ is always defined. We test the two stationary points and a point on either side:

x	-1	$-1/\sqrt{2}$	$1/\sqrt{2}$	1
$f(x)$	-1	-1.17	1.17	1

The graph decreases until $x = -1/\sqrt{2}$, increases until $x = 1/\sqrt{2}$, then decreases again.

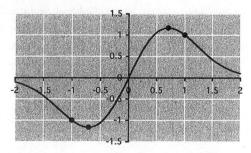

Thus, there is an absolute max at $(1/\sqrt{2}, \sqrt{e/2})$ and an absolute min at $(-1/\sqrt{2}, -\sqrt{e/2})$.

49. $y' = 2x - \dfrac{1}{(x-2)^2}$. The graphs of f and of f' look like this:

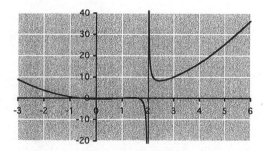

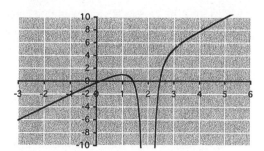

Looking closely at the graph of f' we can see three places where it crosses the x axis. Zooming in we can locate these points approximately as $x \approx 0.15$, $x \approx 1.40$, and $x \approx 2.45$. Thus, we have relative minima at $(0.15, -0.52)$ and $(2.45, 8.22)$ and a relative maximum at $(1.40, 0.29)$.

51. $f'(x) = 2(x-5)(x+4)(x-2) + (x-5)^2 \cdot (x-2) + (x-5)^2(x+4) = (x-5)[2(x+4) \cdot (x-2) + (x-5)(x-2) + (x-5)(x+4)] = (x-5)(4x^2 - 4x - 26)$. The graphs of f and f' look like this:

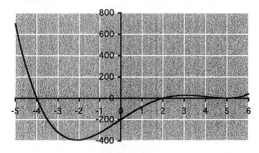

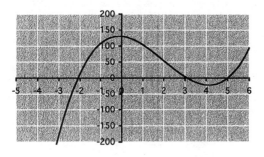

The graph of f' crosses the x axis in three places, hence f has three stationary points. Zooming in we can approximate these as $x \approx -2.10$, $x \approx 3.10$, and $x = 5$. Substituting these and the endpoints -5 and 6 into f, we find that we have an absolute maximum at $(-5, 700)$, relative maxima at $(3.10, 28.19)$ and $(6, 40)$, an absolute minimum at $(-2.10, -392.69)$, and a relative minimum at $(5, 0)$.

53.

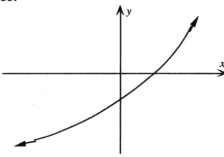

55.

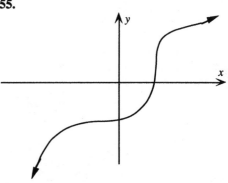

57. Not necessarily; it could be neither a relative maximum nor a relative minimum, as in the graph of $y = x^3$ at the origin.

59. Answers will vary.

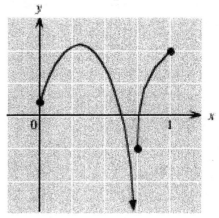

5.2

1. Solve for $y = 10 - x$ and substitute: $P = x(10 - x) = 10x - x^2$. Differentiate: $P'(x) = 10 - 2x$; $P'(x) = 0$ when $x = 5$. Since the graph of $P(x)$ is a parabola opening downward, $x = 5$ must be its vertex, hence gives the maximum. The corresponding value of y is $y = 10 - x = 5$. Hence, $x = y = 5$ and $P = 25$.

3. Solve for $y = 9/x$ and substitute: $S = x + 9/x$ with domain $x > 0$. Differentiate: $S'(x) = 1 - 9/x^2$; $S'(x) = 0$ when $x = 3$ ($x = -3$ is not in the domain). Testing points on either side of $x = 3$ we see that we have the minimum. Hence, $x = y = 3$ and $S = 6$.

5. Solve for $x = 10 - 2y$ and substitute: $F(y) = (10 - 2y)^2 + y^2 = 100 - 40y + 5y^2$. $F'(y) = -40 + 10y$; $F'(y) = 0$ when $y = 4$. Since the graph of F is a parabola opening upward, $y = 4$ must be its vertex, hence gives the minimum. Hence, $x = 2$, $y = 4$, and $F = 20$.

7. Since y appears in both constraints, we can solve for the other two variables in terms of y: $x = 30 - y$ and $z = 30 - y$. Substitute: $P = (30 - y) \cdot y \cdot (30 - y) = y(30 - y)^2$. Since all the variables must be nonnegative, we have $0 \le y \le 30$ as the domain. Differentiate: $P'(y) = (30 - y)^2 - 2y(30 - y) = (30 - y)(30 - y - 2y) = (30 - y)(30 - 3y)$; $P'(y) = 0$ when $y = 10$ or $y = 30$. Substituting these values and the other endpoint $y = 0$ into P, we find that the maximum occurs when $y = 10$. Thus, $x = 20$, $y = 10$, $z = 20$, and $P = 4000$.

9. Let x and y be the dimensions. Then we want to maximize $A = xy$ subject to $2x + 2y = 20$. Solve for $y = 10 - x$ and substitute: $A = x(10 - x) = 10x - x^2$. Differentiate: $A'(x) = 10 - 2x$; $A'(x) = 0$ when $x = 5$. Since the graph of $A(x)$ is a parabola opening downward, $x = 5$ must be its vertex, hence gives the maximum. The corresponding value of y is $y = 10 - x = 5$. Hence, the rectangle should have dimensions 5×5.

11. Let x be the length of the east and west sides and let y be the length of the north and south sides. The area is $A = xy$ and the cost of the fence is $2 \cdot 4x + 2 \cdot 2y = 8x + 4y$. (We multiply by 2 because there are two sides of length x and two sides of length y.) So, our problem is to maximize $A = xy$ subject to $8x + 4y = 80$. Solve for $y = 20 - 2x$ and substitute: $A(x) = x(20 - 2x) = 20x - 2x^2$. Since x and y must both be nonnegative, we have $0 \le x \le 10$. Differentiate: $A'(x) = 20 - 4x$; $A'(x) = 0$ when $20 - 4x = 0$, $x = 5$; $A'(x)$ is always defined. Testing the endpoints 0 and 10 as well as the stationary point 5, we see that the maximum area occurs when $x = 5$. The corresponding value of y is $y = 10$, so the largest area possible is $5 \times 10 = 50$ square feet.

13. $R = pq = p(200{,}000 - 10{,}000p) = 200{,}000p - 10{,}000p^2$. For p and q to both be nonnegative, we must have $0 \le p \le 20$. Differentiate: $R'(p) = 200{,}000 - 20{,}000p$; $R'(p) = 0$ when $p = 10$; $R'(p)$ is always defined. Testing the endpoints 0 and 20 as well as the stationary point 10, we see that the maximum revenue occurs when the price is $p = \$10$.

15. $R = pq = p\left(-\frac{4}{3}p + 80\right) = -\frac{4}{3}p^2 + 80p$.

For p and q to both be nonnegative, we must have $0 \le p \le 60$. Differentiate: $R'(p) = -\frac{8}{3}p + 80$;

105

Section 5.2

$R'(p) = 0$ when $p = 30$; $R'(p)$ is always defined. Testing the endpoints 0 and 60 as well as the stationary point 30, we see that the maximum revenue occurs when the price is $p = \$30$.

17. (a) $R = pq = \dfrac{500{,}000}{q^{1.5}} \cdot q = \dfrac{500{,}000}{q^{0.5}} = 500{,}000q^{-0.5}$. We are given that $q \geq 5000$. Differentiate: $R'(q) = -250{,}000q^{-1.5}$; $R'(q)$ is never 0; $R'(q)$ is defined for all $q > 0$. We test the endpoint $q = 5000$ and a point to the right: $R(5000) = 7071.07$ and $R(10{,}000) = 5000$. Thus, the revenue is decreasing and its maximum value occurs at the endpoint $q = 5000$. The corresponding price is $p = \$1.41$ per pound. **(b)** As found in part (a), the maximum occurs at $q = 5000$ pounds. **(c)** The maximum revenue is $R(5000) = \$7071.07$ per month.

19. We are given two points: $(p, q) = (25, 22)$ and $(14, 27.5)$. The equation of the line through these two points is $q = -p/2 + 34.5$. $R = pq = p(-p/2 + 34.5) = -p^2/2 + 34.5p$. For p and q to both be nonnegative we must have $0 \leq p \leq 69$. Differentiate: $R'(p) = -p + 34.5$; $R'(p) = 0$ when $p = 34.5$; $R'(p)$ is defined for all p. Testing the endpoints 0 and 69 as well as the stationary point 34.5, we find that the revenue is maximized when the price is $p = 34.5$¢ per pound, for an annual (per capita) revenue of $\$5.95$.

21. $R = pq$ and $C = 25q$, so $P = R - C = pq - 25q = p\left(-\dfrac{4}{3}p + 80\right) - 25\left(-\dfrac{4}{3}p + 80\right) = -\dfrac{4}{3}p^2 + \dfrac{340}{3}p - 2000$. For p and q to both be nonnegative, we must have $0 \leq p \leq 60$. Differentiate: $P'(p) = -\dfrac{8}{3}p + \dfrac{340}{3}$; $P'(p) = 0$ when $p = 42.5$; $P'(p)$ is defined for all p. Testing the endpoints 0 and 60 as well as the stationary point 42.5, we find that the profit is maximized when the price is $p = \$42.50$ per ruby, for a weekly profit of $P(42.5) = \$408.33$.

23. (a) $R = pq$ and $C = 100q$, so $P = R - C = pq - 100q = \dfrac{1000}{q^{0.3}} \cdot q - 100q = 1000q^{0.7} - 100q$. For p and q to be defined and nonnegative we need $q > 0$. Differentiate: $P'(q) = 700q^{-0.3} - 100$; $P'(q) = 0$ when $q = 7^{1/0.3} \approx 656$; $P'(q)$ is defined for all $q > 0$. Testing the stationary point at approximately 656 and points on either side, we see that the profit is maximized when you sell 656 headsets, for a profit of $P(656) \approx \$28{,}120$.
(b) The corresponding price is $p \approx \$143$ per headset.

25. $\bar{C}(x) = C(x)/x = \dfrac{150{,}000}{x} + 20 + \dfrac{x}{10{,}000}$, with domain $x > 0$. Differentiate: $\bar{C}'(x) = -\dfrac{150{,}000}{x^2} + \dfrac{1}{10{,}000}$; $\bar{C}'(x) = 0$ when $x = \sqrt{150{,}000 \cdot 10{,}000} \approx 38{,}730$; $\bar{C}'(x)$ is defined for all $x > 0$. Testing the stationary point at 38,730 and points on either side, we see that the average cost is minimized when we manufacture $x = 38{,}730$ CD players, giving an average cost of $\bar{C}(38{,}730) \approx \$28$ per CD player.

27. Net cost is $N = C(q) - 500q = 4000 + 100q^2 - 500q$, with $q \geq 0$. Differentiate: $N'(q) = 200q - 500$; $N'(q) = 0$ when $q = 2.5$; $N'(q)$ is defined for all q. Testing the endpoint $q = 0$, the stationary point $q = 2.5$, and one more point to the right of 2.5, we see that the net cost is minimized when $q = 2.5$ pounds of pollutant per day.

29. Let x be the length of one side of the square cut out of each corner, as in the figure:

106

Section 5.2

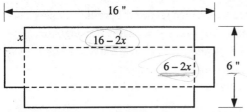

When the sides are folded up, the resulting box will have volume $V = x(16 - 2x)(6 - 2x) = 4x^3 - 44x^2 + 96x$. For the sides to have nonnegative lengths, we must have $0 \le x \le 3$. Differentiate: $V'(x) = 12x^2 - 88x + 96 = 2(x - 6)(6x - 8)$; $V'(x) = 0$ when $x = 4/3$ or $x = 6$, but $x = 6$ is outside of the domain. Testing the endpoints 0 and 3 and the stationary point 4/3, we find that the largest volume occurs when $x = 4/3$". The box with the largest volume has dimensions $13\ 1/3" \times 3\ 1/3" \times 1\ 1/3"$ and it has volume $V(4/3) = 1600/27 \approx 59$ cubic inches.

31. Let l = length, w = width, and h = height. We want to maximize the volume $V = lwh$ but we are restricted by $l + w + h \le 62$. Since we're looking for the largest volume, we can assume that $l + w + h = 62$. We are also told that $h = w$. Thus, our problem is to maximize $V = lwh$ subject to $l + w + h = 62$ and $h = w$. Substitute $h = w$ in the other constraint to get $l + 2w = 62$, then solve for $l = 62 - 2w$. Substitute to get $V = (62 - 2w)w^2 = 62w^2 - 2w^3$. For all dimensions to be nonnegative we need $0 \le w \le 31$. Differentiate: $V'(w) = 124 - 6w^2$; $V'(w) = 0$ when $124w - 6w^2 = 0$, $2w(62 - 3w) = 0$, $w = 0$ or $w = 62/3$; $V'(w)$ is defined for all w. Testing the endpoints 0 and 31 as well as the (other) stationary point 62/3, we see that the volume is maximized when $w = 62/3$. The corresponding values of the other dimensions are $l = h = 62/3$. Thus, the largest volume bag has dimensions $l = w = h \approx 20.67$ in, and volume $V \approx 8827$ in^3.

33. Let l = length, w = width, and h = height. We want to maximize the volume $V = lwh$ but we have the constraints $l + w = 45$ and $w + h = 45$. Solve for $l = 45 - w$ and $h = 45 - w$. Substitute to get $V = w(45 - w)^2$. For all dimensions to be nonnegative we need $0 \le w \le 45$. Differentiate: $V'(w) = (45 - w)^2 - 2w(45 - w) = (45 - w)(45 - w - 2w) = (45 - w)(45 - 3w)$; $V'(w) = 0$ when $w = 15$ or $w = 45$; $V'(w)$ is defined for all w. Testing the endpoints 0 and 45 as well as the stationary point 15, we see that the volume is maximized when $w = 15$. The corresponding values of the other dimensions are $l = h = 30$. Thus, the largest volume bag has dimensions $l = 30$ in, $w = 15$ in, and $h = 30$ in.

35. Let l = length, w = width, and h = height. We want to maximize the volume $V = lwh$ but we are restricted by $l + 2(w + h) \le 108$. Since we're looking for the largest volume, we can assume that $l + 2(w + h) = 108$. We are also told that $w = h$. Substitute to get $l + 2(2w) = 108$, or $l + 4w = 108$. Solve for $l = 108 - 4w$ and substitute to get $V = w^2(108 - 4w) = 108w^2 - 4w^3$. For all the dimensions to be nonnegative we need $0 \le w \le 27$. Differentiate: $V'(w) = 216w - 12w^2 = 12w(18 - w)$; $V'(w) = 0$ when $w = 0$ or $w = 18$; $V'(w)$ is defined for all w. Testing the endpoints 0 and 27 as well as the stationary point 18, we see that the volume is maximized when $w = 18$. The corresponding values of the other dimensions are $h = 18$ and $l = 36$. Thus, the largest volume package has dimensions $l = 36$ in and $w = h = 18$ in, and has volume $V = 11{,}664$ in^3.

37. (a) $R'(t) = 350(39)e^{-0.3t} - 0.3(350)(39t + 68)e^{-0.3t} = 350(39 - 11.7t - 20.4)e^{-0.3t} = 350(18.6 - 11.7t)e^{-0.3t}$. $R'(t) = 0$ when $t \approx 1.6$;

$R'(t)$ is defined for all t. Testing points on either side of the stationary point, we see that the maximum revenue occurs at $t \approx 1.6$ years, or year 2001.6. **(b)** The maximum revenue is $R(1.6) \approx$ \$28,241 million.

39. Let $C(t) = \dfrac{D(t)}{S(t)} = \dfrac{10 + t}{2.5e^{0.08t}} = 0.4(10 + t)e^{-0.08t}$.
$C'(t) = 0.4e^{-0.08t} - 0.032(10 + t)e^{-0.08t} = (0.4 - 0.32 - 0.032t)e^{-0.08t} = (0.08 - 0.032t)e^{-0.08t}$;
$C'(t) = 0$ when $t = 2.5$; $C'(t)$ is defined for all t. Testing points on either side of 2.5 we see that the maximum revenue occurs at $t = 2.5$ or midway through 1972, when $D(2.5)/S(2.5) \approx 4.09$. Midway through 1972 the number of new (approved) drugs per \$1 billion dollars of spending on research and development reached a high of around 4 approved drugs per \$1 billion.

41. $p = ve^{-0.05t} = (300{,}000 + 1000t^2)e^{-0.05t}$. $p' = 2000te^{-0.05t} - 0.05(300{,}000 + 1000t^2)e^{-0.05t} = (-15{,}000 + 2000t - 50t^2)e^{-0.05t}$; $p' = 0$ when $-15{,}000 + 2000t - 50t^2 = 0$, $-50(x - 30)(x - 10) = 0$, $x = 10$ or $x = 30$; p' is always defined. Testing $x = 0$, 10, 30, and 40, we see that the maximum discounted value occurs $t = 30$ years from now.

43. $R = (100 + 2t)(400{,}000 - 2500t) = 40{,}000{,}000 + 550{,}000t - 5000t^2$. $R' = 550{,}000 - 10{,}000t$; $R' = 0$ when $t = 55$; R' is defined for all t. Testing $t = 55$ and points on either side of it (or recognizing that the graph of R is a parabola), we see that the release should be delayed for 55 days.

45. $R = 500\sqrt{x}$, $C = 10{,}000 + 2x$, so the total profit is $P = R - C = 500\sqrt{x} - (10{,}000 + 2x)$

and the average profit per copy is $\bar{P} = 500x^{-1/2} - 10{,}000x^{-1} + 2$. $\bar{P}' = -250x^{-3/2} + 10{,}000x^{-2}$; $\bar{P}' = 0$ when $x = (10{,}000/250)^2 = 1600$; $\bar{P}'$ is defined for all $x > 0$. Testing the stationary point 1600 and a point on either side, we see that the average profit is maximized at $x = 1600$ copies. For this many copies, the average profit is $\bar{P}(1600) = \$8.25$/copy. Since $P' = 250x^{-1/2} + 2$, the marginal profit is $P'(1600) = \$8.25$/copy also. At this value of x, average profit equals marginal profit; beyond this the marginal profit is smaller than the average and so the average declines.

47. We are being asked to find the extreme values of the derivative, $N'(t)$. Call this function $M(t) = N'(t) = 0.084702t^2 - 2.1844t + 13.029$. Then $M'(t) = 0.169404t - 2.1844$; $M'(t) = 0$ when $t \approx 13$. Testing the endpoints 0 and 39 as well as the stationary point 13, we see that M has an absolute minimum of $M(13) \approx -1.05$ and an absolute maximum of $M(39) \approx 56.7$. Hence, N was decreasing most rapidly in 1963 and increasing most rapidly in 1989.

49. Let $r(t) = c'(t) = -0.00813t^2 + 0.274t - 0.892$. $r'(t) = -0.01626t + 0.274$; $r'(t) = 0$ when $t \approx 17$. Testing the endpoints 8 and 30 as well as the stationary point 17, we see that $c'(t)$ has its maximum when $t = 17$ days. This means that the embryo's oxygen consumption is increasing most rapidly 17 days after the egg is laid.

51. We want to minimize $S = \pi r^2 + 2\pi rh$ subject to $\pi r^2 h = 5000$. Solve for $h = 5000/(\pi r^2)$ and substitute to get $S = \pi r^2 + 2\pi r \dfrac{5000}{\pi r^2} = \pi r^2 + \dfrac{10{,}000}{r}$. For the dimensions to be defined and nonnegative we need $r > 0$. $S'(r) = 2\pi r - \dfrac{10{,}000}{r^2}$; $S'(r) = 0$ when $r = \sqrt[3]{\dfrac{5000}{\pi}} = 10\sqrt[3]{\dfrac{5}{\pi}} \approx 11.7$. Testing the stationary point and a point on either side, we see that $S(r)$ has an absolute minimum at $r \approx 11.7$. The corresponding value of h is $h = 10\sqrt[3]{\dfrac{5}{\pi}}$ also. Hence, the bucket using the least plastic has dimensions $h = r \approx 11.7$ cm.

53. Let $y(x)$ be the annual yield per tree when there are x trees; $y(x) = 100 - (x - 50) = 150 - x$. If $Y(x)$ is the total annual yield from x trees, then $Y(x) = xy(x) = x(150 - x) = 150x - x^2$. $Y'(x) = 150 - 2x$; $Y'(x) = 0$ when $x = 75$. Since the graph of Y is a parabola opening downward, we know that this gives the maximum value of Y. Hence, the total annual yield is largest when there are 75 trees, or 25 additional trees beyond the 50 we already have.

55. Let $R(t) = N'(t) = \dfrac{82.8(21.8)(\ln 7.14)(7.14)^{-t}}{[1 + 21.8(7.14)^{-t}]^2} \approx \dfrac{3548(7.14)^{-t}}{[1 + 21.8(7.14)^{-t}]^2}$. $R'(t) =$

$\dfrac{-6974(7.14)^{-t}[1 + 21.8(7.14)^{-t}]^2 + 304{,}000(7.14)^{-t}[1 + 21.8(7.14)^{-t}](7.14)^{-t}}{[1 + 21.8(7.14)^{-t}]^4} =$

$\dfrac{(7.14)^{-t}[-6974 - 152{,}000(7.14)^{-t} + 304{,}000(7.14)^{-t}]}{[1 + 21.8(7.14)^{-t}]^3} = \dfrac{(7.14)^{-t}[-6974 + 152{,}000(7.14)^{-t}]}{[1 + 21.8(7.14)^{-t}]^3}$; $R'(t) = 0$

when $t = -\ln(6974/152{,}000)/\ln(7.14) \approx 1.6$. Testing the endpoints 0 and 3 and the stationary point 1.6, we see that $R(t)$ has an absolute maximum when $t \approx 1.6$. Hence, book sales were increasing most rapidly in the third quarter of 1998 ($t \approx 1.6$) and the rate of increase was $N'(1.6) \approx 40.7$ million books per year.

57. $\dfrac{d}{dt}\left(\dfrac{S(t)}{W(t)}\right) = \dfrac{S'(t)W(t) - S(t)W'(t)}{W(t)^2} =$
$\dfrac{0.0000013t^4 - 0.00006t^3 + 0.0004967t^2 + 0.0014336t - 0.19068}{(0.001t^3 - 0.028t^2 + 0.28t - 0.70)^2}$
Graph:

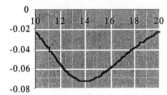

From this graph, or from the graph of its derivative, we see that $\dfrac{d}{dt}\left(\dfrac{S(t)}{W(t)}\right)$ had a relative minimum in 1994 ($t \approx 14$), when $\dfrac{d}{dt}\left(\dfrac{S(t)}{W(t)}\right)_{t=14} \approx -0.072$. The fraction of bottled water sales due to sparkling water was decreasing most rapidly in 1994. At that time, it was decreasing at a rate of 7.2 percentage points per year.

59. $p = v(1.05)^{-t} = \dfrac{10{,}000(1.05)^{-t}}{1 + 500e^{-0.5t}}$.

Section 5.2

$$p' = 10{,}000 \, \frac{-\ln(1.05)(1.05)^{-t}(1 + 500e^{-0.5t}) + 500(0.5)(1.05)^{-t}e^{-0.5t}}{(1 + 500e^{-0.5t})^2} =$$

$$10{,}000 \, \frac{(1.05)^{-t}\{-\ln(1.05) + [250 - 500\ln(1.05)]e^{-0.5t}\}}{(1 + 500e^{-0.5t})^2}; \, p' = 0 \text{ when } e^{-0.5t} = \frac{\ln(1.05)}{250 - 500\ln(1.05)},$$

$$t = -2\ln\left(\frac{\ln(1.05)}{250 - 500\ln(1.05)}\right) \approx 17.$$ Testing the stationary point at 17 and a point on either side, we see that the maximum value of p occurs when $t \approx 17$. So, you should sell them in 17 years, when they will be worth approximately $3960.

61. We want to minimize $C = 20{,}000x + 365y$ subject to $x^{0.4}y^{0.6} = 1000$. Solve for $y = (1000x^{-0.4})^{1/0.6} = 1000^{5/3}x^{-2/3}$ and substitute: $C = 20{,}000x + 365(1000^{5/3})x^{-2/3}$. The domain is $x > 0$. Differentiate: $C'(x) = 20{,}000 - \frac{730}{3}1000^{5/3}x^{-5/3}$;

$C'(x) = 0$ when $x = \left(\frac{60{,}000}{730 \cdot 1000^{5/3}}\right)^{-3/5} \approx 71$.

Testing the stationary point at 71 and a point on either side, we see that C has its minimum at $x \approx 71$. So, you should hire 71 employees.

63. $\frac{d(TR)}{dQ} = b - 2cQ$; $\frac{d(TR)}{dQ} = 0$ when $Q = \frac{b}{2c}$, which is (D).

65. The problem is uninteresting because the company can accomplish the objective by cutting away the entire sheet of cardboard, resulting in a box with surface area zero. Put another way, if it doesn't cut away everything, it can make the surface area be as close to zero as you like.

67. Not all absolute extrema occur at stationary points; some may occur at an end-point or singular point of the domain, as in Exercises 17, 18, 47 and 48.

69. The minimum of dq/dp is the fastest that the demand is dropping in response to increasing price.

5.3

1. $\frac{dy}{dx} = 6x;\ \frac{d^2y}{dx^2} = 6$

3. $\frac{dy}{dx} = -\frac{2}{x^2};\ \frac{d^2y}{dx^2} = \frac{4}{x^3}$

5. $\frac{dy}{dx} = 1.6x^{-0.6} - 1;\ \frac{d^2y}{dx^2} = -0.96x^{-1.6}$

7. $\frac{dy}{dx} = -e^{-(x-1)} - 1;\ \frac{d^2y}{dx^2} = e^{-(x-1)}$

9. $\frac{dy}{dx} = -\frac{1}{x^2} - \frac{1}{x};\ \frac{d^2y}{dx^2} = \frac{2}{x^3} + \frac{1}{x^2}$

11. (a) $v = 3 - 32t;\ a = -32$ ft/sec² (b) $a(2) = -32$ ft/sec²

13. (a) $v = -\frac{1}{t^2} - \frac{2}{t^3};\ a = \frac{2}{t^3} + \frac{6}{t^4}$ ft/sec²
(b) $a(1) = 8$ ft/sec²

15. (a) $v = \frac{1}{2}t^{-1/2} + 2t;\ a = -\frac{1}{4}t^{-3/2} + 2$ ft/sec²
(b) $a(4) = \frac{63}{32}$ ft/sec²

17. (1, 0): changes from concave down to concave up

19. (1, 0): changes from concave down to concave up

21. None: always concave down

23. (−1, 0): changes from concave up to concave down; (1, 1): changes from concave down to concave up

For exercises 25 & 27, remember that a point of inflection of f corresponds to a relative extreme point of f' that is internal, not an endpoint.

25. Points of inflection at $x = -1$ (relative max of f') and $x = 1$ (relative min of f')

27. One point of inflection, at $x = -2$ (relative min of f'). Note that f' has a stationary point at $x = 1$ but not a relative extreme point there.

For exercises 29–33, remember that a point of inflection of f corresponds to a point at which f'' changes sign, from positive to negative or vice versa.

29. Points of inflection at $x = -2$, $x = 0$, and $x = 2$

31. One point of inflection, at $x = 0$. Note that f'' is not defined at $x = 0$, but does change from negative to positive there.

33. Points of inflection at $x = -2$ and $x = 2$

35. $f'(x) = 2x + 2;\ f'(x) = 0$ when $x = -1;\ f'(x)$ is always defined. $f''(x) = 2;\ f''(x)$ is never 0 and is always defined.

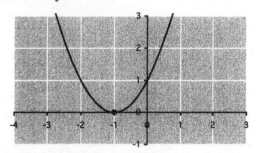

f has an absolute min at $(-1, 0)$ and no points of inflection.

37. $f'(x) = 6x^2 + 6x - 12 = 6(x^2 + x - 2);$

$f'(x) = 0$ when $x^2 + x - 2 = 0$, $(x + 2)(x - 1) = 0$, $x = -2$ or $x = 1$; $f'(x)$ is always defined.
$f''(x) = 12x + 6$; $f''(x) = 0$ when $x = -1/2$; $f''(x)$ is always defined.

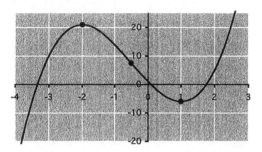

f has a relative max at $(-2, 21)$, a relative min at $(1, -6)$, and a point of inflection at $(-1/2, 15/2)$.

39. $g'(x) = 3x^2 - 12$; $g'(x) = 0$ when $x = \pm 2$; $g'(x)$ is always defined. $g''(x) = 6x$; $g''(x) = 0$ when $x = 0$; $g''(x)$ is always defined.

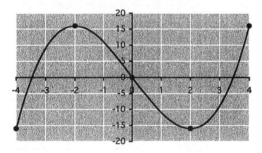

g has absolute mins at $(-4, -16)$ and $(2, -16)$, absolute maxes at $(-2, 16)$ and $(4, 16)$, and a point of inflection at $(0, 0)$.

41. $g'(t) = t^3 - 2t^2 + t = t(t^2 - 2t + 1)$; $g'(t) = 0$ when $t = 0$ or $t = 1$; $g'(t)$ is always defined. $g''(t) = 3t^2 - 4t + 1$; $g''(t) = 0$ when $3t^2 - 4t + 1 = 0$, $(3t - 1)(t - 1) = 0$, $t = 1/3$ or $t = 1$.

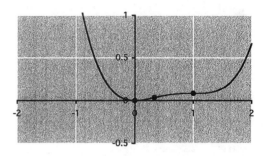

Note that g increases from the stationary point at $t = 0$ to the one at $t = 1$, then continues to increase to the right (as is clear from the graph or from a test point to the right of $t = 1$). So, g has an absolute min at $(0, 0)$ and points of inflection at $(1/3, 11/324)$ and $(1, 1/12)$.

43. Notice that $t = \pm 1$ are not in the domain of f.
$$f'(t) = \frac{2t(t^2 - 1) - (t^2 + 1)(2t)}{(t^2 - 1)^2} = \frac{-4t}{(t^2 - 1)^2};$$
$f'(t) = 0$ when $t = 0$; $f'(t)$ is defined for all t in the domain of f. $f''(t) =$
$$\frac{-4(t^2 - 1)^2 + 4t(2)(t^2 - 1)(2t)}{(t^2 - 1)^4} =$$
$$\frac{(t^2 - 1)[-4(t^2 - 1) + 16t^2]}{(t^2 - 1)^4} = \frac{12t^2 + 4}{(t^2 - 1)^3}; f''(t)$$ is never 0; $f''(t)$ is defined for all t in the domain of f.

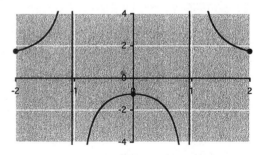

f has relative mins at $(-2, 5/3)$ and $(2, 5/3)$, a relative max at $(0, -1)$, and vertical asymptotes at $x = \pm 1$.

45. $g'(x) = \dfrac{3x^2(x^2 + 3) - x^3(2x)}{(x^2 + 3)^2} = \dfrac{x^4 + 9x^2}{(x^2 + 3)^2}$;

$g'(x) = 0$ when $x^4 + 9x^2 = 0$, $x^2(x^2 + 9) = 0$, $x = 0$; $g'(x)$ is defined for all x. $g''(x) =$
$\dfrac{(4x^3 + 18x)(x^2 + 3)^2 - (x^4 + 9x^2)(2)(x^2 + 3)(2x)}{(x^2 + 3)^4}$
$= \dfrac{(x^2 + 3)(-6x^3 + 54x)}{(x^2 + 3)^4} = \dfrac{-6x(x^2 - 9)}{(x^2 + 3)^3}$;

$g''(x) = 0$ when $x = 0$ or $x = \pm 3$; $g''(x)$ is always defined.

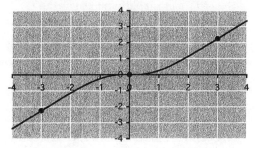

It is difficult to tell from the graph, but the points at $x = \pm 3$ are points of inflection. We can tell by computing, for example, $g''(2) = 60/7^3 > 0$ and $g''(4) = -168/19^3 < 0$, which shows that the concavity changes at $x = 3$. So, g has no extrema and points of inflection at $(0, 0)$, $(-3, -9/4)$, and $(3, 9/4)$.

47. Notice that the domain of f includes all numbers except 0. $f'(x) = 1 - \dfrac{1}{x^2}$; $f'(x) = 0$ when $x^2 = 1$, $x = \pm 1$; $f'(x)$ is defined for all x in the domain of f. $f''(x) = \dfrac{2}{x^3}$; $f''(x)$ is never 0; $f''(x)$ is defined for all x in the domain of f.

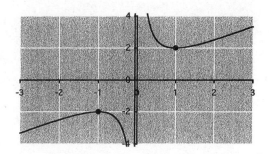

f has a relative min at $(1, 2)$, a relative max at $(-1, -2)$, no points of inflection, and a vertical asymptote at $y = 0$.

49. The domain of g is $x \geq 0$. $g'(x) = \sqrt{x} +$
$(x - 3)\dfrac{1}{2\sqrt{x}} = \dfrac{3\sqrt{x}}{2} - \dfrac{3}{2\sqrt{x}}$; $g'(x) = 0$ when
$\sqrt{x} = \dfrac{1}{\sqrt{x}}$, $x = 1$; $g'(x)$ is not defined when $x = 0$. $g''(x) = \dfrac{3}{4\sqrt{x}} + \dfrac{3}{4x^{3/2}}$; $g''(x)$ is never 0; $g''(x)$ is not defined when $x = 0$.

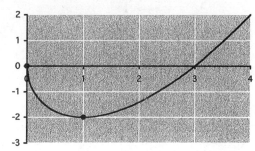

g has a relative max at $(0, 0)$ and an absolute min at $(1, -2)$.

51. $f'(x) = 1 - \dfrac{1}{x}$; $f'(x) = 0$ when $x = 1$; $f'(x)$ is defined for all x in the domain of f. $f''(x) = \dfrac{1}{x^2}$; $f''(x)$ is never 0; $f''(x)$ is defined for all x in the domain of f.

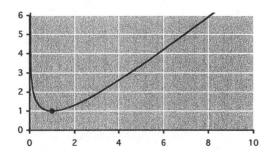

f has an absolute min at $(1, 1)$ and a vertical asymptote at $x = 0$.

53. The domain of f is all numbers except 0. $f'(x) = 2x + \frac{2x}{x^2} = 2x + \frac{2}{x}$; $f'(x)$ is never 0; $f'(x)$ is defined for all x in the domain of f. $f''(x) = 2 - \frac{2}{x^2}$; $f'(x) = 0$ when $x^2 = 1$, $x = \pm 1$; $f''(x)$ is defined for all x in the domain of f.

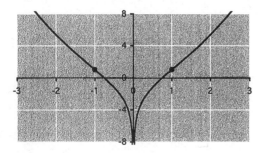

f has no relative extrema, points of inflection at $(1, 1)$ and $(-1, 1)$, and a vertical asymptote at $x = 0$.

55. $g'(t) = e^t - 1$; $g'(t) = 0$ when $e^t = 1$, $t = \ln 1 = 0$; $g'(t)$ is always defined. $g''(t) = e^t$; $g''(t)$ is never 0; $g''(t)$ is always defined.

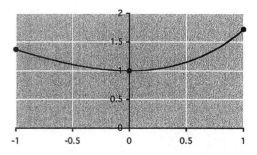

g has an absolute min at $(0, 1)$, an absolute max at $(1, e - 1)$, and a relative max at $(-1, e^{-1} + 1)$.

57. $f'(x) = 4x^3 - 6x^2 + 2x - 2$; by graphing $f'(x)$ we see that $f'(x) = 0$ for $x \approx 1.40$; $f'(x)$ is always defined. $f''(x) = 12x^2 - 12x + 2$; $f''(x) = 0$ for $x = \frac{1}{2} \pm \frac{\sqrt{3}}{6}$ (by the quadratic formula), $x \approx 0.21$ or $x \approx 0.79$; $f''(x)$ is always defined.

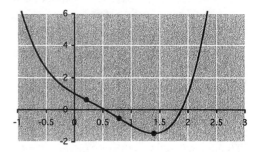

f has an absolute min at $(1.40, -1.49)$ and points of inflection at $(0.21, 0.61)$ and $(0.79, -0.55)$.

59. $f'(x) = e^x - 3x^2$; by graphing $f'(x)$ we see that $f'(x) = 0$ for $x \approx -0.46$, $x \approx 0.91$, and $x \approx 3.73$; $f'(x)$ is always defined. $f''(x) = e^x - 6x$; by graphing $f''(x)$ we see that $f''(x) = 0$ for $x \approx 0.20$ and $x \approx 2.83$; $f''(x)$ is always defined.

Section 5.3

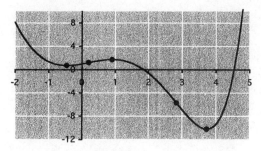

f has a relative min at $(-0.46, 0.73)$, a relative max at $(0.91, 1.73)$, an absolute min at $(3.73, -10.22)$, and points of inflection at $(0.20, 1.22)$ and $(2.83, -5.74)$.

61. (a) Where the graph is steepest: 2 years into the epidemic. (b) At the point at inflection 2 years into the epidemic: There the steepness stops increasing and starts to decline, so the rate of new infections starts to drop.

63. (a) 2000: the point where the graph is increasing and steepest. (b) 2002: the point where the graph is decreasing and steepest. (c) 1998: where the graph changed from concave down to concave up.

65. (a) $p(15) = 465,000$ prisoners (b) $p'(t) = 0.0030t - 0.0035$; $p'(15) = 0.0415$. So, the population was increasing at a rate of 41,500 prisoners per year. (c) $p''(t) = 0.0030$. So, the population was accelerating at a rate of 3000 prisoners per year per year.

67. $W'(t) = 0.003t^2 - 0.056t + 0.28$; $W''(t) = 0.006t - 0.056$; $W''(t) = 0$ when $t = 9.333$. $W''(t)$ changes from negative to positive there, so there is a point of inflection at $(9.333, 0.2873)$. Here is the graph of W:

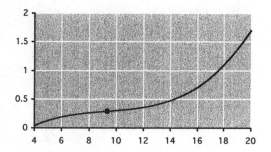

Annual sales of bottled water were increasing at the slowest rate at $t = 9.333$ (that is, around October, 1989) at which time approximately 287.3 million gallons were being sold per year.

69. (a) $S'(n) \approx -1757(n - 180)^{-2.325}$; $S''(n) \approx 4085(n - 180)^{-3.325}$; $S''(n)$ is never zero and is always defined for $n > 180$. So, there are no points of inflection in the graph of S.
(b) Since the graph is concave up ($S''(n) > 0$ for $n > 180$), the derivative of S is increasing, and so the rate of *decrease* of SAT scores with increasing numbers of prisoners is diminishing. In other words, the apparent effect on SAT scores of increasing numbers of prisoners is diminishing.

71. (a) $\dfrac{dn}{ds} = 144.42 - 47.72s + 4.371s^2$; $\dfrac{d^2n}{ds^2} = -47.72 + 8.742s$; $\left.\dfrac{d^2n}{ds^2}\right|_{s=3} = -21.494$. Thus, for a firm with annual sales of $3 million, the rate at which new patents are produced decreases with increasing firm size. This means that the returns (as measured in the number of new patents per increase of $1 million in sales) are diminishing as the firm size increases.

(b) $\left.\dfrac{d^2n}{ds^2}\right|_{s=7} = 13.474$. Thus, for a firm with annual sales of $7 million, the rate at which new patents are produced increases with increasing

115

firm size by 13.474 new patents per $1 million increase in annual sales.
(c) There is a point of inflection when $s \approx 5.4587$, so that in a firm with sales of $5,458,700 per year, the number of new patents produced per additional $1 million in sales is a minimum.

73. In Exercise 59 in section 5.2 we calculated $p' = 10,000 \dfrac{(1.05)^{-t}\{-\ln(1.05) + [250 - 500\ln(1.05)]e^{-0.5t}\}}{(1 + 500e^{-0.5t})^2} \approx 10,000 \dfrac{(1.05)^{-t}(-0.04879 + 225.6e^{-0.5t})}{(1 + 500e^{-0.5t})^2}$.

Now we calculate $p'' \approx 10,000 \dfrac{(1.05)^{-t}(0.002380 - 152e^{-0.5t} + 48,396e^{-t})}{(1 + 500e^{-0.5t})^3}$. Graphing this expression, we see that it is 0 when $t \approx 12$. [Alternatively, we could use technology to compute numerical approximations to p' and p'' and then see where p'' is (approximately) zero.] Graphing p', we can see that this zero gives a maximum of p'. Thus, the greatest rate of increase of the present value occurs after about 12 years, when $p' \approx \$570$ per year.

75. $p = (300,000 + 1000t^2)e^{-0.05t}$; $p' = (-15,000 + 2000t - 50t^2)e^{-0.05t}$ from Exercise 41 in section 5.2; $p'' = (2000 - 100t)e^{-0.05t} - 0.05(-15,000 + 2000t - 50t^2)e^{-0.05t} = (2750 - 200t + 2.5t^2)e^{-0.05t}$. $p'' = 0$ when $2750 - 200t + 2.5t^2 = 0$, $t \approx 17.64$ or $t \approx 62.34$. Here is the graph of p':

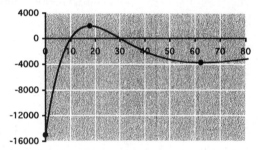

From the graph we can see that the absolute minimum of p' occurs at $t = 0$ while the absolute maximum occurs at $t \approx 17.64$. Therefore, the value p is increasing most rapidly in 17.64 years and decreasing most rapidly now (at $t = 0$).

77. (a) $\lim\limits_{t \to +\infty} M(t)/B(t) = 2.48/15,700 \approx 0.000158$ manatees per boat, or 15.8 manatees per 100,000 boats.

Section 5.3

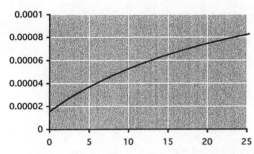

(b) $M(t)/B(t)$ represents the number of manatees killed in year t per registered boat. $\lim_{t \to +\infty} M(t)/B(t)$ represents the number of manatees killed per registered boat in the long-term. (c) From the graph, it appear to be concave down. We can verify this by looking at the second derivative: $\dfrac{d}{dt}\left(\dfrac{M(t)}{B(t)}\right) = \dfrac{2.48(15{,}700t + 444{,}000) - 15{,}700(2.48t + 6.87)}{(15{,}700t + 444{,}000)^2} = \dfrac{993{,}000}{(15{,}700t + 444{,}000)^2}$; $\dfrac{d^2}{dt^2}\left(\dfrac{M(t)}{B(t)}\right) = \dfrac{-3.12 \times 10^{10}}{(15{,}700t + 444{,}000)^3}$ is negative for all $t \geq 0$. The correct choice is (C) because the number of manatees killed per boat is increasing, but the rate of increase is going down.

79. nonnegative

81.

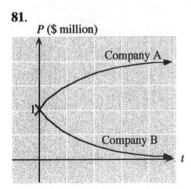

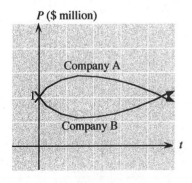

83. Sales were decreasing most rapidly in June 2002.

85. At a point of inflection, the graph of a function changes either from concave up to concave down, or vice-versa. If it changes from concave up to concave down, then the derivative changes from increasing to decreasing, and hence has a relative maximum. Similarly, if it changes from concave down to concave up, the derivative has a relative minimum.

Section 5.4

5.4

1. $P = 10{,}000$ and $\dfrac{dP}{dt} = 1000$.

3. Let R be the annual revenue of my company and let q be annual sales. $R = 7000$ and $\dfrac{dR}{dt} = -700$. Find $\dfrac{dq}{dt}$.

5. Let p be the price of a pair of shoes and let q be the demand for shoes. $\dfrac{dp}{dt} = 5$. Find $\dfrac{dq}{dt}$.

7. Let T be the average global temperature and let q be the number of Bermuda shorts sold per year. $T = 60$ and $\dfrac{dT}{dt} = 0.1$. Find $\dfrac{dq}{dt}$.

9. (a) Let r = the radius and A = the area. $\dfrac{dA}{dt} = 1200$. Find $\dfrac{dr}{dt}$ when $r = 10{,}000$. $A = \pi r^2$, so $\dfrac{dA}{dt} = 2\pi r \dfrac{dr}{dt}$. Substitute: $1200 = 2\pi(10{,}000)\dfrac{dr}{dt}$, so $\dfrac{dr}{dt} = 6/(100\pi) \approx 0.019$ km/sec. **(b)** This time the problem is to find $\dfrac{dr}{dt}$ when $A = 640{,}000$. Since r appears in the equation but not A, we need to find r from $A = \pi r^2$: $640{,}000 = \pi r^2$, $r = \sqrt{640{,}000/\pi} = 800\sqrt{\pi}$. So, $1200 = 2\pi(800\sqrt{\pi})\dfrac{dr}{dt}$, $\dfrac{dr}{dt} = 6/(8\sqrt{\pi}) \approx 0.4231$ km/sec.

11. Let b = the distance of the base of the ladder from the wall and h = the height of the top of the ladder. $\dfrac{db}{dt} = 10$. Find $\dfrac{dh}{dt}$ when $b = 30$. $b^2 + h^2 = 50^2$, so $2b\dfrac{db}{dt} + 2h\dfrac{dh}{dt} = 0$. We need the value of h: $30^2 + h^2 = 50^2$, $h = \sqrt{2500 - 900} = 40$. Substitute: $2(30)(10) + 2(40)\dfrac{dh}{dt} = 0$, $\dfrac{dh}{dt} = -600/80 = -7.5$, so the top of the ladder is sliding down at 7.5 ft/sec.

13. $x = 3000$ and $\dfrac{dx}{dt} = 100$. Find $\dfrac{d\bar{C}}{dt}$. $\dfrac{d\bar{C}}{dt} = -150{,}000 x^{-2}\dfrac{dx}{dt} + 0.0001 \dfrac{dx}{dt}$. Substitute: $\dfrac{d\bar{C}}{dt} = -150{,}000(3000)^{-2}(100) + 0.0001(100) = -1.66$. The average cost is decreasing at a rate of $1.66 per player per week.

15. $p = 15$ and $\dfrac{dp}{dt} = 2$. Find $\dfrac{dq}{dt}$. $\dfrac{dq}{dt} = -50p^{-0.5}\dfrac{dp}{dt}$. Substitute: $\dfrac{dq}{dt} = -50(15)^{-0.5}(2) \approx -26$. Monthly sales will drop at a rate of 26 T-shirts per month.

17. Let p be the price, let q be the weekly demand, and let R be the weekly revenue. $q = 50$, $p = 30¢$, and $\dfrac{dq}{dt} = -5$. Find $\dfrac{dp}{dt}$ if $\dfrac{dR}{dt} = 0$. $R = pq$, so $\dfrac{dR}{dt} = \dfrac{dp}{dt}q + p\dfrac{dq}{dt}$. Substitute: $0 = \dfrac{dp}{dt}(50) + (30)(-5)$, $\dfrac{dp}{dt} = 150/50 = 3$. You must raise the price by 3¢ per week.

19. $q = 900$ and $\dfrac{dq}{dt} = 100$. Find $\dfrac{dp}{dt}$. From $pq^{1.5} = 50{,}000$ we get $\dfrac{dp}{dt}q^{1.5} + 1.5pq^{0.5}\dfrac{dq}{dt} = 0$. We will also need the value of p: $p(900)^{1.5} = 50{,}000$, $p = 50{,}000/(900)^{1.5}$. Substitute: $\dfrac{dp}{dt}(900)^{1.5} + 1.5[50{,}000/(900)^{1.5}](900)^{0.5}(100) = 0$, $\dfrac{dp}{dt} = -(75{,}000/9)/(900)^{1.5} \approx -0.31$. The price is decreasing at a rate of approximately 31¢ per pound per month.

Section 5.4

21. $\frac{dV}{dt} = 3$. Find $\frac{dr}{dt}$ when $r = 1$. From $V = \frac{4}{3}\pi r^3$ we get $\frac{dV}{dt} = 4\pi r^2 \frac{dr}{dt}$. Substitute: $3 = 4\pi(1)^2 \frac{dr}{dt}$, $\frac{dr}{dt} = \frac{3}{4\pi} \approx 0.24$ ft/min.

23. Let x be the distance of the Mona Lisa from Montauk and let y be the distance of the Dreadnaught from Montauk. Let z be the distance between the two ships. $x = 50$, $\frac{dx}{dt} = 30$, $y = 40$, and $\frac{dy}{dt} = 20$. Find $\frac{dz}{dt}$. $z^2 = x^2 + y^2$, so $2z\frac{dz}{dt} = 2x\frac{dx}{dt} + 2y\frac{dy}{dt}$. We need the value of z. $z^2 = 50^2 + 40^2 = 4100$, so $z = \sqrt{4100}$. Substitute: $2\sqrt{4100}\frac{dz}{dt} = 2(50)(30) + 2(40)(20) = 4600$, $\frac{dz}{dt} = \frac{2300}{\sqrt{4100}} \approx 36$ mph.

25. $\frac{dx}{dt} = 4$ and $y = 2$. Find $\frac{dy}{dt}$. From $y = x^{-1}$ we get $\frac{dy}{dt} = -x^{-2}\frac{dx}{dt}$. We need the value of x: $2 = 1/x$, $x = 1/2$. Substitute: $\frac{dy}{dt} = -(1/2)^{-2}(4) = -16$. The y coordinate is decreasing at a rate of 16 units per second.

27. P is constant at 1000. $x = 150$ and $\frac{dx}{dt} = 10$. Find $\frac{dy}{dt}$. From $P = 10x^{0.3}y^{0.7}$ we get $0 = 3x^{-0.7}y^{0.7}\frac{dx}{dt} + 7x^{0.3}y^{-0.3}\frac{dy}{dt}$. This problem is a bit simpler if we solve for $\frac{dy}{dt}$ before substituting values: $\frac{dy}{dt} = -\frac{3y\,dx}{7x\,dt}$. We need the value of y: $1000 = 10(150)^{0.3}y^{0.7}$, $y = (100/150^{0.3})^{1/0.7} = 100^{10/7}/150^{3/7}$. Substitute:

$\frac{dy}{dt} = -\frac{3(100^{10/7})/150^{3/7}}{7(150)}(10) = -\frac{30(100^{10/7})}{7(150)^{10/7}} \approx -2.40$. The daily operating budget is dropping at a rate of \$2.40 per year.

29. $n = 13$ and $\frac{dn}{dt} = \frac{1}{3}$. Find $\frac{dI}{dn}$. From the formula for $I(n)$ we get $\frac{dI}{dt} = 8786.4n^2\frac{dn}{dt} - 231{,}720n\frac{dn}{dt} + 1{,}532{,}900\frac{dn}{dt}$. Substitute: $\frac{dI}{dt} = 8786.4(13)^2(1/3) - 231{,}720(13)(1/3) + 1{,}532{,}900(1/3) \approx \$1814$ per year.

31. V is constant at 200. $g = 3.0$ and $\frac{dg}{dt} = -0.2$. Find $\frac{de}{dt}$. From $V = 3e^2 + 5g^3$ we get $0 = 6e\frac{de}{dt} + 15g^2\frac{dg}{dt}$. We need the value of e: $200 = 3e^2 + 5(3.0)^3$, $e = \sqrt{65/3} \approx 4.655$. Substitute: $0 = 6(4.655)\frac{de}{dt} + 15(3.0)^2(-0.2)$, $\frac{de}{dt} = \frac{27}{27.93} \approx 0.97$. Their prior experience must increase at a rate of approximately 0.97 years every year.

33. $\frac{dV}{dt} = 100$ and $V = 200\pi$. Find $\frac{dh}{dt}$. We also need to know something about r. Looking at the vessel from the side, we can see that, for any given value of h, the corresponding radius r must satisfy $r/h = 30/50$, the ratio at the brim of the vessel. So, $r = \frac{3}{5}h$. Substituting into $V = \frac{1}{3}\pi r^2 h$ we get $V = \frac{3}{25}\pi h^3$. So, $\frac{dV}{dt} = \frac{9}{25}\pi h^2 \frac{dh}{dt}$. We need the value of h: $200\pi = \frac{3}{25}\pi h^3$, $h = (5000/3)^{1/3}$. Substitute: $100 = \frac{9\pi}{25}\left(\frac{5000}{3}\right)^{2/3}\frac{dh}{dt}$, $\frac{dh}{dt} = \frac{2500}{9\pi}\left(\frac{3}{5000}\right)^{2/3} \approx 0.63$ m/sec.

Section 5.4

35. r is constant at 2. $V = 4t^2 - t$ and $h = 2$. Find $\frac{dh}{dt}$. From $V = \pi r^2 h = 4\pi h$ we get $h = \frac{V}{4\pi}$, so $\frac{dh}{dt} = \frac{1}{4\pi}\frac{dV}{dt} = \frac{1}{4\pi}(8t - 1)$. We need the value of t when $h = 2$. We first find $V = 4\pi h = 8\pi$, so $8\pi = 4t^2 - t$, $4t^2 - t - 8\pi = 0$, $t = \frac{1 \pm \sqrt{1 + 128\pi}}{8}$ by the quadratic formula. We take the positive solution $t = \frac{1 + \sqrt{1 + 128\pi}}{8}$, where the volume is rising. Substituting: $\frac{dh}{dt} = \frac{\sqrt{1 + 128\pi}}{4\pi} \approx 1.6$ cm/sec.

37. $x = 30{,}000$ and $\frac{dx}{dt} = 2000$. Find q and $\frac{dq}{dt}$. From the formula, $q = 0.3454 \ln(30{,}000) - 3.047 \approx 0.5137$. $\frac{dq}{dt} = \frac{0.3454}{x}\frac{dx}{dt} = \frac{0.3454}{30{,}000}(2000) \approx 0.0230$. So, there are approximately 0.5137 computers per household, increasing at a rate of 0.0230 computers per household per year.

39. $n = 475$ and $\frac{dn}{dt} = 35$. Find S and $\frac{dS}{dt}$. $S(475) \approx 904.71$ from the formula. $\frac{dS}{dt} = 1326(-1.325)(n - 180)^{-2.325}\frac{dn}{dt}$. Substitute: $\frac{dS}{dt} = -1756.95(475 - 180)^{-2.325}(35) \approx -0.11$. The average SAT score was 904.71, decreasing at a rate of 0.11 per year.

41. $r = 1.1$ and $\frac{dr}{dt} = 0.05$. Find $\frac{d}{dt} d(r)$. Note that $r = 1.1$ is in the range where $d(r) = -40r + 74$ and that we stay in that range because we are interested in the slope at that point. Thus, $\frac{d}{dt} d(r) = -40\frac{dr}{dt} = -40(0.05) = -2$. The divorce rate is decreasing by 2 percentage points per year.

43. The section is called "related rates" because the goal is to compute the rate of change of a quantity based on a knowledge of the rate of change of a related quantity. The relationship between the quantities gives a relationship between their rates of change.

45. Answers may vary. For example: A rectangular solid has dimensions 2 cm $\times$ 5 cm $\times$ 10 cm and each side is expanding at a rate of 3 cm/second. How fast is the volume increasing?

47. $R = pq$, so $R' = p'q + pq'$, where the derivatives are with respect to time t. Divide by $R = pq$ to get $\frac{R'}{R} = \frac{p'q}{pq} + \frac{pq'}{pq} = \frac{p'}{p} + \frac{q'}{q}$ as claimed.

49. The derived equation is linear in the unknown rate X. This follows from the chain rule, since if Q is a quantity and $f(Q)$ is any expression in Q, we have $\frac{d}{dt} f(Q) = f'(Q) \frac{dQ}{dt}$, which is linear in the derivative $\frac{dQ}{dt}$. The presence of other variables may add terms not containing $\frac{dQ}{dt}$ but those maintain the linearity.

51. Let $x = $ my grades and $y = $ your grades. If $dx/dt = 2\, dy/dt$, then $dy/dt = (1/2)\, dx/dt$.

5.5

1. $E = -(-20)\dfrac{p}{1000 - 20p} = \dfrac{20p}{1000 - 20p}$.
When $p = 30$, $E = 1.5$: The demand is going down 1.5% per 1% increase in price at that price level. $E = 1$ when $20p = 1000 - 20p$, $p = 25$: Revenue is maximized when $p = \$25$; weekly revenue at that price is $R = pq = 25(1000 - 20 \cdot 25) = \$12{,}500$.

3. (a) $E = -(-2)(100 - p)\dfrac{p}{(100 - p)^2} = \dfrac{2p}{100 - p}$.
When $p = 30$, $E = 6/7$. The demand is going down 6% per 7% increase in price at that price level. Thus, a price increase is in order. **(b)** $E = 1$ when $2p = 100 - p$, $p = 100/3$: Revenue is maximized when $p = 100/3 \approx \$33.33$.
(c) Demand would be $(100 - 100/3)^2 = (200/3)^2 \approx 4444$ cases per week.

5. (a) $E = -(-2.17)\dfrac{p}{9859.39 - 2.17p} = \dfrac{2.17p}{9859.39 - 2.17p}$. When $p = 2867$, $E = 1.71$.
Thus, the demand is elastic at the given tuition level, showing that a decrease in tuition will result in an increase in revenue. **(b)** $E = 1$ when $2.17p = 9859.39 - 2.17p$, $p \approx 2271.75$: They should charge an average of $\$2271.75$ per student, and this will result in an enrollment of about $q = 4930$ students, giving a revenue of about $pq = \$11{,}199{,}000$.

7. (a) $E = -m\dfrac{p}{mp + b} = -\dfrac{mp}{mp + b}$ **(b)** $E = 1$ when $-mp = mp + b$, $p = -\dfrac{b}{2m}$

9. (a) $E = -(-r)\dfrac{k}{p^{r+1}}\dfrac{p}{k/p^r} = r$ **(b)** E is independent of p. **(c)** If $r = 1$ the revenue is not affected by the price. If $r > 1$ the revenue is always elastic, whereas if $r < 1$ the revenue is always inelastic. This is an unrealistic model because there should be a price at which the revenue is a maximum.

11. (a) $E = -(-6p + 1)100e^{-3p^2+p}\dfrac{p}{100e^{-3p^2+p}} = p(6p - 1)$. When $p = 3$, $E = 51$: The demand is going down 51% per 1% increase in price at that price level. Thus, a large price decrease is advised. **(b)** $E = 1$ when $p(6p - 1) = 1$, $6p^2 - p - 1 = 0$, $(3p + 1)(2p - 1) = 0$, $p = 1/2$. (We reject the solution $p = -1/3$ because we must have $p > 0$.) Revenue is maximized when $p = ¥0.50$. **(c)** Demand would be $q = 100e^{-3/4+1/2} \approx 78$ paint-by-number sets per month.

13. (a) We have two data points: $(p, q) = (2.00, 3000)$ and $(4.00, 0)$. The line through these two points is $q = -1500p + 6000$. **(b)** $E = -(-1500)\dfrac{p}{-1500p + 6000} = \dfrac{1500p}{-1500p + 6000}$.
$E = 1$ when $1500p = -1500p + 6000$, $p = \$2$ per hamburger. This gives a total weekly revenue of $R = pq = 2(-1500 \cdot 2 + 6000) = \$6000$.

15. $E = \dfrac{dq}{dx} \cdot \dfrac{x}{q} = \dfrac{0.01(-0.0156x + 1.5) \cdot x}{0.01(-0.0078x^2 + 1.5x + 4.1)} = \dfrac{-0.0156x^2 + 1.5x}{-0.0078x^2 + 1.5x + 4.1}$. When $x = 20$, $E \approx 0.77$: At a family income level of $\$20{,}000$, the fraction of children attending a live theatrical performance is increasing by 0.77% per 1% increase in household income.

Section 5.5

17. (a) $E = \dfrac{0.3454}{x} \cdot \dfrac{x}{0.3454 \ln x - 3.047} = \dfrac{0.3454}{0.3454 \ln x - 3.047}$. When $x = 60{,}000$, $E \approx 0.46$. The demand for computers is increasing by 0.46% per 1% increase in household income. **(b)** E decreases as income increases because the denominator of E gets larger. **(c)** Unreliable; it predicts a likelihood greater than 1 at incomes of \$123,000 and above. ($0.3454 \ln x - 3.047 = 1$ when $x = e^{4.047/0.3454} \approx 123{,}000$) In a model appropriate for large incomes, one would expect the curve to level off at or below 1. **(d)** E approaches 0 as x goes to infinity, so for very large x we have $E \approx 0$.

19. The income elasticity of demand is $\dfrac{dQ}{dY} \cdot \dfrac{Y}{Q} = \alpha\beta P^\alpha Y^{\beta-1} \dfrac{Y}{aP^\alpha Y^\beta} = \beta$. An increase in income of $x\%$ will result in an increase in demand of $\beta x\%$.

21. (a) The data points $(p, q) = (3.00, 407)$ and $(5.00, 223)$ give us the exponential function $q = 1000e^{-0.3p}$. **(b)** $E = 300e^{-0.3p} \dfrac{p}{1000e^{-0.3p}} = 0.3p$. At $p = \$3$, $E = 0.9$; at $p = \$4$, $E = 1.2$; and at $p = \$5$, $E = 1.5$. **(c)** $E = 1$ when $0.3p = 1$, $p = \$3.33$. **(d)** We first find the price that produces a demand of 200 pounds: $1000e^{-0.3p} = 200$, $p = -(\ln 0.2)/0.3 = \$5.36$. Selling at a lower price would increase demand, but you cannot sell more than 200 pounds anyway so your revenue would go down. On the other hand, if you set the price higher than \$5.36 the decrease in sales will outweigh the increase in price, which we know because the elasticity at $p = 5.36$ is $E \approx 1.6 > 1$. You should therefore set the price at \$5.36 per pound.

23. the price is lowered

25. Start with $R = pq$ and differentiate with respect to p to obtain $\dfrac{dR}{dp} = q + p\dfrac{dq}{dp}$. For a stationary point, $dR/dp = 0$, so $q + p\dfrac{dq}{dp} = 0$. Rearranging gives $p\dfrac{dq}{dp} = -q$, and hence $-\dfrac{dq}{dp} \cdot \dfrac{p}{q} = 1$, or $E = 1$, showing that stationary points of R correspond to points of unit elasticity.

27. The distinction is best illustrated by an example. Suppose that q is measured in weekly sales and p is the unit price in dollars. Then the quantity $-dq/dp$ measures the drop in weekly sales per \$1 increase in price. The elasticity of demand E, on the other hand, measures the *percentage* drop in sales per *one percent* increase in price. Thus, $-dq/dp$ measures absolute change, while E measures fractional, or percentage, change.

Chapter 5 Review Test

1. (a) $f'(x) = 6x^2 - 6$; $f'(x) = 0$ when $6x^2 - 6 = 0$, $6(x^2 - 1) = 0$, $x = \pm 1$; $f'(x)$ is defined for all x. Thus, we have stationary points at $x = \pm 1$ and the endpoint $x = -2$. We test a point to the right of 1 as well:

x	−2	−1	1	2
$f(x)$	−3	5	−3	5

The graph must increase from $x = -2$ to $x = -1$, decrease to $x = 1$, and increase from then on.

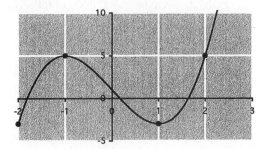

This gives absolute mins at $(-2, -3)$ and $(1, -3)$ and a relative max at $(-1, 5)$.

(b) $f'(x) = \dfrac{(x-1)^2 - 2(x+1)(x-1)}{(x-1)^4} = \dfrac{x - 1 - 2(x+1)}{(x-1)^3} = \dfrac{-x - 3}{(x-1)^3}$; $f'(x) = 0$ when $x = -3$, which is outside of the domain; $f'(x)$ is undefined at $x = 1$, which is not in the domain of f. Thus, there are no critical points in the domain of f. We test the endpoints and another point on each side of the point $x = 1$:

x	−2	−1	1.5	2
$f(x)$	−0.11	0	10	3

The graph must increase from $x = -2$ until the vertical asymptote at $x = 1$. It then decreases on the other side of $x = 1$ until $x = 2$.

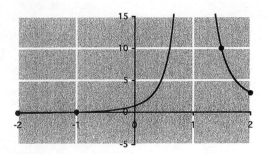

This gives an absolute min at $(-2, -1/9)$ and a relative min at $(2, 3)$.

(c) The domain of g is all real numbers. $g'(x) = \dfrac{2}{3}(x-1)^{-1/3}$; $g'(x)$ is never 0; $g'(x)$ is not defined at $x = 1$. Thus, we have a critical point at $x = 1$. We test a point on either side:

x	0	1	2
$g(x)$	1	0	1

The graph must decrease until $x = 1$ and then increase.

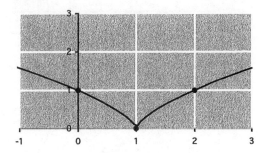

This gives an absolute min at $(1, 0)$.

(d) $g'(x) = 2x + 1/x$; $g'(x) = 0$ when $2x + 1/x = 0$, which cannot happen for $x > 0$; $g'(x)$ is defined for all x in the domain of g. Thus, g has no critical points in its domain.

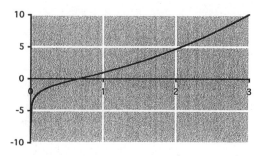

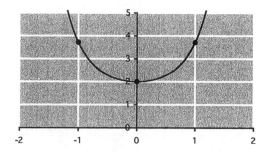

g has no extrema.

(e) The domain of h is all $x \neq 0$. $h'(x) = -\dfrac{1}{x^2} - \dfrac{2}{x^3}$; $h'(x) = 0$ when $2x^2 = -x^3$, $x^3 + 2x^2 = 0$, $x^2(x+2) = 0$, $x = -2$ ($x = 0$ is not in the domain); $h'(x)$ is defined for all x in the domain of h. Thus, h has a stationary point at $x = -2$. We test points on either side of -2 and points to the right of $x = 0$:

x	-3	-2	-1	1	2
$h(x)$	-0.22	-0.25	0	2	0.75

The graph decreases to $x = -2$ then increases approaching the vertical asymptote at $x = 0$. On the other side of the asymptote it decreases.

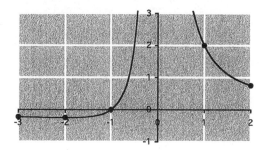

h has an absolute min at $(-2, -1/4)$.

(f) $h'(x) = 2xe^{x^2}$; $h'(x) = 0$ when $x = 0$; $h'(x)$ is always defined. We test points on either side of the stationary point at $x = 0$:

x	-1	0	1
$h(x)$	3.72	2	3.72

The graph decreases until $x = 0$ then increases.

h has an absolute min at $(0, 2)$.

2. (a) Relative max at $x = 1$, point of inflection at $x = -1$. **(b)** Relative min at $x = 2$, point of inflection at $x = -1$. **(c)** Relative max at $x = -2$: the derivative goes from positive to negative, so the f must go from increasing to decreasing; relative min at $x = 1$: f goes from decreasing to increasing; point of inflection at $x = -1$: a min of f' gives a point of inflection of f. **(d)** Relative min at $x = -2.5$ and $x = 1.5$: in both cases f goes from decreasing to increasing; relative max at $x = 3$: f goes from increasing to decreasing; point of inflection at $x = 2$: a max of f' gives a point of inflection of f.

3. (a) $f'(x) = 3x^2 - 12$; $f'(x) = 0$ when $3(x^2 - 4) = 0$, $x = \pm 2$; $f'(x)$ is always defined. $f''(x) = 6x$; $f''(x) = 0$ when $x = 0$; $f''(x)$ is always defined.

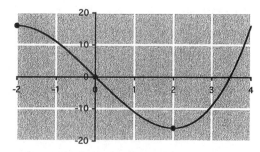

f has a relative max at $(-2, 16)$, an absolute min at $(2, -16)$, and a point of inflection at $(0, 0)$. It has no horizontal or vertical asymptotes.

(b) The domain of f includes all numbers except 0. $f'(x) = \dfrac{2x \cdot x^3 - (x^2 - 3)(3x^2)}{x^6} = \dfrac{-x^4 + 9x^2}{x^6} =$

$\dfrac{-x^2 + 9}{x^4}$; $f'(x) = 0$ when $x = \pm 3$; $f'(x)$ is defined for all x in the domain of f. $f''(x) =$
$\dfrac{-2x \cdot x^4 - (-x^2 + 9)(4x^3)}{x^8} = \dfrac{2x^5 - 36x^3}{x^8} =$

$\dfrac{2(x^2 - 18)}{x^5}$; $f''(x) = 0$ when $x = \pm\sqrt{18} = \pm 3\sqrt{2}$;

$f''(x)$ is defined for all x in the domain of f.

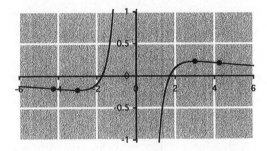

f has a relative min at $(-3, -2/9)$, a relative max at $(3, 2/9)$, and points of inflection at $(-3\sqrt{2}, -5\sqrt{2}/36)$ and $(3\sqrt{2}, 5\sqrt{2}/36)$. f has a vertical asymptote at $x = 0$ and a horizontal asymptote at $y = 0$.

(c) $f(x)$ is defined for all x. $f'(x) = \dfrac{2}{3}(x - 1)^{-1/3} +$ $\dfrac{2}{3}$; $f'(x) = 0$ when $(x - 1)^{-1/3} = -1$, $x - 1 = (-1)^{-3} = -1$, $x = 0$; $f'(x)$ is not defined when $x = 1$. $f''(x) = -\dfrac{2}{9}(x - 1)^{-4/3}$; $f''(x)$ is never 0; $f''(x)$ is not defined when $x = 1$.

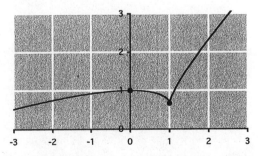

f has a relative max at $(0, 1)$ and a relative min at $(1, 2/3)$. It has no points of inflection and no horizontal or vertical asymptotes.

4. (a) $E = -(-2p + 33)\dfrac{p}{-p^2 + 33p + 9} =$
$\dfrac{2p^2 - 33p}{-p^2 + 33p + 9}$ **(b)** $E(20) \approx 0.52$ and $E(25) \approx 2.03$. When the price is $20, demand is dropping at a rate of 0.52% per 1% increase in the price; when the price is $25, demand is dropping at a rate of 2.03% per 1% increase in the price.
(c) $E = 1$ when $2p^2 - 33p = -p^2 + 33p + 9$, $3p^2 - 66p - 9 = 0$, $p^2 - 22p - 3 = 0$, $p \approx \$22.14$ per book (using the quadratic formula; the other solution is negative so we reject it).

5. (a) Profit $P = R - C = pq - (9q + 100) =$
$p(-p^2 + 33p + 9) - 9(-p^2 + 33p + 9) - 100 =$
$-p^3 + 42p^2 - 288p - 181$ **(b)** $P' = -3p^2 + 84p - 288 = -3(p^2 - 28p + 96) =$
$-3(p - 24)(p - 4)$; $P' = 0$ when $p = 4$ or 24. To see which, if either, gives us the maximum profit, we test points on either side:

p	0	4	24	30
$P(p)$	-181	-725	3275	1979

P decreases to a low at $p = 4$ then increases to a maximum at $p = 24$, after which it decreases again. So, the company should charge $24 per copy. **(c)** $P(24) = \$3275$ **(d)** For maximum revenue, the company should charge $22.14 per copy. At this price, the cost is decreasing at a linear rate with increasing price, while the revenue is not decreasing (its derivative is zero). Thus, the profit is increasing with increasing price, suggesting that the maximum profit will occur at a higher price. This is, in fact, what we just found.

6. (a) $s'(t) = \dfrac{2460.7 \, e^{-0.55(t-4.8)}}{(1 + e^{-0.55(t-4.8)})^2}$ Graph:

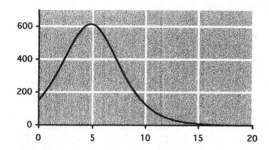

Sales were growing fastest in week 5. **(b)** The maximum of s' corresponds to a point of inflection of s. **(c)** As t goes to infinity the exponential term goes to 0, so $s(t)$ approaches $6053 + 4474 = 10{,}527$. If weekly sales continue as predicted by the model, they will level off at around 10,500 books per week in the long term. **(d)** Again, the exponential terms go to 0, so s' goes to 0. If weekly sales continue as predicted by the model, the rate of change of sales approaches zero in the long term.

7. The combined area is $A = hw + \frac{1}{4}\pi h^2$. Given that $w = 1$, $\frac{dw}{dt} = 0.5$, and $h = 3$, we want to find $\frac{dh}{dt}$. Using the fact that A is constant, we differentiate: $0 = \frac{dh}{dt}w + h\frac{dw}{dt} + \frac{1}{2}\pi h \frac{dh}{dt}$. Substitute: $0 = \frac{dh}{dt} + 3(0.5) + \frac{1}{2}\pi(3)\frac{dh}{dt} = 1.5 + (1 + 3\pi/2)\frac{dh}{dt}$, $\frac{dh}{dt} = -1.5/(1 + 3\pi/2) \approx -0.263$ inches per second.

Section 6.1

Chapter 6
6.1

1. $\dfrac{x^6}{6} + C$

3. $6x + C$

5. $\dfrac{x^2}{2} + C$

7. $\dfrac{x^3}{3} - \dfrac{x^2}{2} + C$

9. $x + \dfrac{x^2}{2} + C$

11. $-\dfrac{x^{-4}}{4} + C$

13. $\dfrac{x^{3.3}}{3.3} - \dfrac{x^{-0.3}}{0.3} + C$

15. $\dfrac{u^3}{3} - \ln|u| + C$

17. $\int x^{1/2}\, dx = \dfrac{x^{3/2}}{3/2} + C = \dfrac{2x^{3/2}}{3} + C$

19. $\dfrac{3x^5}{5} + 2x^{-1} - \dfrac{x^{-4}}{4} + 4x + C$

21. $\int (x^{-1} + 2x^{-2} - x^{-3})\, dx = \ln|x| + 2\dfrac{x^{-1}}{-1} - \dfrac{x^{-2}}{-2} + C = \ln|x| - \dfrac{2}{x} + \dfrac{1}{2x^2} + C$

23. $\dfrac{3x^{1.1}}{1.1} - \dfrac{x^{5.3}}{5.3} - 4.1x + C$

25. $\int (3x^{-0.1} - 4x^{-1.1})\, dx = 3\dfrac{x^{0.9}}{0.9} - 4\dfrac{x^{-0.1}}{-0.1} + C = \dfrac{x^{0.9}}{0.3} + \dfrac{40}{x^{0.1}} + C$

27. $2e^x + 5\ln|x| + \dfrac{x}{4} + C$

29. $\int \left(6.1x^{-0.5} + \dfrac{x^{0.5}}{6} - e^x\right) dx = 6.1\dfrac{x^{0.5}}{0.5} + \dfrac{x^{1.5}}{6 \cdot 1.5} - e^x + C = 12.2x^{0.5} + \dfrac{x^{1.5}}{9} - e^x + C$

31. $\dfrac{2^x}{\ln 2} - \dfrac{3^x}{\ln 3} + C$

33. $\dfrac{100(1.1^x)}{\ln(1.1)} + C$

35. $\int \left(\dfrac{1}{x^2} + \dfrac{2}{x^3}\right) dx = \int (x^{-2} + 2x^{-3})\, dx = -x^{-1} - x^{-2} = -\dfrac{1}{x} - \dfrac{1}{x^2} + C$

37. $f'(x) = x$, so $f(x) = \int x\, dx = \dfrac{x^2}{2} + C$. $f(0) = 1$, so $\dfrac{0^2}{2} + C = 1$, $C = 1$. So, $f(x) = \dfrac{x^2}{2} + 1$.

39. $f'(x) = e^x - 1$, so $f(x) = \int (e^x - 1)\, dx = e^x - x + C$. $f(0) = 0$, so $e^0 - 0 + C = 0$, $C = -1$. So, $f(x) = e^x - x - 1$.

41. $C(x) = \int \left(5 - \dfrac{x}{10{,}000}\right) dx = 5x - \dfrac{x^2}{20{,}000} + K$. $C(0) = 20{,}000$, so $0 - 0 + K = 20{,}000$.

Section 6.1

$C(x) = 5x - \dfrac{x^2}{20{,}000} + 20{,}000$.

43. $C(x) = \displaystyle\int \left(5 + 2x + \dfrac{1}{x}\right) dx = 5x + x^2 +$ $\ln x + K$. (Note that $x > 0$.) $C(1) = 1000$, so $5 + 1 + \ln 1 + K = 1000$, $K = 994$. $C(x) = 5x + x^2 + \ln x + 994$.

45. $I'(t) = -0.7$ and $I(2) = 6$. $I(t) = \displaystyle\int(-0.7)\,dt$ $= -0.7t + C$. $-0.7(2) + C = 6$, $C = 7.4$. $I(t) = -0.7t + 7.4$ and $I(8) = 1.8$.

47. $S(t) = \displaystyle\int(0.0166t^2 - 0.050t + 0.27)\,dt =$ $0.00533t^3 - 0.025t^2 + 0.27t + C$. $S(0) = 0$ because the total sales since 1990 are 0 at the beginning of 1990. So, $S(t) = 0.00553t^3 - 0.025t^2 + 0.27t$ billion gallons.

49. (a) We have two data points: $(t, C'(t)) =$ $(0, 3.7)$ and $(7, 3)$. The line through these two points is $C'(t) = -0.1t + 3.7$ (billion dollars per year). **(b)** $C(t) = \displaystyle\int(-0.1t + 3.7)\,dt = -0.05t^2 +$ $3.7t + K$. $C(0) = 135$, so $0 + 0 + K = 135$. $C(t) = -0.05t^2 + 3.7t + 135$ billion dollars.

51. (a) $s(t) = \displaystyle\int v(t)\,dt = \int(t^2 + 1)\,dt = \dfrac{t^3}{3} + t +$ C **(b)** $0 + 0 + C = 1$ so $s = \dfrac{t^3}{3} + t + 1$

53. $a(t) = -32$, $v(t) = \displaystyle\int a(t)\,dt = \int(-32)\,dt =$ $-32t + C$. $v(0) = 0$ is given, so $C = 0$ and $v(t) =$ $-32t$. After 10 seconds, $v(10) = -320$, so the stone is traveling 320 ft/s downward.

55. $a(t) = -32$, $v(t) = \displaystyle\int a(t)\,dt = \int(-32)\,dt =$ $-32t + C$. $v(0) = v_0$ is given, so $C = v_0$, $v(t) = -32t + v_0$. The projectile has zero velocity when $-32t + v_0 = 0$, $t = v_0/32$. This is when it reaches its highest point.

57. $v_0^2/64 = 20$, $v_0 = (1280)^{1/2} \approx 35.78$ ft/s

59. (a) $v_0^2/64 = 100$, $v_0 = 80$ ft/s **(b)** As in Exercise 56, $s(t) = -16t^2 + v_0 t = -16t^2 + 100t$. The projectile strikes the ceiling when $s(t) = 100$, $-16t^2 + 100t = 100$, $16t^2 - 100t + 100 = 0$, $4(4t - 5)(t - 5) = 0$, $t = 1.25$ or 5. We take the first solution, which is the first time it strikes the ceiling. Now, $v(t) = -32t + v_0 = -32t + 100$, so the velocity when it strikes the ceiling is $v(1.25) = 60$ ft/s. **(c)** Start with $s(0) = 100$ and $v(0) = -60$. As before, $v(t) = -32t + v_0 = -32t - 60$. Now, $s(t) = \displaystyle\int v(t)\,dt = \int(-32t - 60)\,dt =$ $-16t^2 - 60t + C$. $0 - 0 + C = 100$, so $s(t) = -16t^2 - 60t + 100$. Now we find when $s(t) = 0$: $-16t^2 - 60t + 100 = 0$, $-4(4t - 5)(t + 5) = 0$, $t = 1.25$ or -5. We take the positive solution and say that it takes 1.25 seconds to hit the ground.

61. Let v_0 be the speed at which Prof. Strong throws and let w_0 be the speed at which Prof. Weak throws. We have $v_0^2/64 = 2w_0^2/64$, so $v_0 = w_0\sqrt{2}$. Thus, Prof. Strong throws $\sqrt{2} \approx 1.414$ times as fast as Prof. Weak.

63. They differ by a constant, $G(x) - F(x) =$ Constant.

Section 6.1

65. Antiderivative; marginal

67. $\int f(x)\, dx$ represents the total cost of manufacturing x items. The units of $\int f(x)\, dx$ are the product of the units of $f(x)$ and the units of x.

69. $\int [f(x) + g(x)]\, dx$ is, by definition, an antiderivative of $f(x) + g(x)$. Let $F(x)$ be an antiderivative of $f(x)$ and let $G(x)$ be an antiderivative of $g(x)$. Then, since the derivative of $F(x) + G(x)$ is $f(x) + g(x)$ (by the rule for sums of derivatives), $F(x) + G(x)$ is an antiderivative of $f(x) + g(x)$. In symbols, $\int [f(x) + g(x)]\, dx = F(x) + G(x) + C = \int f(x)\, dx + \int g(x)\, dx$, the sum of the indefinite integrals.

71. $\int x \cdot 1\, dx = \int x\, dx = \dfrac{x^2}{2} + C$, whereas $\int x\, dx \cdot \int 1\, dx = \left(\dfrac{x^2}{2} + D\right) \cdot (x + E)$, which is not the same as $\dfrac{x^2}{2} + C$, no matter what values we choose for the constants C, D and E.

73. derivative; indefinite integral; indefinite integral; derivative

Section 6.2

6.2

1. $u = 3x + 1$, $du = 3dx$, $dx = \frac{1}{3} du$;

$$\int (3x+1)^5 \, dx = \int u^5 \frac{1}{3} du = \frac{u^6}{18} + C =$$

$$\frac{(3x+1)^6}{18} + C$$

3. $u = -2x + 2$, $du = -2 \, dx$, $dx = -\frac{1}{2} du$;

$$\int (-2x+2)^{-2} \, dx = -\int u^{-2} \frac{1}{2} du = \frac{u^{-1}}{2} + C =$$

$$\frac{(-2x+2)^{-1}}{2} + C$$

5. $u = -x$, $du = -dx$, $dx = -du$; $\int e^{-x} \, dx =$

$$-\int e^u \, du = -e^u + C = -e^{-x} + C$$

7. $u = 3x - 4$, $du = 3 \, dx$, $dx = \frac{1}{3} du$;

$$\int 7.2\sqrt{3x-4} \, dx = \int 7.2\sqrt{u} \frac{1}{3} du = 2.4 \int u^{1/2} \, du$$

$$= 2.4 \frac{u^{3/2}}{3/2} + C = 1.6(3x-4)^{3/2} + C$$

9. $u = 0.6x + 2$, $du = 0.6 \, dx$, $dx = \frac{1}{0.6} du$;

$$\int 1.2 e^{(0.6x+2)} \, dx = \int 1.2 e^u \frac{1}{0.6} du = 2 e^u + C =$$

$$2 e^{(0.6x+2)} + C$$

11. $u = 3x^2 + 3$, $du = 6x \, dx$, $dx = \frac{1}{6x} du$;

$$\int x(3x^2+3)^3 \, dx = \int xu^3 \frac{1}{6x} du = \frac{1}{6} \int u^3 \, du =$$

$$\frac{u^4}{12} + C = \frac{(3x^2+3)^4}{24} + C$$

13. $u = x^2 + 1$, $du = 2x \, dx$, $dx = \frac{1}{2x} du$;

$$\int x(x^2+1)^{1.3} \, dx = \int xu^{1.3} \frac{1}{2x} du = \frac{1}{2} \int u^{1.3} \, du =$$

$$\frac{u^{2.3}}{4.6} + C = \frac{(x^2+1)^{2.3}}{4.6} + C$$

15. $u = 3.1x - 2$, $du = 3.1 \, dx$, $dx = \frac{1}{3.1} du$;

$$\int (1 + 9.3 e^{3.1x-2}) \, dx = \int dx + \int 9.3 e^{3.1x-2} \, dx =$$

$$x + \int 9.3 e^u \frac{1}{3.1} du = x + 3 e^u + C = x + 3 e^{3.1x-2} + C$$

17. $\int \frac{e^x + e^{-x}}{2} dx = \int \frac{e^x}{2} + \int \frac{e^{-x}}{2} dx$; let $u = -x$,

$du = -dx$, $dx = -du$; $\frac{e^x}{2} + \int \frac{e^{-x}}{2} dx = \frac{e^x}{2} - \int \frac{e^u}{2} du$

$$= \frac{e^x}{2} - \frac{e^u}{2} + C = \frac{e^x - e^{-x}}{2} + C$$

19. $u = 3x^2 - 1$, $du = 6x \, dx$, $dx = \frac{1}{6x} du$;

$$\int 2x\sqrt{3x^2-1} \, dx = \int 2x\sqrt{u} \frac{1}{6x} du = \frac{1}{3} \int u^{1/2} \, du$$

$$= \frac{u^{3/2}}{9/2} + C = \frac{2}{9}(3x^2-1)^{3/2} + C$$

21. $u = -x^2 + 1$, $du = -2x \, dx$, $dx = -\frac{1}{2x} du$;

$$\int x e^{-x^2+1} \, dx = -\int x e^u \frac{1}{2x} du = -\frac{1}{2} \int e^u \, du =$$

$$-\frac{1}{2} e^u + C = -\frac{1}{2} e^{-x^2+1} + C$$

23. $u = -(x^2 + 2x)$, $du = -(2x+2) \, dx$, $dx =$

Section 6.2

$-\dfrac{1}{2(x+1)}\,du;\ \int(x+1)e^{-(x^2+2x)}\,dx =$

$-\int(x+1)e^u\,\dfrac{1}{2(x+1)}\,du = -\dfrac{1}{2}\int e^u\,du =$

$-\dfrac{1}{2}e^u + C = -\dfrac{1}{2}e^{-(x^2+2x)} + C$

25. $u = x^2 + x + 1,\ du = (2x+1)\,dx,\ dx =$

$\dfrac{1}{2x+1}\,du;\ \int\dfrac{-2x-1}{(x^2+x+1)^3}\,dx =$

$\int\dfrac{-2x-1}{u^3}\cdot\dfrac{1}{2x+1}\,du = -\int u^{-3}\,du = -\dfrac{u^{-2}}{-2} + C$

$= \dfrac{(x^2+x+1)^{-2}}{2} + C$

27. $u = 2x^3 + x^6 - 5,\ du = (6x^2+6x^5)\,dx,\ dx =$

$\dfrac{1}{6(x^2+x^5)}\,du;\ \int\dfrac{x^2+x^5}{\sqrt{2x^3+x^6-5}}\,dx =$

$\int\dfrac{x^2+x^5}{\sqrt{u}}\cdot\dfrac{1}{6(x^2+x^5)}\,du = \dfrac{1}{6}\int u^{-1/2}\,du =$

$\dfrac{1}{6}\cdot\dfrac{u^{1/2}}{1/2} + C = \dfrac{1}{3}(2x^3+x^6-5)^{1/2} + C =$

$\dfrac{1}{3}\sqrt{2x^3+x^6-5} + C$

29. $u = x-2,\ du = dx;\ x = u+2;\ \int x(x-2)^5\,dx$

$= \int(u+2)u^5\,du = \int(u^6+2u^5)\,du = \dfrac{1}{7}u^7 +$

$\dfrac{1}{3}u^6 + C = \dfrac{1}{7}(x-2)^7 + \dfrac{1}{3}(x-2)^6 + C$

31. $u = x+1,\ du = dx;\ x = u-1;\ \int 2x\sqrt{x+1}\,dx$

$= \int 2(u-1)\sqrt{u}\,du = 2\int(u^{3/2}-u^{1/2})\,du =$

$2\dfrac{u^{5/2}}{5/2} - 2\dfrac{u^{3/2}}{3/2} + C = \dfrac{4}{5}(x+1)^{5/2} -$

$\dfrac{4}{3}(x+1)^{3/2} + C$

33. $u = -\dfrac{1}{x},\ du = \dfrac{1}{x^2}\,dx,\ dx = x^2\,du;\ \int\dfrac{3e^{-1/x}}{x^2}\,dx$

$= \int\dfrac{3e^u}{x^2}\,x^2\,du = \int 3e^u\,du = 3e^u + C = 3e^{-1/x} + C$

35. $u = 1 - e^{-0.05x},\ du = 0.05e^{-0.05x}\,dx,\ dx =$

$\dfrac{1}{0.05e^{-0.05x}}\,du;\ \int\dfrac{e^{-0.05x}}{1-e^{-0.05x}}\,dx =$

$\int\dfrac{e^{-0.05x}}{u}\cdot\dfrac{1}{0.05e^{-0.05x}}\,du = 20\int\dfrac{1}{u}\,du = 20\ln|u| + C$

$= 20\ln|1-e^{-0.05x}| + C$

37. $u = e^x + e^{-x},\ du = (e^x - e^{-x})\,dx,\ dx =$

$\dfrac{1}{e^x-e^{-x}}\,du;\ \int\dfrac{e^x-e^{-x}}{e^x+e^{-x}}\,dx =$

$\int\dfrac{e^x-e^{-x}}{u}\cdot\dfrac{1}{e^x-e^{-x}}\,du = \int\dfrac{1}{u}\,du = \ln|u| + C =$

$\ln|e^x + e^{-x}| + C$

39. $\int[(2x-1)e^{2x^2-2x} + xe^{x^2}]\,dx =$

$\int(2x-1)e^{2x^2-2x}\,dx + \int xe^{x^2}\,dx$. For the first

integral, let $u = 2x^2 - 2x,\ du = (4x-2)\,dx$,

$dx = \dfrac{1}{2(2x-1)}\,du;\ \int(2x-1)e^{2x^2-2x}\,dx =$

$\int(2x-1)e^u\,\dfrac{1}{2(2x-1)}\,du = \dfrac{1}{2}\int e^u\,du = \dfrac{1}{2}e^u +$

$C = \dfrac{1}{2}e^{2x^2-2x} + C.$ For the second integral, let

131

Section 6.2

$u = x^2$, $du = 2x\,dx$, $dx = \dfrac{1}{2x}du$; $\int xe^{x^2}\,dx =$

$\int xe^u \dfrac{1}{2x}du = \dfrac{1}{2}\int e^u\,du = \dfrac{1}{2}e^u + C = \dfrac{1}{2}e^{x^2} + C$.

So, $\int [(2x-1)e^{2x^2-2x} + xe^{x^2}]\,dx = \dfrac{1}{2}e^{2x^2-2x} + \dfrac{1}{2}e^{x^2} + C$.

41. Let $u = ax + b$, $du = a\,dx$, $dx = \dfrac{1}{a}du$;

$\int (ax+b)^n\,dx = \int u^n \dfrac{1}{a}du = \dfrac{u^{n+1}}{a(n+1)} + C =$

$\dfrac{(ax+b)^{n+1}}{a(n+1)} + C$ (if $n \neq -1$)

43. Let $u = ax + b$, $du = a\,dx$, $dx = \dfrac{1}{a}du$;

$\int \sqrt{ax+b}\,dx = \int \sqrt{u} \dfrac{1}{a}du = \dfrac{u^{3/2}}{(3/2)a} + C =$

$\dfrac{2}{3a}(ax+b)^{3/2} + C$

45. $-e^{-x} + C$

47. $\dfrac{1}{2}e^{2x-1} + C$

49. $\dfrac{1}{6}(2x+4)^3 + C$

51. $\dfrac{1}{5}\ln|5x-1| + C$

53. $\dfrac{1}{6}(1.5x)^4 + C$

55. $\dfrac{1.5^{3x}}{3\ln(1.5)} + C$

57. $\dfrac{2^{3x+4} - 2^{-3x+4}}{3\ln 2} + C$

59. $f(x) = \int x(x^2+1)^3\,dx$; $u = x^2 + 1$, $du =$

$2x\,dx$, $dx = \dfrac{1}{2x}du$; $\int x(x^2+1)^3\,dx = \int xu^3 \dfrac{1}{2x}du$

$= \dfrac{1}{2}\int u^3\,du = \dfrac{1}{8}u^4 + C = \dfrac{1}{8}(x^2+1)^4 + C$.

$f(0) = 0$, so $\dfrac{1}{8}(0+1)^4 + C = 0$, $C = -\dfrac{1}{8}$.

$f(x) = \dfrac{1}{8}(x^2+1)^4 - \dfrac{1}{8}$

61. $f(x) = \int xe^{x^2-1}\,dx$; $u = x^2 - 1$, $du = 2x\,dx$,

$dx = \dfrac{1}{2x}du$; $\int xe^{x^2-1}\,dx = \int xe^u \dfrac{1}{2x}du = \dfrac{1}{2}\int e^u\,du$

$= \dfrac{1}{2}e^u + C = \dfrac{1}{2}e^{x^2-1} + C$. $f(1) = 1/2$, so $\dfrac{1}{2}e^{1-1} +$

$C = 1/2$, $C = 0$. $f(x) = \dfrac{1}{2}e^{x^2-1}$

63. $C(x) = \int [5 + (x+1)^{-2}]\,dx = 5x +$

$\int (x+1)^{-2}\,dx$; $u = x + 1$, $du = dx$;

$\int (x+1)^{-2}\,dx = \int u^{-2}\,du = -u^{-1} + K =$

$-(x+1)^{-1} + K$; $\int [5 + (x+1)^{-2}]\,dx = 5x -$

$(x+1)^{-1} + K$. $C(1) = 1000$, so $5 - 2^{-1} + C =$

1000, $C = 995.5$. $C(x) = 5x - \dfrac{1}{(x+1)} + 995.5$

65. $S(t) = \int D(t)\,dt = \int [0.0166(t-1990)^2 -$

$0.05(t-1990) + 0.27]\,dt =$

$0.00553(t-1990)^3 - 0.025(t-1990)^2 +$

132

Section 6.2

$0.27(t - 1990) + K$ (using the shortcut formulas). $S(1990) = 0$, so $K = 0$. Thus, $S(t) = 0.00553(t - 1990)^3 - 0.025(t - 1990)^2 + 0.27(t - 1990)$ billion gallons.

67. (a) $s = \int v(t)\, dt = \int [t(t^2 + 1)^4 + t]\, dt =$

$\int t(t^2 + 1)^4\, dt + \frac{1}{2}t^2;\ u = t^2 + 1;\ du = 2t\, dt;$

$dt = \frac{1}{2t}\, du;\ \int t(t^2 + 1)^4\, dt = \int tu^4 \frac{1}{2t}\, du =$

$\frac{1}{2}\int u^4\, du = \frac{1}{10} u^5 + C = \frac{1}{10}(t^2 + 1)^5 + C.$

$s = \frac{1}{10}(t^2 + 1)^5 + \frac{1}{2}t^2 + C$ (b) $s(0) = 1,$

$\frac{1}{10}(0 + 1)^5 + 0 + C = 1,\ C = \frac{9}{10};$

$s = \frac{1}{10}(t^2 + 1)^5 + \frac{1}{2}t^2 + \frac{9}{10}$

69. None; the substitution $u = x$ simply replaces the letter x throughout by the letter u, and thus does not change the integral at all. For instance, the integral $\int x(3x^2 + 1)\, dx$ becomes

$\int u(3u^2 + 1)\, du$ if we substitute $u = x$.

71. The integral $\int (2x + 1)(x^2 + x)\, dx$ can be calculated by expanding the integrand, or by using the substitution $u = x^2 + x$, as follows. Expanding the integrand: $\int (2x + 1)(x^2 + x)\, dx =$

$\int (2x^3 + 3x^2 + x)\, dx = \frac{1}{2}x^4 + x^3 + \frac{1}{2}x^2 + C.$

Substituting $u = x^2 + x$: $\int (2x + 1)(x^2 + x)\, dx =$

$\int u\, du = \frac{1}{2}u^2 + C = \frac{1}{2}(x^2 + x)^2 + C.$ Expanding

the second answer gives the first, so both methods result in the same solution.

73. The purpose of substitution is to introduce a new variable that is defined in terms of the variable of integration. One cannot say $u = u^2 + 1$, since u is not a new variable. Instead, define $w = u^2 + 1$ (or any other letter different from u).

75. The substitution $u = -x$ leads to $\int e^{-u^2}\, du$, which is just the original integral. The substitution $u = x^2$ leads to $\int e^{-u} \frac{1}{2x}\, du = \int e^{-u} \frac{1}{2\sqrt{u}}\, du$ which is no easier to evaluate. The substitution $u = -x^2$ is similar.

133

6.3

1. With $f(x) = 4x - 1$, $[f(0) + f(0.5) + f(1) + f(1.5)](0.5) = 4$

3. With $f(x) = x^2$, $[f(-2) + f(-1) + f(0) + f(1)](1) = 6$

5. With $f(x) = 1/(1 + x)$, $[f(0) + f(0.2) + f(0.4) + f(0.6) + f(0.8)](0.2) \approx 0.7456$

7. With $f(x) = e^{-x}$, $[f(0) + f(2) + f(4) + f(6) + f(8)](2) \approx 2.3129$

9. With $f(x) = e^{-x^2}$, $[f(0) + f(2.5) + f(5) + f(7.5)](2.5) \approx 2.5048$

11. $n = 10$: 3.3045; $n = 100$: 3.1604; $n = 1000$: 3.1436

13. $n = 10$: 0.0275; $n = 100$: 0.0258; $n = 1000$: 0.0256

15. $\int_0^5 C'(x)\, dx \approx [C'(0) + C'(1) + C'(2) + C'(3) + C'(4)](1) = \$99.95$

17. $\int_5^{10} D(t)\, dt \approx [D(5) + D(6) + D(7) + D(8) + D(9)](1) \approx 3.8$ billion gallons

19. $\int_5^{10} W(t)\, dt \approx [W(5) + W(6) + W(7) + W(8) + W(9)](1) = 1058 + 1058 + 1149 + 1186 + 1350 = 5801$. This represents the total number of wiretaps authorized by U.S. courts from 1995 through 1999.

21. $\int_0^2 v(t)\, dt \approx [v(0) + v(0.2) + v(0.4) + \ldots + v(1.8)](0.2) = 91.2$ ft

23. $\int_0^4 R(t)\, dt \approx 198$. Approximately 198 million books were sold online during the period 1997 through 2000.

25. (a) $p(x) = \dfrac{1}{5.2\sqrt{2\pi}} e^{-(x-72.6)^2/54.08}$, $\int_{60}^{100} p(x)\, dx \approx 0.994$, so approximately 99.4% of students obtained between 60 and 100. (b) $\int_0^{30} p(x)\, dx \approx 0$ (to at least 15 decimal places)

27. increases: The left sum underestimates the function, by less as n increases.

29. The total cost is $c(1) + c(2) + \ldots + c(60)$, which is represented by the Riemann sum approximation of $\int_1^{61} c(t)\, dt$ with $n = 60$.

31. $[f(x_1) + f(x_2) + \ldots + f(x_n)]\Delta x = \sum_{k=1}^{n} f(x_k)\Delta x$

33. If increasing n by a factor of 10 does not change the value of the answer when rounded to three decimal places, then the answer is (likely) accurate to three decimal places.

6.4

1. The area is a square of height 1 and width 1, so has area $1 \times 1 = 1$.

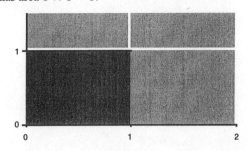

3. The area is a triangle of height 1 and base 1, so has area $\frac{1}{2}(1 \times 1) = \frac{1}{2}$.

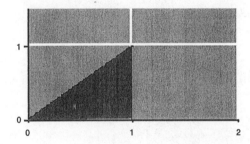

5. The area is a triangle of height $\frac{1}{2}$ and base 1, so has area $\frac{1}{2}\left(\frac{1}{2} \times 1\right) = \frac{1}{4}$.

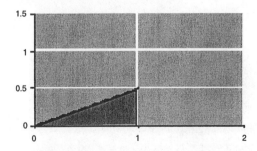

7. The area is a triangle of height 2 and base 2, so has area $\frac{1}{2}(2 \times 2) = 2$.

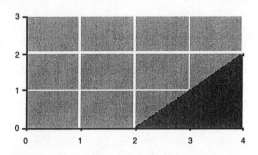

9. Since the are below the x axis is the same as the area above, the integral is 0.

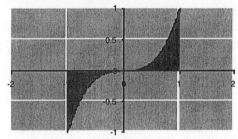

11. Compute $\int_0^1 x \, dx$. $n = 10$: 0.45; $n = 100$:

0.495

13. Compute $\int_0^4 (x^2 - 1) \, dx$. $n = 10$: 14.24; $n = 100$:

17.0144

15. Compute $\int_0^1 e^{x^2} \, dx$. $n = 10$: 1.3813; $n = 100$:

1.4541

17. The area is $2650 + 2800 + 2900 + 3000 + 3400 + 3500 = 18{,}250$ million. If $S(t)$ is the rate of sales, in millions of SUVs per year, this area is the definite integral of $S(t)$, so represents the total number of SUVs sold in the U.S. from 1998 through 2003.

19. The are is $-220 - 170 - 170 + 5 - 25 = -\$580$ million. If $I(t)$ is Amazon's net income rate, in millions of dollars per quarter (t measured in quarters since the start of 1999), this area is the definite integral of $I(t)$ from 8 to 13, so represents the total net income for the first quarter of 2001 through the first quarter of 2002.

21. Yes. The Riemann sum gives an estimated area of $(0 + 15 + 18 + 8 + 7 + 16 + 20)(5) = 420$ square feet.

23. (a) The area below represents the total oil revenue earned by Saudi Arabia in 1999.

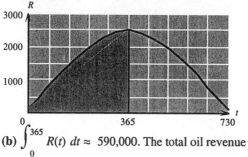

(b) $\int_0^{365} R(t)\,dt \approx 590{,}000$. The total oil revenue earned by Saudi Arabia in 1999 was approximately \$590 billion.

25. Answers may vary. For example: Take $P(t)$ to be the rate of change of profit at time t. If $P(t)$ is negative, then the profit is decreasing, so the change in profit, represented by the definite integral of $P(t)$, is negative.

27. Answers may vary. For example:

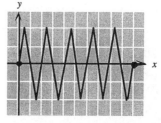

29. Answers may vary. For example:

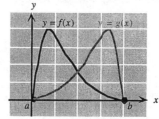

31. Here is a sketch of a typical example:

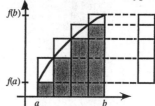

The difference between the right and left sums for $\int_a^b f(x)\,dx$ is the area of the unshaded regions between the shaded rectangles (giving the left Riemann sum) and the larger rectangles extending over the graph (giving the right Riemann sum). When stacked up, the rectangles combine to give the rectangle to the right of the sketch. This rectangle has height $f(b) - f(a)$ and width Δx, hence its area, which is the difference between the right and left Riemann sums, is $[f(b) - f(a)]\Delta x$. As $n \to \infty$, $f(b)$ and $f(a)$ remain fixed while $\Delta x \to 0$, so the product $[f(b) - f(a)]\Delta x$ goes to 0.

6.5

1. $A(x)$ is the area of a rectangle of height 1 and width x, so $A(x) = x$.

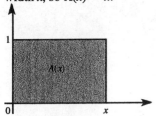

3. $A(x)$ is the difference between the area of a triangle of width and height x and one of width and height 2, so $A(x) = x^2/2 - 2$

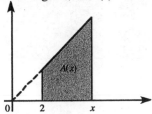

5. $A(x)$ is the area of a rectangle of height 4 and width $x - a$, so $A(x) = 4(x - a)$.

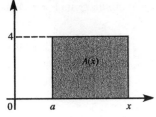

7. $A(x) = \int_0^x t\, dt = \left[\frac{t^2}{2}\right]_0^x = \frac{x^2}{2} - \frac{0}{2} = \frac{x^2}{2} \cdot f(x)$:

$A(x)$:

9. $A(x) = \int_0^x t^2\, dt = \left[\frac{t^3}{3}\right]_0^x = \frac{x^3}{3} - \frac{0}{3} = \frac{x^3}{3} \cdot f(x)$:

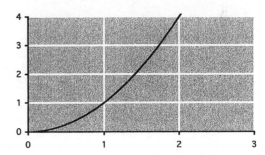

$A(x)$:

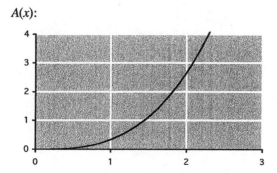

11. $A(x) = \int_0^x e^t \, dt = [e^t]_0^x = e^x - 1$. $f(x)$:

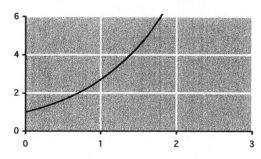

$A(x)$:

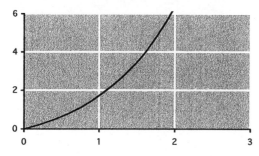

13. $A(x) = \int_1^x \frac{1}{t} \, dt = [\ln t]_1^x = \ln x - \ln 1 = \ln x$.

$f(x)$:

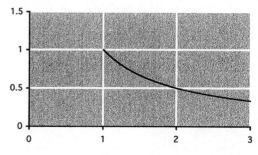

$A(x)$:

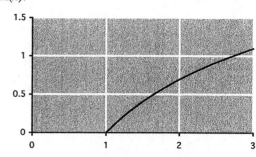

15. For $0 \leq x \leq 1$, $A(x) = \int_0^x f(t) \, dt = \int_0^x t \, dt = \left[\frac{t^2}{2}\right]_0^x = \frac{x^2}{2} - \frac{0}{2} = \frac{x^2}{2}$. In particular, $A(1) = \frac{1}{2}$. For $x > 1$, $A(x) = A(1) + \int_1^x f(t) \, dt = \frac{1}{2} + \int_1^x 2 \, dt = \frac{1}{2} + [2t]_1^x = \frac{1}{2} + 2x - 2 = 2x - \frac{3}{2}$. Thus,

$$A(x) = \begin{cases} \dfrac{x^2}{2} & \text{if } 0 \leq x \leq 1 \\ 2x - \dfrac{3}{2} & \text{if } x > 1 \end{cases}.$$

Section 6.5

$f(x)$:

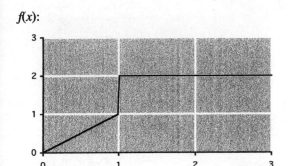

$A(x)$:

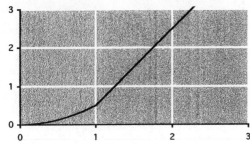

17. $\int_{-1}^{1}(x^2 + 2)\, dx = \left[\dfrac{x^3}{3} + 2x\right]_{-1}^{1} =$

$\dfrac{1}{3} + 2 - \left(\dfrac{-1}{3} - 2\right) = \dfrac{14}{3}$

19. $\int_{0}^{1}(12x^5 + 5x^4 - 6x^2 + 4)\, dx = [2x^6 + x^5 -$

$2x^3 + 4x]_0^1 = 2 + 1 - 2 + 4 - (0) = 5$

21. $\int_{-2}^{2}(x^3 - 2x)\, dx = \left[\dfrac{x^4}{4} - x^2\right]_{-2}^{2} =$

$\dfrac{16}{4} - 4 - \left(\dfrac{16}{4} - 4\right) = 0$

23. $\int_{1}^{3}\left(\dfrac{2}{x^2} + 3x\right) dx = \int_{1}^{3}(2x^{-2} + 3x)\, dx =$

$\left[-2x^{-1} + \dfrac{3}{2}x^2\right]_1^3 = -\dfrac{2}{3} + \dfrac{27}{2} - \left(-2 + \dfrac{3}{2}\right) = \dfrac{40}{3}$

25. $\int_{0}^{1}(2.1x - 4.3x^{1.2})\, dx = \left[1.05x^2 - \dfrac{4.3}{2.2}x^{2.2}\right]_0^1 =$

$1.05 - \dfrac{4.3}{2.2} - (0) \approx -0.9045$

27. $\int_{0}^{1} 2e^x\, dx = [2e^x]_0^1 = 2e - 2 = 2(e - 1)$

29. $\int_{0}^{1}\sqrt{x}\, dx = \int_{0}^{1} x^{1/2}\, dx = \left[\dfrac{2}{3}x^{3/2}\right]_0^1 = \dfrac{2}{3} - 0 = \dfrac{2}{3}$

31. $\int_{0}^{1} 2^x\, dx = \left[\dfrac{2^x}{\ln 2}\right]_0^1 = \dfrac{2}{\ln 2} - \dfrac{1}{\ln 2} = \dfrac{1}{\ln 2}$

33. Let $u = 3x + 1$, $du = 3dx$, $dx = \dfrac{1}{3}du$; when $x = 0$, $u = 1$; when $x = 1$, $u = 4$;

$\int_{0}^{1} 18(3x + 1)^5\, dx = \int_{1}^{4} 18u^5 \dfrac{1}{3}du = [u^6]_1^4 = 4^6 - 1$

$= 4095$

In Exercises 35–39, we use the shortcut integration formulas rather than substitution.

35. $\int_{-1}^{1} e^{2x-1}\, dx = \left[\dfrac{1}{2}e^{2x-1}\right]_{-1}^{1} = \dfrac{1}{2}e^1 - \dfrac{1}{2}e^{-3} =$

$\dfrac{1}{2}(e - e^{-3})$

Section 6.5

37. $\int_0^2 2^{-x+1} \, dx = \left[-\dfrac{2^{-x+1}}{\ln 2}\right]_0^2 = -\dfrac{2^{-1}}{\ln 2} - \left(-\dfrac{2}{\ln 2}\right)$

$= \dfrac{3}{2 \ln 2}$

39. $\int_0^{50} e^{-0.02x-1} \, dx = \left[-\dfrac{e^{-0.02x-1}}{0.02}\right]_0^{50} = -50e^{-2} -$

$(-50e^{-1}) = 50(e^{-1} - e^{-2})$

41. $\int_{-1.1}^{1.1} e^{x+1} \, dx = [e^{x+1}]_{-1.1}^{1.1} = e^{2.1} - e^{-0.1}$

43. Let $u = 2x^2 + 1$, $du = 4x \, dx$, $dx = \dfrac{1}{4x} du$;

when $x = -\sqrt{2}$, $u = 5$; when $x = \sqrt{2}$, $u = 5$;

$\int_{-\sqrt{2}}^{\sqrt{2}} 3x\sqrt{2x^2+1} \, dx = \int_5^5 3x\sqrt{u} \cdot \dfrac{1}{4x} du = \dfrac{3}{4} \int_0^0 u^{1/2} \, du$

$= \left[\dfrac{1}{2} u^{3/2}\right]_0^0 = 0 - 0 = 0$

45. Let $u = x^2 + 2$, $du = 2x \, dx$, $dx = \dfrac{1}{2x} du$;

when $x = 0$, $u = 2$; when $x = 1$, $u = 3$;

$\int_0^1 5xe^{x^2+2} \, dx = \int_2^3 5xe^u \cdot \dfrac{1}{2x} du = \dfrac{5}{2} \int_2^3 e^u \, du =$

$\left[\dfrac{5}{2} e^u\right]_2^3 = \dfrac{5}{2} (e^3 - e^2)$

47. Let $u = x^3 - 1$, $du = 3x^2 \, dx$, $dx = \dfrac{1}{3x^2} du$;

when $x = 2$, $u = 7$; when $x = 3$, $u = 26$;

$\int_2^3 \dfrac{x^2}{x^3 - 1} \, dx = \int_7^{26} \dfrac{x^2}{u} \cdot \dfrac{1}{3x^2} du = \dfrac{1}{3} \int_7^{26} \dfrac{1}{u} \, du =$

$\left[\dfrac{1}{3} \ln|u|\right]_7^{26} = \dfrac{1}{3} (\ln 26 - \ln 7) = \dfrac{1}{3} \ln\left(\dfrac{26}{7}\right)$

49. Let $u = -x^2$, $du = -2x \, dx$, $dx = -\dfrac{1}{2x} du$;

when $x = 0$, $u = 0$; when $x = 1$, $u = -1$;

$\int_0^1 x(1.1)^{-x^2} \, dx = -\int_0^{-1} x(1.1)^u \dfrac{1}{2x} du =$

$-\dfrac{1}{2} \int_0^{-1} (1.1)^u \, du = \left[-\dfrac{(1.1)^u}{2 \ln 1.1}\right]_0^{-1} =$

$-\dfrac{(1.1)^{-1}}{2 \ln 1.1} - \left(-\dfrac{1}{2 \ln 1.1}\right) = \dfrac{0.1}{2.2 \ln 1.1}$

51. Let $u = x^{-1}$, $du = -x^{-2} \, dx$, $dx = -x^2 \, du$;

when $x = 1$, $u = 1$; when $x = 2$, $u = 1/2$;

$\int_1^2 \dfrac{e^{1/x}}{x^2} \, dx = -\int_1^{1/2} \dfrac{e^u}{x^2} x^2 \, du = -\int_1^{1/2} e^u \, du = [-e^u]_1^{1/2} =$

$-e^{1/2} - (-e) = e - e^{1/2}$

53. Let $u = x + 1$, $du = dx$; when $x = 0$, $u = 1$;

when $x = 2$, $u = 3$; $x = u - 1$; $\int_0^2 \dfrac{x}{x+1} \, dx =$

$\int_1^3 \dfrac{x}{u} \, du = \int_1^3 \dfrac{u-1}{u} \, du = \int_1^3 \left(1 - \dfrac{1}{u}\right) du = [u -$

$\ln|u|]_1^3 = 3 - \ln 3 - (1 - \ln 1) = 2 - \ln 3$

55. Let $u = x - 2$, $du = dx$; when $x = 1$, $u = -1$; when $x = 2$, $u = 0$; $x = u + 2$;

Section 6.5

$$\int_1^2 x(x-2)^5\,dx = \int_{-1}^0 xu^5\,du = \int_{-1}^0 (u+2)u^5\,du =$$

$$\int_{-1}^0 (u^6 + 2u^5)\,du = \left[\frac{u^7}{7} + \frac{u^6}{3}\right]_{-1}^0 = 0 + 0 -$$

$$\left(\frac{-1}{7} + \frac{1}{3}\right) = -\frac{4}{21}$$

57. Let $u = 2x + 1$, $du = 2\,dx$, $dx = \frac{1}{2}du$; when $x = 0$, $u = 1$; when $x = 1$, $u = 3$; $x = \frac{1}{2}(u-1)$;

$$\int_0^1 x\sqrt{2x+1}\,dx = \int_1^3 x\sqrt{u}\,\frac{1}{2}du =$$

$$\frac{1}{2}\int_1^3 \frac{1}{2}(u-1)u^{1/2}\,du = \frac{1}{4}\int_1^3 (u^{3/2} - u^{1/2})\,du =$$

$$\left[\frac{1}{10}u^{5/2} - \frac{1}{6}u^{3/2}\right]_1^3 = \frac{3^{5/2}}{10} - \frac{3^{3/2}}{6} - \left(\frac{1}{10} - \frac{1}{6}\right) =$$

$$\frac{3^{5/2}}{10} - \frac{3^{3/2}}{6} + \frac{1}{15}$$

59. The line $y = x$ crosses the x axis at $x = 0$, so we can calculate the area as $\int_0^1 x\,dx = \left[\frac{x^2}{2}\right]_0^1 = \frac{1}{2}$

61. The curve $y = \sqrt{x}$ touches the x axis at $x = 0$ only, so we can calculate the area as $\int_0^4 \sqrt{x}\,dx =$

$$\int_0^4 x^{1/2}\,dx = \left[\frac{2}{3}x^{3/2}\right]_0^4 = \frac{16}{3} - 0 = \frac{16}{3}$$

63. The curve $y = x^2 - 1$ crosses the x axis where $x^2 - 1 = 0$, so at $x = \pm 1$. We compute two

integrals: $\int_0^1 (x^2 - 1)\,dx = \left[\frac{x^3}{3} - x\right]_0^1 = \frac{1}{3} - 1 - 0$

$$= -\frac{2}{3} \text{ and } \int_1^4 (x^2 - 1)\,dx = \left[\frac{x^3}{3} - x\right]_1^4 =$$

$\frac{64}{3} - 4 - \left(\frac{1}{3} - 1\right) = \frac{54}{3}$. The total area is the sum of the absolute values, so $\frac{2}{3} + \frac{54}{3} = \frac{56}{3}$.

65. The curve $y = xe^{x^2}$ crosses the x axis at $x = 0$, so we can calculate the area as $\int_0^{(\ln 2)^{1/2}} xe^{x^2}\,dx$. Let

$u = x^2$, $du = 2x\,dx$, $dx = \frac{1}{2x}du$; when $x = 0$,

$u = 0$; when $x = (\ln 2)^{1/2}$, $u = \ln 2$. $\int_0^{(\ln 2)^{1/2}} xe^{x^2}\,dx =$

$$\int_0^{\ln 2} xe^u \frac{1}{2x}\,du = \frac{1}{2}\int_0^{\ln 2} e^u\,du = \left[\frac{1}{2}e^u\right]_0^{\ln 2} = \frac{1}{2}e^{\ln 2} - \frac{1}{2}$$

$$= \frac{1}{2}\cdot 2 - \frac{1}{2} = \frac{1}{2}$$

67. $\int_{10}^{100}\left(5 + \frac{x^2}{1000}\right)dx = \left[5x + \frac{x^3}{3000}\right]_{10}^{100} = 500 +$

$$\frac{1000}{3} - \left(50 + \frac{1}{3}\right) = \$783$$

69. $\int_5^{10}(0.0166t^2 - 0.050t + 0.27)\,dt =$

$[0.005533t^3 - 0.025t^2 + 0.27t]_5^{10} \approx 5.73 -$
$1.42 \approx 4.3$ billion gallons

Section 6.5

71. $\int_{14}^{26} 5.4e^{0.06t}\, dt = [90e^{0.06t}]_{14}^{26} \approx 428.3 - 208.5 \approx$ $220 billion

73. $\int_{1}^{6}(60 - e^{-t/10})\, dt = [60t + 10e^{-t/10}]_{1}^{6} = 360 + 10e^{-6/10} - (60 + 10e^{-1/10}) \approx 296$ miles

75. (a) We have two data points: $(t, C'(t)) = (0, 3.7)$ and $(7, 3)$. The line through these two points is $C'(t) = -0.1t + 3.7$ billion dollars per year. (b) $\int_{0}^{5}(-0.1t + 3.7)\, dt = [-0.05t^2 + 3.7t]_{0}^{5}$
$= -1.25 + 18.5 - (0) = \$17.25$ billion dollars.

77. Since $c(t)$ measures oxygen consumption in milliliters *per hour*, we need to multiply by 24 to convert to a function that measures consumption in milliliters per day:

$24\int_{8}^{10}(-0.0027t^3 + 0.137t^2 - 0.892t + 0.149)\, dt$
$= 24[-0.000675t^4 + 0.04567t^3 - 0.446t^2 + 0.149t]_{8}^{10} = 24[-6.75 + 45.67 - 44.6 + 1.49 - (-2.7648 + 23.383 - 28.544 + 1.192)] \approx$ 61 milliliters.

79. If $S(t)$ is the weekly sales rate, we have $S(t) = 50(1 - 0.05)^t$ T-shirts per week after t weeks. The total sales over the coming year will be
$\int_{0}^{52} 50(0.95)^t\, dt = \left[\dfrac{50(0.95)^t}{\ln 0.95}\right]_{0}^{52} =$
$\dfrac{50}{\ln 0.95}[(0.95)^{52} - 1] \approx 907$ T-shirts.

81. $\int_{0}^{10}(1 - e^{-t})\, dt = [t + e^{-t}]_{0}^{10} = 10 + e^{-10} - (0 + e^0) \approx 9$ gallons

83. (a) Let $u = 0.77 + e^{0.15t}$, $du = 0.15e^{0.15t}\, dt$, $dt = \dfrac{1}{0.15e^{0.15t}}\, du$; when $t = 0$, $u = 1.77$; when $t = 10$, $u \approx 5.252$. $\int_{0}^{10}\dfrac{1500e^{0.15t}}{0.77 + e^{0.15t}}\, dt =$

$\int_{1.77}^{5.252}\dfrac{1500e^{0.15t}}{u} \cdot \dfrac{1}{0.15e^{0.15t}}\, du = 10,000 \int_{1.77}^{5.252}\dfrac{1}{u}\, du =$

$10,000[\ln u]_{1.77}^{5.252} = 10,000(\ln 5.252 - \ln 1.77) \approx$ 11,000 wiretaps. (b) To estimate the actual number of wiretaps over the given period from the chart, take half of the 1990 and 2000 numbers and add the numbers from 1991 through 1999. This gives 10,737, which agrees with the answer in part (a) when we round to two significant digits. Therefore, the integral in part (a) does give an accurate estimate.

85. (a) We use the two data points $(t, s(t)) = (0, 330)$ and $(12, 940)$ to construct the exponential model $s(t) = 330e^{0.0872t}$.

(b) $\int_{0}^{14} 330e^{0.0872t}\, dt = \dfrac{330}{0.0872}[e^{0.0872t}]_{0}^{14} =$
$\dfrac{330}{0.0872}(e^{1.2208} - 1) \approx \$9044$ billion, compared to $8225 billion computed from the graph.

(c) $\int_{14}^{19} 330e^{0.0872t}\, dt = \dfrac{330}{0.0872}[e^{0.0872t}]_{14}^{19} =$
$\dfrac{330}{0.0872}(e^{1.6568} - e^{1.2208}) \approx \$7011$ billion

Section 6.5

87. $\int_0^x m(t)\, dt = C(x) - C(0)$ by the FTC, so

$C(x) = C(0) + \int_0^x m(t)\, dt$. $C(0)$ is the *fixed cost*.

89. (a) $\dfrac{N}{1 + Ab^{-x}} = \dfrac{Nb^x}{(1 + Ab^{-x})b^x} = \dfrac{Nb^x}{b^x + A} = \dfrac{Nb^x}{A + b^x}$ **(b)** Let $u = A + b^x$, $du = (\ln b)b^x\, dx$,

$dx = \dfrac{1}{(\ln b)b^x}\, du$. $\int \dfrac{N}{1 + Ab^{-x}}\, dx = \int \dfrac{Nb^x}{A + b^x}\, dx =$

$\int \dfrac{Nb^x}{u} \cdot \dfrac{1}{(\ln b)b^x}\, du = \dfrac{N}{\ln b} \int \dfrac{1}{u}\, du = \dfrac{N \ln u}{\ln b} + C =$

$\dfrac{N \ln(A + b^x)}{\ln b} + C$ **(c)** $\int_0^4 \dfrac{82.8}{1 + 21.8(7.14)^{-t}}\, dt =$

$\left[\dfrac{82.8 \ln(21.8 + 7.14^t)}{\ln 7.14} \right]_0^4 =$

$\dfrac{82.8 \ln(21.8 + 7.14^4)}{\ln 7.14} - \dfrac{82.8 \ln(21.8 + 7.14^0)}{\ln 7.14} \approx$

200 million books

91. $W = \int_{v_0}^{v_1} v \dfrac{d}{dv}(p)\, dv = \int_{v_0}^{v_1} v \dfrac{d}{dv}(mv)\, dv =$

$\int_{v_0}^{v_1} v \cdot m\, dv = \left[\tfrac{1}{2} mv^2 \right]_{v_0}^{v_1} = \tfrac{1}{2} mv_1^2 - \tfrac{1}{2} mv_0^2$

93. (a) According to the first part of the FTC,

$\int_a^x \dfrac{1}{t}\, dt$ is an antiderivative of $\dfrac{1}{x}$ for any fixed a.

(b) $M(x) = \int_1^x \dfrac{1}{t}\, dt$. $M(1) = \int_1^1 \dfrac{1}{t}\, dt = 0$; $M'(x) =$

$\dfrac{1}{x}$ by the first part of the FTC. **(c)** $\dfrac{d}{dx}[M(ax)] =$

$\dfrac{1}{ax} \cdot a = \dfrac{1}{x}$. **(d)** If $F(x) = M(ax) - M(x)$, $F'(x) =$

$\dfrac{1}{x} - \dfrac{1}{x} = 0$. Hence, $F(x)$ is constant. **(e)** $F(1) =$

$M(a) - M(1) = M(a)$. Hence, $M(ax) - M(x) =$

$M(a)$ for all x, that is, $M(ax) = M(a) + M(x)$.

[Hence, $M(x)$ has the characteristic property of the logarithm: It converts multiplication to addition. The other properties of the logarithm can be deduced in a similar way.]

95. They are related by the Fundamental Theorem of Calculus, which (briefly) states that the definite integral of a suitably nice function can be calculated by evaluating the indefinite integral at the two endpoints and subtracting.

97. An example is $v(t) = t - 5$.

99. An example is $f(x) = e^{-x}$.

101. Two things: (1) calculate definite integrals by using antiderivatives and (2) obtain formulas for an antiderivative of a function in terms of a definite integral, or area.

103. $F(x) = \begin{cases} 0 & \text{if } x < 0; \\ x & \text{if } x \geq 0. \end{cases}$ $F'(x) = f(x)$ for

every $x \neq 0$ (since $F(x)$ is not differentiable at $x = 0$).

Chapter 6 Review Test

1. (a) $\frac{x^3}{3} - 5x^2 + 2x + C$

(b) $e^x + \frac{2}{3}x^{3/2} + C$

(c) Let $u = x^2 + 4$, $du = 2x\,dx$, $dx = \frac{1}{2x}du$;

$$\int x(x^2 + 4)^{10}\,dx = \int xu^{10}\frac{1}{2x}du = \frac{1}{2}\int u^{10}\,du =$$

$$\frac{1}{22}u^{11} + C = \frac{1}{22}(x^2 + 4)^{11} + C$$

(d) Let $u = x^3 + 3x + 2$, $du = (3x^2 + 3)\,dx = 3(x^2 + 1)\,dx$, $dx = \frac{1}{3(x^2+1)}du$;

$$\int \frac{x^2 + 1}{(x^3 + 3x + 2)^2}\,dx = \int \frac{x^2+1}{u^2}\cdot\frac{1}{3(x^2+1)}\,du =$$

$$\frac{1}{3}\int u^{-2}\,du = -\frac{1}{3}u^{-1} + C =$$

$$-\frac{1}{3(x^3 + 3x + 2)} + C$$

(e) $-\frac{5}{2}e^{-2x} + C$

(f) Let $u = -x^2/2$, $du = -x\,dx$, $dx = -\frac{1}{x}du$;

$$\int xe^{-x^2/2}\,dx = -\int xe^u\frac{1}{x}\,du = -\int e^u\,du = -e^u + C = -e^{-x^2/2} + C$$

2. (a) $\int_0^1(x - x^3)\,dx = \left[\frac{x^2}{2} - \frac{x^4}{4}\right]_0^1 = \frac{1}{2} - \frac{1}{4} - (0)$

$= \frac{1}{4}$

(b) Let $u = x + 1$, $du = dx$; when $x = 0$, $u = 1$; when $x = 9$, $u = 10$; $\int_0^9\frac{1}{x+1}\,dx = \int_1^{10}\frac{1}{u}\,du =$

$[\ln u]_1^{10} = \ln 10 - \ln 1 = \ln 10 \approx 2.303$

(c) Let $u = x^3 + 1$, $du = 3x^2\,dx$, $dx = \frac{1}{3x^2}$; when $x = 0$, $u = 1$; when $x = 2$, $u = 9$; $\int_0^2 x^2\sqrt{x^3 + 1}\,dx$

$= \int_1^9 x^2\sqrt{u}\frac{1}{3x^2}\,du = \frac{1}{3}\int_1^9 u^{1/2}\,du = \frac{2}{9}[u^{3/2}]_1^9 =$

$\frac{2}{9}(27 - 1) = \frac{52}{9}$

(d) $\int_{-1}^1 3^{2x-2}\,dx = \frac{1}{2\ln 3}[3^{2x-2}]_{-1}^1 =$

$\frac{1}{2\ln 3}(3^0 - 3^{-4}) = \frac{40}{81\ln 3} \approx 0.4495$

3. (a) $y = 4 - x^2$ crosses the x axis when $x = \pm 2$, so we can compute the area as $\int_{-2}^2(4 - x^2)\,dx =$

$\left[4x - \frac{x^3}{3}\right]_{-2}^2 = 8 - \frac{8}{3} - \left(-8 + \frac{8}{3}\right) = \frac{32}{3}$.

(b) $y = 4 - x^2$ crosses the x axis when $x = \pm 2$, so we compute two integrals: $\int_0^2(4 - x^2)\,dx =$

$\left[4x - \frac{x^3}{3}\right]_0^2 = 8 - \frac{8}{3} - (0) = \frac{16}{3}$ and

$\int_2^5(4 - x^2)\,dx = \left[4x - \frac{x^3}{3}\right]_2^5 = 20 - \frac{125}{3} - \left(8 - \frac{8}{3}\right)$

$= -27$. Adding the absolute values gives a total area of $\frac{16}{3} + 27 = \frac{97}{3}$.

(c) $y = xe^{-x^2}$ crosses the x axis at $x = 0$, so we can compute the area as $\int_0^5 xe^{-x^2}\,dx$. Let $u = -x^2$,

144

Chapter 6 Review Test

$du = -2x\,dx$, $dx = -\dfrac{1}{2x}du$; when $x = 0$, $u = 0$;

when $x = 5$, $u = -25$. So the area is

$-\displaystyle\int_0^{-25} xe^u \dfrac{1}{2x}\,du = \dfrac{1}{2}\displaystyle\int_0^{-25} e^u\,du = -\dfrac{1}{2}[e^u]_0^{-25} =$

$\dfrac{1-e^{-25}}{2}$.

(d) Since $y = -2x$ when $x < 0$ and $y = 2x$ when $x \geq 0$, we compute the area using two integrals:

$\displaystyle\int_{-1}^0 (-2x)\,dx = [-x^2]_{-1}^0 = 0 - (-1) = 1$ and

$\displaystyle\int_0^1 2x\,dx = [x^2]_0^1 = 1 - 0 = 1$. Adding these

together gives $1 + 1 = 2$.

4. (a) $n = 10$: 0.77781682, $n = 100$: 0.7499786, $n = 1000$: 0.74714013
(b) $n = 10$: 0.78690744, $n = 100$: 0.69719004, $n = 1000$: 0.68849294

5. (a) $q = \displaystyle\int (-20p)\,dp = -10p^2 + C$; $q = 100{,}000$ when $p = 0$, so $0 + C = 100{,}000$, $C = 100{,}000$, $q = 100{,}000 - 10p^2$ **(b)** $q = 0$ when $100{,}000 - 10p^2 = 0$, $p^2 = 10{,}000$, $p = \$100$.

6. (a) If $S(t)$ is the weekly sales in week t, they were estimating that $S(t) = 6400(2)^{t/2}$. The total sales over the first five weeks would be

$\displaystyle\int_0^5 S(t)\,dt = \displaystyle\int_0^5 6400(2)^{t/2}\,dt = \dfrac{12{,}800}{\ln 2}[2^{t/2}]_0^5 =$

$\dfrac{12{,}800}{\ln 2}[2^{5/2} - 2^0] \approx 86{,}000$ books. **(b)** Let $u = e^{0.55t} + 14.01$, $du = 0.55e^{0.55t}\,dt$, $dt =$

$\dfrac{1}{0.55e^{0.55t}}\,du$; when $t = 0$, $u = 15.01$; when $t = 5$,

$u \approx 29.653$; $\displaystyle\int_0^5 \left(6053 + \dfrac{4474e^{0.55t}}{e^{0.55t}+14.01}\right)dt =$

$[6053t]_0^5 + \displaystyle\int_{15.01}^{29.653} \dfrac{4474e^{0.55t}}{u} \cdot \dfrac{1}{0.55e^{0.55t}}\,du =$

$30{,}265 + 8135\displaystyle\int_{15.01}^{29.653} \dfrac{1}{u}\,du = 30{,}265 +$

$8135[\ln u]_{15.01}^{29.653} \approx 35{,}800$ books.

7. (a) Let $s(t)$ be the height function, $v(t)$ the velocity, and $a(t)$ the acceleration. We know that $a(t) = -32$ ft/s^2; we are told that $v(0) = 60$ ft/s and $s(0) = 100$ ft. So, $v(t) = \displaystyle\int (-32)\,dt =$

$-32t + C$; $60 = -32(0) + C$, $C = 60$; $v(t) = -32t + 60$. Thus, $s(t) = \displaystyle\int (-32t + 60)\,dt =$

$-16t^2 + 60t + C$; $100 = -16(0) + 60(0) + C$, $C = 100$; $s(t) = -16t^2 + 60t + 100$. The book will hit the ground when $s(t) = 0$: $-16t^2 + 60t + 100 = 0$, $t = 5$ seconds. (The other solution is $t = -1.25$ s, which we reject because it is negative.)
(b) $v(5) = -100$ ft/s, so the book is traveling 100 ft/s when it hits the ground. **(c)** The maximum value of $s(t)$ occurs when $v(t) = 0$, so when

$-32t + 60 = 0$, $t = 1.875$ s. The height at that time is $s(1.875) = 156.25$ ft above ground level.

Chapter 6 Review Test

Section 7.1

Chapter 7
7.1

1.

	D	I
+	$2x$	e^x
−	2	e^x
+∫	0 →	e^x

$\int 2xe^x\, dx = 2xe^x - 2e^x + C = 2(x-1)e^x + C$

3.

	D	I
+	$3x - 1$	e^{-x}
−	3	$-e^{-x}$
+∫	0 →	e^{-x}

$\int (3x-1)e^{-x}\, dx = -(3x-1)e^{-x} - 3e^{-x} + C =$

$-(3x+2)e^{-x} + C$

5.

	D	I
+	$x^2 - 1$	e^{2x}
−	$2x$	$\frac{1}{2}e^{2x}$
+	2	$\frac{1}{4}e^{2x}$
−∫	0 →	$\frac{1}{8}e^{2x}$

$\int (x^2 - 1)e^{2x}\, dx = \frac{1}{2}(x^2 - 1)e^{2x} - \frac{1}{2}xe^{2x} +$

$\frac{1}{4}e^{2x} + C = \frac{1}{4}(2x^2 - 2x - 1)e^{2x} + C$

7.

	D	I
+	$x^2 + 1$	e^{-2x+4}
−	$2x$	$-\frac{1}{2}e^{-2x+4}$
+	2	$\frac{1}{4}e^{-2x+4}$
−∫	0 →	$-\frac{1}{8}e^{-2x+4}$

$\int (x^2 + 1)e^{-2x+4}\, dx = -\frac{1}{2}(x^2 + 1)e^{-2x+4} -$

$\frac{1}{2}xe^{-2x+4} - \frac{1}{4}e^{-2x+4} + C =$

$-\frac{1}{4}(2x^2 + 2x + 3)e^{-2x+4} + C$

9.

	D	I
+	$2 - x$	2^x
−	-1	$2^x/\ln 2$
+∫	0 →	$2^x/(\ln 2)^2$

$\int (2 - x)2^x\, dx = \frac{1}{\ln 2}(2 - x)2^x + \frac{1}{(\ln 2)^2}2^x + C$

$= \left[\frac{2-x}{\ln 2} + \frac{1}{(\ln 2)^2}\right]2^x + C$

147

Section 7.1

11.

D	I
+ $\quad x^2 - 1$	3^{-x}
$-\quad 2x$	$-3^{-x}/\ln 3$
+ $\quad 2$	$3^{-x}/(\ln 3)^2$
$-\int\quad 0$	$-3^{-x}/(\ln 3)^3$

$$\int (x^2 - 1)3^{-x}\, dx = -\frac{1}{\ln 3}(x^2 - 1)3^{-x} -$$

$$\frac{2}{(\ln 3)^2}x3^{-x} - \frac{2}{(\ln 3)^3}3^{-x} + C =$$

$$-\left[\frac{x^2 - 1}{\ln 3} + \frac{2x}{(\ln 3)^2} + \frac{2}{(\ln 3)^3}\right]3^{-x} + C$$

13.

D	I
+ $\quad x^2 - x$	e^{-x}
$-\quad 2x - 1$	$-e^{-x}$
+ $\quad 2$	e^{-x}
$-\int\quad 0$	$-e^{-x}$

$$\int \frac{x^2 - x}{e^x}\, dx = \int (x^2 - x)e^{-x}\, dx = -(x^2 - x)e^{-x} -$$

$$(2x - 1)e^{-x} - 2e^{-x} + C = -(x^2 + x + 1)e^{-x} + C$$

15.

D	I
+ $\quad x$	$(x + 2)^6$
$-\quad 1$	$(x + 2)^7/7$
$+\int\quad 0$	$(x + 2)^8/56$

$$\int x(x + 2)^6\, dx = \frac{1}{7}x(x + 2)^7 - \frac{1}{56}(x + 2)^8 + C$$

17.

D	I
+ $\quad x$	$(x - 2)^{-3}$
$-\quad 1$	$-(x - 2)^{-2}/2$
$+\int\quad 0$	$(x - 2)^{-1}/2$

$$\int \frac{x}{(x - 2)^3}\, dx = \int x(x - 2)^{-3}\, dx =$$

$$-\frac{1}{2}x(x - 2)^{-2} - \frac{1}{2}(x - 2)^{-1} + C =$$

$$-\frac{x}{2(x - 2)^2} - \frac{1}{2(x - 2)} + C$$

19.

D	I
+ $\quad \ln x$	x^3
$-\int\quad 1/x$	$x^4/4$

$$\int x^3 \ln x\, dx = \frac{1}{4}x^4 \ln x - \int \frac{1}{4}x^3\, dx =$$

$$\frac{1}{4}x^4 \ln x - \frac{1}{16}x^4 + C$$

148

Section 7.1

21.

	D	I
+	$\ln(2t)$	$t^2 + 1$
	↘	
$-\int$	$1/t$ → $t^3/3 + t$	

$\int (t^2 + 1)\ln(2t)\, dt = \left(\frac{1}{3}t^3 + t\right)\ln(2t) -$

$\int \left(\frac{1}{3}t^2 + 1\right) dt = \left(\frac{1}{3}t^3 + t\right)\ln(2t) - \frac{1}{9}t^3 - t + C$

23.

	D	I
+	$\ln t$	$t^{1/3}$
	↘	
$-\int$	$1/t$ → $3t^{4/3}/4$	

$\int t^{1/3} \ln t\, dt = \frac{3}{4} t^{4/3} \ln t - \int \frac{3}{4} t^{1/3}\, dt =$

$\frac{3}{4} t^{4/3} \ln t - \frac{9}{16} t^{4/3} + C = \frac{3}{4} t^{4/3} \left(\ln t - \frac{3}{4}\right) + C$

25.

	D	I
+	$\log_3 x$	1
	↘	
$-\int$	$1/(x \ln 3)$ → x	

$\int \log_3 x\, dx = x \log_3 x - \int \frac{1}{\ln 3}\, dx =$

$x \log_3 x - \dfrac{x}{\ln 3} + C$

27. $\int (xe^{2x} - 4e^{3x})\, dx = \int xe^{2x}\, dx - \int 4e^{3x}\, dx =$

$\int xe^{2x}\, dx - \frac{4}{3} e^{3x}$. To evaluate the remaining integral we use integration by parts:

	D	I
+	x	e^{2x}
	↘	
$-$	1	$\frac{1}{2} e^{2x}$
	↘	
$+\int$	0 → $\frac{1}{4} e^{2x}$	

$\int xe^{2x}\, dx - \frac{4}{3} e^{3x} = \frac{1}{2} xe^{2x} - \frac{1}{4} e^{2x} - \frac{4}{3} e^{3x} + C =$

$\left(\frac{1}{2} x - \frac{1}{4}\right) e^{2x} - \frac{4}{3} e^{3x} + C$

29. $\int (x^2 e^x - xe^{x^2})\, dx = \int x^2 e^x\, dx - \int xe^{x^2}\, dx$.

To evaluate the first integral we use integration by parts:

	D	I
+	x^2	e^x
	↘	
$-$	$2x$	e^x
	↘	
+	2	e^x
	↘	
$-\int$	0 → e^x	

$\int x^2 e^x\, dx = x^2 e^x - 2xe^x + 2e^x + C = (x^2 - 2x +$

$2)e^x + C$. To evaluate the second integral we use substitution: $u = x^2$, $du = 2x\, dx$, $dx = \dfrac{1}{2x} du$.

$\int xe^{x^2}\, dx = \int xe^u \dfrac{1}{2x}\, du = \frac{1}{2} \int e^u\, du = \frac{1}{2} e^u + C$

Section 7.1

$= \frac{1}{2} e^{x^2} + C$. Combining the two integrals we get

$\int (x^2 e^x - x e^{x^2}) \, dx = (x^2 - 2x + 2) e^x - \frac{1}{2} e^{x^2} + C$

31.

	D	I
+	$x + 1$	e^x
−	1	e^x
+∫	0 →	e^x

$\int_0^1 (x+1) e^x \, dx = [(x+1) e^x - e^x]_0^1 = [x e^x]_0^1 =$

$e - 0 = e$

33.

	D	I
+	x^2	$(x+1)^{10}$
−	$2x$	$(x+1)^{11}/11$
+	2	$(x+1)^{12}/132$
−∫	0 →	$(x+1)^{13}/1716$

$\int_0^1 x^2 (x+1)^{10} \, dx =$

$\left[\frac{1}{11} x^2 (x+1)^{11} - \frac{1}{66} x (x+1)^{12} + \frac{1}{858} (x+1)^{13} \right]_0^1$

$= \frac{1}{11} 2^{11} - \frac{1}{66} 2^{12} + \frac{1}{858} 2^{13} - \frac{1}{858} = \frac{38,229}{286}$

35.

	D	I
+	$\ln(2x)$	x
−∫	$1/x$ →	$x^2/2$

$\int_1^2 x \ln(2x) \, dx = \left[\frac{1}{2} x^2 \ln(2x) \right]_1^2 - \int_1^2 \frac{1}{2} x \, dx =$

$\left[\frac{1}{2} x^2 \ln(2x) \right]_1^2 - \left[\frac{1}{4} x^2 \right]_1^2 = 2 \ln 4 - \frac{1}{2} \ln 2 -$

$\left(1 - \frac{1}{4} \right) = 4 \ln 2 - \frac{1}{2} \ln 2 - \frac{3}{4} = \frac{7}{2} \ln 2 - \frac{3}{4}$

37.

	D	I
+	$\ln(x+1)$	x
−∫	$1/(x+1)$ →	$x^2/2$

$\int_0^1 x \ln(x+1) \, dx = \left[\frac{1}{2} x^2 \ln(x+1) \right]_0^1 -$

$\int_0^1 \frac{x^2}{2(x+1)} \, dx = \frac{1}{2} \ln 2 - \int_0^1 \frac{x^2}{2(x+1)} \, dx$. We

evaluate this integral using a substitution: $u = x + 1$, $du = dx$; $x = u - 1$; when $x = 0$, $u = 1$; when $x = 1$, $u = 2$. $\int_0^1 \frac{x^2}{2(x+1)} \, dx = \int_1^2 \frac{x^2}{2u} \, du =$

$\int_1^2 \frac{(u-1)^2}{2u} \, du = \frac{1}{2} \int_1^2 \frac{u^2 - 2u + 1}{u} \, du =$

$\frac{1}{2} \int_1^2 \left(u - 2 + \frac{1}{u} \right) du = \left[\frac{1}{4} u^2 - u + \frac{1}{2} \ln u \right]_1^2 =$

Section 7.1

$1 - 2 + \frac{1}{2}\ln 2 - \left(\frac{1}{4} - 1 + 0\right) = -\frac{1}{4} + \frac{1}{2}\ln 2$.

Combining with our earlier calculation we get

$\int_0^1 x \ln(x + 1)\, dx = \frac{1}{2}\ln 2 - \left(-\frac{1}{4} + \frac{1}{2}\ln 2\right) = \frac{1}{4}$

39. We calculate the area using $\int_0^{10} xe^{-x}\, dx$. To evaluate this integral we use integration by parts:

	D	I
+	x	e^{-x}
−	1	$-e^{-x}$
$+\int$	0	e^{-x}

$\int_0^{10} xe^{-x}\, dx = [-xe^{-x} - e^{-x}]_0^{10} = -10e^{-10} - e^{-10} -$
$(0 - e^0) = 1 - 11e^{-10}$

41. We calculate the area using $\int_1^2 (x + 1) \ln x\, dx$.

To evaluate this integral we use integration by parts:

	D	I
+	$\ln x$	$x + 1$
$-\int$	$1/x$	$x^2/2 + x$

$\int_1^2 (x + 1) \ln x\, dx = \left[\left(\frac{1}{2}x^2 + x\right)\ln x\right]_1^2 -$

$\int_1^2 \left(\frac{1}{2}x + 1\right) dx = \left[\left(\frac{1}{2}x^2 + x\right)\ln x\right]_1^2 -$

$\left[\frac{1}{4}x^2 + x\right]_1^2 = 4\ln 2 - 0 - \left[1 + 2 - \left(\frac{1}{4} + 1\right)\right] =$
$4\ln 2 - \frac{7}{4}$

43. We compute the displacement by integrating the velocity over the first two minutes, or 120 seconds: $\int_0^{120} 2000te^{-t/120}\, dt$. We evaluate this integral using integration by parts:

	D	I
+	t	$e^{-t/120}$
−	1	$-120e^{-t/120}$
$+\int$	0	$14{,}400e^{-t/120}$

$\int_0^{120} 2000te^{-t/120}\, dt = 2000[-120te^{-t/120} -$
$14{,}400e^{-t/120}]_0^{120} = 2000[-14{,}400e^{-1} -$
$14{,}400e^{-1} - (-14{,}400e^0)] =$
$28{,}800{,}000(1 - 2e^{-1})$ ft $\approx 7{,}610{,}000$ ft

45. We are given $C'(x) = 10 + \dfrac{\ln(x + 1)}{(x + 1)^2}$ and $C(0) = 5000$. So, $C(x) = \int \left[10 + \dfrac{\ln(x + 1)}{(x + 1)^2}\right] dx$

$= 10x + \int (x + 1)^{-2} \ln(x + 1)\, dx$. To evaluate this integral we use integration by parts:

	D	I
+	$\ln(x + 1)$	$(x + 1)^{-2}$
$-\int$	$1/(x + 1)$	$-(x + 1)^{-1}$

151

Section 7.1

$C(x) = 10x - (x + 1)^{-1} \ln(x + 1) + \int (x + 1)^{-2} \, dx$

$= 10x - (x + 1)^{-1} \ln(x + 1) - (x + 1)^{-1} + K$. To determine K we substitute $C(0) = 5000$: $5000 = -\ln 1 - 1 + K$, $K = 5001$. So, $C(x) = 10x - \dfrac{\ln(x + 1)}{x + 1} - \dfrac{1}{x + 1} + 5001$.

47. If $p(t)$ is the price in week t, we are told that $p(t) = 10 + 0.5t$. If $q(t)$ is the weekly sales, we are told that $q(t) = 50e^{-0.02t}$. The weekly revenue is therefore $R(t) = p(t)q(t) = 50(10 + 0.5t)e^{-0.02t}$ and the total revenue over the coming year is

$\displaystyle\int_0^{52} 50(10 + 0.5t)e^{-0.02t} \, dt = \int_0^{52} 500e^{-0.02t} \, dt +$

$25\displaystyle\int_0^{52} te^{-0.02t} \, dt$. We evaluate each integral:

$\displaystyle\int_0^{52} 500e^{-0.02t} \, dt = [-25{,}000e^{-0.02t}]_0^{52} =$

$-25{,}000e^{-1.04} - (-25{,}000e^0) = 25{,}000 - 25{,}000e^{-1.04}$. We evaluate the second integral using integration by parts:

D	I
$+$ t	$e^{-0.02t}$
$-$ 1	$-50e^{-0.02t}$
$+\int$ 0 $\to$	$2500e^{-0.02t}$

$\displaystyle\int_0^{52} te^{-0.02t} \, dt = [-50te^{-0.02t} - 2500e^{-0.02t}]_0^{52} =$

$-2600e^{-1.04} - 2500e^{-1.04} - (-2500e^0) = 2500 - 5100e^{-1.04}$. So, the total revenue is $25{,}000 - 25{,}000e^{-1.04} + 25(2500 - 5100e^{-1.04}) \approx \$33{,}598$.

49. If $p(t)$ is the price t years after 1982 and $q(t)$ is the annual sales, we are told that $p(t) = 2e^{0.05t}$ and $q(t) = 70.5 - 2.5t$ million magazines. The annual revenue is therefore $R(t) = p(t)q(t) = 2(70.5 - 2.5t)e^{0.05t}$, so the total revenue from 1982 to 1992 is $\displaystyle\int_0^{10} 2(70.5 - 2.5t)e^{0.05t} \, dt = \int_0^{10} 141e^{0.05t} \, dt -$

$5\displaystyle\int_0^{10} te^{0.05t} \, dt$. We evaluate each integral:

$\displaystyle\int_0^{10} 141e^{0.05t} \, dt = [2820e^{0.05t}]_0^{10} = 2820e^{0.5} - 2820$.

We evaluate the second integral using integration by parts:

D	I
$+$ t	$e^{0.05t}$
$-$ 1	$20e^{0.05t}$
$+\int$ 0 $\to$	$400e^{0.05t}$

$\displaystyle\int_0^{10} te^{0.05t} \, dt = [20te^{0.05t} - 400e^{0.05t}]_0^{10} = 200e^{0.5} -$

$400e^{0.5} - (-400e^0) = 400 - 200e^{0.5}$. So, the total revenue is $2820e^{0.5} - 2820 - 5(400 - 200e^{0.5}) \approx \$1478$ million.

51. Answers will vary. Examples are xe^{x^2} and $e^{x^2} = 1 \cdot e^{x^2}$

53. $n+1$ times

55. If $f(x)$ is a polynomial of degree n, then

Section 7.1

$f^{(n+1)}(x) = 0$. Using integration by parts to evaluate $\int_0^b f(x)e^{-x}\,dx$ we get the following table:

D	I
$+\quad f(x)$	e^{-x}
$-\quad f'(x)$	$-e^{-x}$
$+\quad f''(x)$	e^{-x}
$\cdots$	
$\pm\quad f^{(n)}(x)$	$\pm e^{-x}$
$\mp\quad 0$	$\mp e^{-x}$

So, $\int_0^b f(x)e^{-x}\,dx =$

$[-f(x)e^{-x} - f'(x)e^{-x} - \ldots - f^{(n)}(x)e^{-x}]_0^b =$
$-[f(b) + f'(b) + \ldots + f^{(n)}(b)]e^{-b} + [f(0) + f'(0) + \ldots + f^{(n)}(0)]e^0 = F(0) - F(b)e^{-b}$.

Section 7.2

7.2

In these solutions we always take the integral of $f(x) - g(x)$ with $f(x) \geq g(x)$. Remember that, if you reverse the order, you will simply get the negative of that integral and should then take the absolute value.

1. We have $x^2 \geq 0 > -1$ for all x, so the two graphs do not cross. The area is
$$\int_{-1}^{1} [x^2 - (-1)] \, dx = \int_{-1}^{1} (x^2 + 1) \, dx = \left[\frac{1}{3}x^3 + x\right]_{-1}^{1}$$
$$= \frac{1}{3} + 1 - \left(-\frac{1}{3} - 1\right) = \frac{8}{3}.$$

3. $-x = x$ when $x = 0$. The area is
$$\int_{0}^{2} [x - (-x)] \, dx = \int_{0}^{2} 2x \, dx = [x^2]_0^2 = 4 - 0 = 4.$$

5. $x = x^2$ when $x^2 - x = 0$, $x(x-1) = 0$, $x = 0$ or $x = 1$. We calculate the area using two integrals: $\int_{-1}^{0} (x^2 - x) \, dx = \left[\frac{1}{3}x^3 - \frac{1}{2}x^2\right]_{-1}^{0} =$
$0 - \left(-\frac{1}{3} - \frac{1}{2}\right) = \frac{5}{6}$ and $\int_{0}^{1} (x - x^2) \, dx =$
$\left[\frac{1}{2}x^2 - \frac{1}{3}x^3\right]_0^1 = \frac{1}{2} - \frac{1}{3} - (0) = \frac{1}{6}$. The total area is therefore $\frac{5}{6} + \frac{1}{6} = 1$.

7. $e^x > x$ for all x. (Examine the graphs, or consider the fact that $e^x - x$ has its minimum value when its derivative, $e^x - 1$, is 0, which occurs when $x = 0$.) The area is $\int_{0}^{1} (e^x - x) \, dx =$
$\left[e^x - \frac{1}{2}x^2\right]_0^1 = e - \frac{1}{2} - (1) = e - \frac{3}{2} \approx 1.218.$

9. $(x-1)^2 \geq 0 \geq -(x-1)^2$, so the area is
$$\int_{0}^{1} [(x-1)^2 + (x-1)^2] \, dx = \int_{0}^{1} 2(x-1)^2 \, dx =$$
$$\left[\frac{2}{3}(x-1)^3\right]_0^1 = 0 - \left(-\frac{2}{3}\right) = \frac{2}{3}.$$

11. $x = x^4$ when $x^4 - x = 0$, $x(x^3 - 1) = 0$, $x = 0$ or $x = 1$. So, the area is $\int_{0}^{1} (x - x^4) \, dx =$
$$\left[\frac{1}{2}x^2 - \frac{1}{5}x^5\right]_0^1 = \frac{1}{2} - \frac{1}{5} = \frac{3}{10}.$$

13. $x^3 = x^4$ when $x^4 - x^3 = 0$, $x^3(x - 1) = 0$, $x = 0$ or $x = 1$. So, the area is $\int_{0}^{1} (x^3 - x^4) \, dx =$
$$\left[\frac{1}{4}x^4 - \frac{1}{5}x^5\right]_0^1 = \frac{1}{4} - \frac{1}{5} - (0) = \frac{1}{20}.$$

15. $x^2 = x^4$ when $x^4 - x^2 = 0$, $x^2(x^2 - 1) = 0$, $x = 0$ or $x = \pm 1$. So, the area is $\int_{-1}^{0} (x^2 - x^4) \, dx +$
$$\int_{0}^{1} (x^2 - x^4) \, dx = \left[\frac{1}{3}x^3 - \frac{1}{5}x^5\right]_{-1}^{0} + \left[\frac{1}{3}x^3 - \frac{1}{5}x^5\right]_0^1$$
$$= 0 - \left(-\frac{1}{3} + \frac{1}{5}\right) + \left(\frac{1}{3} - \frac{1}{5}\right) = \frac{4}{15}.$$ (In fact, since $x^2 \geq x^4$ on all of $[-1, 1]$, we could have used the single integral $\int_{-1}^{1} (x^2 - x^4) \, dx$ to calculate this area.)

17. Here are the graphs of $y = e^x$ and $y = 2$:

154

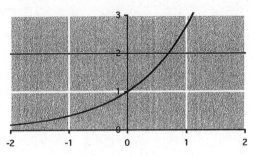

The two graphs intersect where $e^x = 2$, $x = \ln 2$. From the graph we can see that the area we want is the area between these two graphs for $0 \leq x \leq \ln 2$. So we compute $\int_0^{\ln 2} (2 - e^x)\, dx =$

$[2x - e^x]_0^{\ln 2} = 2\ln 2 - e^{\ln 2} - (0 - 1) =$
$2\ln 2 - 2 + 1 = 2\ln 2 - 1$.

19. Here are the graphs of $y = \ln x$ and $y = 2 - \ln x$:

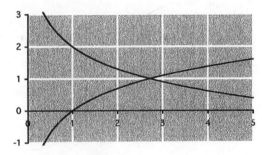

The two graphs intersect where $\ln x = 2 - \ln x$, $2\ln x = 2$, $\ln x = 1$, $x = e$. From the graph we can see that the area we want is the area between these two graphs for $e \leq x \leq 4$. So we compute
$\int_e^4 (2\ln x - 2)\, dx = [2(x\ln x - x) - 2x]_e^4$ (using the antiderivative of $\ln x$ we derived in Section 7.1) $= [2x\ln x - 4x]_e^4 = 8\ln 4 - 16 - (2e\ln e - 4e) = 8\ln 4 + 2e - 16 \approx 0.5269$.

21. Here are the graphs of $y = e^x$ and $y = 2x + 1$:

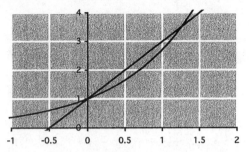

The two graphs intersect at $x = 0$ and at a point somewhere between $x = 1$ and $x = 1.5$. We cannot solve the equation $e^x = 2x + 1$ algebraically but we can use technology to estimate the second intersection point. To four decimal places it is $x \approx 1.2564$. Therefore, the area is approximately $\int_0^{1.2564} (2x + 1 - e^x)\, dx =$

$[x^2 + x - e^x]_0^{1.2564} = (1.2564)^2 + 1.2564 - e^{1.2564} - (-e^0) \approx 0.3222$.

23. Here are the graphs of $y = \ln x$ and $y = \frac{1}{2}x - \frac{1}{2}$:

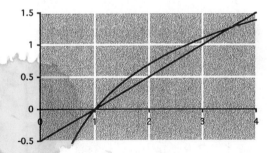

The two graphs intersect at $x = 1$ and at a point somewhere between $x = 3$ and $x = 4$. We cannot solve the equation $\ln x = \frac{1}{2}x - \frac{1}{2}$ algebraically, but we can use technology to estimate the second intersection point. To four decimal places it is $x \approx 3.5129$. Therefore, the area is approximately

155

Section 7.2

$$\int_1^{3.5129}\left(\ln x - \frac{1}{2}x + \frac{1}{2}\right)dx =$$

$$\left[x\ln x - x - \frac{1}{4}x^2 + \frac{1}{2}x\right]_1^{3.5129} =$$

$3.5129 \ln(3.5129) - 3.5129 - \frac{1}{4}(3.5129)^2 + \frac{1}{2}3.5129 - \left(0 - 1 - \frac{1}{4} + \frac{1}{2}\right) \approx 0.3222$. (Why is this the same answer as Exercise 21?)

25. $q = 5 - p/2$, so $\bar{q} = 5 - 5/2 = 5/2$. The consumers' surplus is $\int_0^{5/2}(10 - 2q - 5)\,dq =$

$\int_0^{5/2}(5 - 2q)\,dq = [5q - q^2]_0^{5/2} = 5\left(\frac{5}{2}\right) - \left(\frac{5}{2}\right)^2 - (0) = \$6.25$.

27. $q = (100 - p)^2/9$, so $\bar{q} = (100 - 76)^2/9 = 64$. The consumers' surplus is

$\int_0^{64}(100 - 3\sqrt{q} - 76)\,dq = \int_0^{64}(24 - 3\sqrt{q})\,dq =$

$[24q - 2q^{3/2}]_0^{64} = 24(64) - 2(64)^{3/2} - (0) = \$512$.

29. $q = -\frac{1}{2}\ln(p/500)$, so $\bar{q} = -\frac{1}{2}\ln(1/5) = \frac{1}{2}\ln 5$. The consumers' surplus is

$\int_0^{(\ln 5)/2}(500e^{-2q} - 100)\,dq = [-250e^{-2q} -$

$100q]_0^{(\ln 5)/2} = -250(1/5) - 50\ln 5 - (-250) = \$119.53$.

31. $\bar{q} = 100 - 2(20) = 60$; $p = 50 - q/2$. The consumers' surplus is $\int_0^{60}\left(50 - \frac{1}{2}q - 20\right)dq =$

$\int_0^{60}\left(30 - \frac{1}{2}q\right)dq = \left[30q - \frac{1}{4}q^2\right]_0^{60} = 30(60) - \frac{1}{4}(60)^2 - (0) = \$900$.

33. $\bar{q} = 100 - 0.25(10)^2 = 75$; $p = 2\sqrt{100 - q}$. The consumers' surplus is

$\int_0^{75}(2\sqrt{100 - q} - 10)\,dq =$

$\left[-\frac{4}{3}(100 - q)^{3/2} - 10q\right]_0^{75} = -\frac{4}{3}(100 - 75)^{3/2}$

$- 10(75) - \left(-\frac{4}{3}(100)^{3/2} - 0\right) = \$416.67$.

35. $\bar{q} = 500e^{-0.5} - 50$; $p = -2\ln\left(\frac{1}{500}q + \frac{1}{10}\right)$. The consumers' surplus is

$\int_0^{500e^{-0.5}-50}\left[-2\ln\left(\frac{1}{500}q + \frac{1}{10}\right) - 1\right]dq =$

$-2\int_0^{500e^{-0.5}-50}\ln\left(\frac{1}{500}q + \frac{1}{10}\right)dq - [q]_0^{500e^{-0.5}-50} =$

$-2\int_0^{500e^{-0.5}-50}\ln\left(\frac{1}{500}q + \frac{1}{10}\right)dq - 500e^{-0.5} + 50$. We evaluate the remaining integral using substitution (and the integral of $\ln u$ we found by integration by parts): Let $u = \frac{1}{500}q + \frac{1}{10}$; $du = \frac{1}{500}dq$, $dq = 500\,du$. When $q = 0$, $u = 0.1$; when $q =$

Section 7.2

$500e^{-0.5} - 50$, $u = e^{-0.5}$.

$$\int_0^{500e^{-0.5}-50} \ln\left(\frac{1}{500}q + \frac{1}{10}\right) dq = \int_{0.1}^{e^{-0.5}} 500 \ln u \, du =$$

$500[u \ln u - u]_{0.1}^{e^{-0.5}} = 500[e^{-0.5} \ln(e^{-0.5}) - e^{-0.5} - (0.1 \ln(0.1) - 0.1)] \approx -289.769$. The consumers' surplus is therefore $-2(-289.769) - 500e^{-0.5} + 50 = \$326.27$.

37. $q = p/2 - 5$, so $\bar{q} = 20/2 - 5 = 5$. The producers' surplus is $\int_0^5 [20 - (10 + 2q)] dq =$

$\int_0^5 (10 - 2q) dq = [10q - q^2]_0^5 = 50 - 25 - (0)$

$= \$25$.

39. $q = (p/2 - 5)^3$, so $\bar{q} = (12/2 - 5)^3 = 1$. The producers' surplus is $\int_0^1 [12 - (10 + 2q^{1/3})] dq =$

$\int_0^1 (2 - 2q^{1/3}) dq = \left[2q - \frac{3}{2}q^{4/3}\right]_0^1 = 2 - \frac{3}{2} - (0)$

$= \$0.50$.

41. $q = 2\ln(p/500)$, so $\bar{q} = 2\ln(1000/500) = 2\ln 2$. The producers' surplus is

$\int_0^{2\ln 2} (1000 - 500e^{0.5q}) dq = [1000q -$

$1000e^{0.5q}]_0^{2\ln 2} = 2000 \ln 2 - 2000 - (-1000) = 2000 \ln 2 - 1000 = \$386.29$.

43. $\bar{q} = 2(40) - 50 = 30$; $p = q/2 + 25$. The producers' surplus is $\int_0^{30} \left[40 - \left(\frac{1}{2}q + 25\right)\right] dq =$

$\int_0^{30} \left(15 - \frac{1}{2}q\right) dq = \left[15q - \frac{1}{4}q^2\right]_0^{30} = 450 - 225 - (0) = \$225$.

45. $\bar{q} = 0.25(10)^2 - 10 = 15$; $p = \sqrt{4q + 40} = 2\sqrt{q + 10}$. The producers' surplus is

$\int_0^{15} (10 - 2\sqrt{q + 10}) dq = \left[10q - \frac{4}{3}(q + 10)^{3/2}\right]_0^{15}$

$= 150 - \frac{500}{3} - \left(-\frac{4}{3} \cdot 10^{3/2}\right) = \$25.50$.

47. $\bar{q} = 500e^{0.05(10)} - 50 = 500e^{0.5} - 50$; $p = 20 \ln\left(\frac{1}{500}q + \frac{1}{10}\right)$. The producers' surplus is

$\int_0^{500e^{0.5}-50} \left[10 - 20 \ln\left(\frac{1}{500}q + \frac{1}{10}\right)\right] dq = [10q -$

$(20q + 1000) \ln\left(\frac{1}{500}q + \frac{1}{10}\right) +$

$(20q + 1000)]_0^{500e^{0.5}-50}$ (using the substitution $u = \frac{1}{500}q + \frac{1}{10}$ as in Exercise 35) $= 5000e^{0.5} - 500 - 10{,}000e^{0.5} \ln e^{0.5} + 10{,}000e^{0.5} - \left(-1000 \ln\left(\frac{1}{10}\right) + 1000\right) = \$12{,}684.63$.

49. To find the equilibrium tuition set demand equal to supply: $9859.39 - 2.17p = 100 + 0.5p$, so $\bar{p} = \$3655$. The equilibrium supply and demand is $\bar{q} = 9859.39 - 2.17(3655) = 1928$. To find the consumers' surplus we solve for $p = 4543 - q/2.17$ and compute $CS =$

$$\int_0^{1928}\left(4543 - \frac{q}{2.17} - 3655\right)dq =$$

$$\int_0^{1928}\left(888 - \frac{q}{2.17}\right)dq = \left[888q - \frac{q^2}{4.34}\right]_0^{1928} =$$

$888(1928) - \frac{1928^2}{4.34} - (0) \approx \$856{,}000$. To find the producers' surplus we solve for $p = 2q - 200$ and compute $PS = \int_0^{1928}[3655 - (2q - 200)]\,dq =$

$$\int_0^{1928}(3855 - 2q)\,dq = [3855q - q^2]_0^{1928} =$$

$3855(1928) - 1928^2 - (0) \approx \$3{,}715{,}000$. The total social gain is approximately $\$4{,}571{,}000$.

51. $\bar{q} = b - m\bar{p}$, $p = \frac{1}{m}(b - q)$; $CS =$

$$\int_0^{b-m\bar{p}}\left[\frac{1}{m}(b - q) - \bar{p}\right]dq =$$

$$\left[\frac{b}{m}q - \frac{1}{2m}q^2 - \bar{p}q\right]_0^{b-m\bar{p}} = \frac{b}{m}(b - m\bar{p}) -$$

$$\frac{1}{2m}(b - m\bar{p})^2 - \bar{p}(b - m\bar{p}) - (0) =$$

$$\frac{1}{2m}(b - m\bar{p})[2b - (b - m\bar{p}) - 2m\bar{p}] =$$

$$\frac{1}{2m}(b - m\bar{p})^2$$

53. (a) The area represents the accumulated U.S. trade deficit with China (total excess value of imports over exports) for the 9-year period 1989–1998. **(b)** $\int_1^{10}[I(t) - E(t)]\,dt =$

$$\int_1^{10}[(3.0 + 6.2t) - (4.1 + 0.94t)]\,dt =$$

$$\int_1^{10}(5.26t - 1.1)\,dt = [2.63t^2 - 1.1t]_1^{10} =$$

$2.63(100) - 1.1(10) - (2.63 - 1.1) = 250.47$. The U.S. accumulated a $\$250.47$ billion trade deficit with China over the period 1989–1998.

55. (a) The area represents the difference between the total number of above-ground pools and in-ground pools built from 1996 to 2001.

(b) $\int_6^{11}[b(t) - g(t)]\,dt = \int_6^{11}[(0.36t^2 + 4.1t + 140) - (-0.45t^2 + 15t + 54)]\,dt =$

$\int_6^{11}(0.81t^2 - 10.9t + 86)\,dt = [0.27t^3 - 5.45t^2 + 86t]_6^{11} = 645.92 - 378.12 \approx 268$. From 1996 to 2001, 268 thousand more above-ground pools than in-ground pools were built in the U.S.

Section 7.2

57. (a) $\int_{10}^{50} [N(t) - L(t)]\, dt = \int_{10}^{50} (0.38e^{0.017t} - 0.15e^{0.026t})\, dt = [22.35e^{0.017t} - 5.77e^{0.026t}]_{10}^{50} \approx 31 - 19 = 12$ billion cases **(b)** This is the area of the region between the graphs of $N(t)$ and $L(t)$ for $10 \le t \le 50$. **(c)** Since the exponent for $L(t)$ is larger, the fraction of new cases that occur among people 85 and older is projected to rise.

59. (a) We verify, using the formula in the text, that $C'(8) = 270$ million, $C'(10) = 360$ million, and $C'(12) = 780$ million. **(b)** The net marginal cost is $C'(q) - 400{,}000{,}000$ dollars per million tons of reduction. The total cost to remove 12 million tons of sulfur is therefore $\int_0^{12} [C'(q) - 400{,}000{,}000]\, dq =$
$\int_0^{12} [1{,}000{,}000(41.25q^2 - 697.5q + 3210) - 400{,}000{,}000]\, dq = 1{,}000{,}000 \int_0^{12} (41.25q^2 - 697.5q + 2810)\, dq$
$= 1{,}000{,}000[13.75q^3 - 348.75q^2 + 2810q]_0^{12} = 7260 - 0 = \$7260$ million. **(c)** The two equations are $y = 1{,}000{,}000(41.25q^2 - 697.5q + 3210)$ and $y = 400{,}000{,}000$ for $0 \le q \le 12$.

61. The area between the export and import curves represents Canada's accumulated trade surplus (that is, the total excess of exports over imports) from January 1997 to January 2001.

63. (A)

65. The claim is wrong because the area under a curve can only represent income if the curve is a graph of income *per unit time*. The value of a stock price is not income per unit time — the income can only be realized when the stock is sold and it amounts to the current market price. The total net income (per share) from the given investment would be the stock price on the date of sale minus the purchase price of $40.

7.3

1. Average $= \frac{1}{2} \int_0^2 x^3 \, dx = \frac{1}{2} \left[\frac{1}{4} x^4 \right]_0^2 =$

$\frac{1}{2}(4 - 0) = 2$

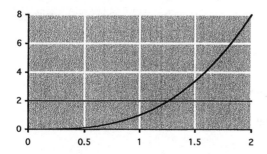

3. Average $= \frac{1}{2} \int_0^2 (x^3 - x) \, dx = \frac{1}{2} \left[\frac{1}{4} x^4 - \frac{1}{2} x^2 \right]_0^2$

$= \frac{1}{2}(4 - 2 - 0) = 1$

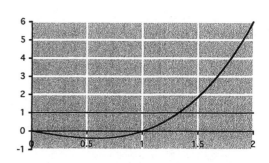

5. Average $= \frac{1}{2} \int_0^2 e^{-x} \, dx = \frac{1}{2} [-e^{-x}]_0^2 =$

$\frac{1}{2}(1 - e^{-2}) \approx 0.43$

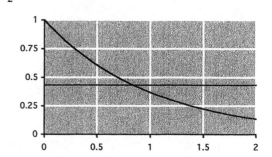

7. $\bar{r}(2) = (3 + 5 + 10)/3 = 6$, and so on.

x	0	1	2	3	4	5	6	7
r(x)	3	5	10	3	2	5	6	7
$\bar{r}(x)$			6	6	5	10/3	13/3	6

9. We must have $(1 + 2 + r(2))/3 = \bar{r}(2) = 3$, so $r(2) = 6$. Working from left to right we fill in the other missing values similarly.

x	0	1	2	3	4	5	6	7
r(x)	1	2	6	7	11	15	10	2
$\bar{r}(x)$			3	5	8	11	12	9

Section 7.3

11. Moving average: $\bar{f}(x) = \frac{1}{5} \int_{x-5}^{x} t^3 \, dt =$

$\frac{1}{5} \left[\frac{1}{4} t^4 \right]_{x-5}^{x} = \frac{1}{20} [x^4 - (x-5)^4] = \frac{1}{20} (20x^3 - 150x^2 + 500x - 625) = x^3 - \frac{15}{2}x^2 + 25x - \frac{125}{4}$

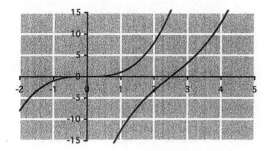

13. Moving average: $\bar{f}(x) = \frac{1}{5} \int_{x-5}^{x} t^{2/3} \, dt =$

$\frac{1}{5} \left[\frac{3}{5} t^{5/3} \right]_{x-5}^{x} = \frac{3}{25} [x^{5/3} - (x-5)^{5/3}]$

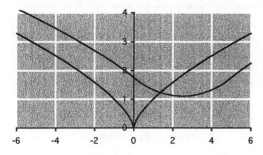

15. Moving average: $\bar{f}(x) = \frac{1}{5} \int_{x-5}^{x} e^{0.5t} \, dt =$

$\frac{1}{5} [2e^{0.5t}]_{x-5}^{x} = \frac{2}{5} [e^{0.5x} - e^{0.5(x-5)}] = \frac{2}{5}(1 - e^{-2.5})e^{0.5x} \approx 0.367 e^{0.5x}$

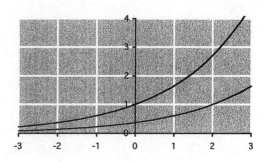

17. Moving average: $\bar{f}(x) = \frac{1}{5} \int_{x-5}^{x} t^{1/2} \, dt =$

$\frac{1}{5} \left[\frac{2}{3} t^{3/2} \right]_{x-5}^{x} = \frac{2}{15} [x^{3/2} - (x-5)^{3/2}]$ (Note that the domain is $x \geq 5$.)

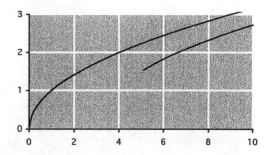

19.

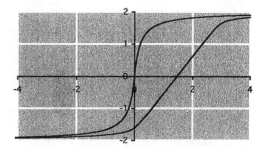

Section 7.3

21.

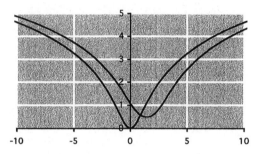

23. $(120 + 123 + 125 + 130 + 132 + 132)/6 = 127$ million people

25. $\frac{1}{3} \int_8^{11} (0.355t - 1.6) \, dt = \frac{1}{3} [0.1755t^2 - 1.6t]_8^{11}$

$= \$1.7345$ million

27. The amount you have in the account at time t is $A(t) = 10{,}000e^{0.08t}$, $0 \leq t \leq 1$. The average amount over the first year is $\int_0^1 10{,}000e^{0.08t} \, dt = [125{,}000e^{0.08t}]_0^1 = \$10{,}410.88$.

29. The amount in the account begins at \$3000 at the beginning of the month and then declines linearly to 0 by the end of the month. So, the amount in the account during the month is $A(t) = 3000 - 3000t$, $0 \leq t \leq 1$. The average over one month is therefore $\int_0^1 (3000 - 3000t) \, dt = [3000t - 1500t^2]_0^1 = \$1500$. Since the average over each month is \$1500, the average over several months is also \$1500.

31.

Year t	5	6	7	8	9	10	11
Employment (millions)	117	120	123	125	130	132	132
Moving average (millions)				121.25	124.5	127.5	129.75

Some changes are larger (for example, $t = 9$ to $t = 10$) and others are smaller (for example, $t = 8$ to $t = 9$).

33. (a)

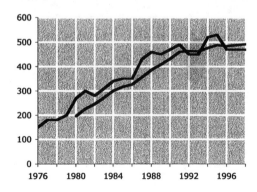

(b) The moving average is 226 in 1981 and 460 in 1991. This gives a rate of change of $(460 - 226)/10 = \$23.4$ million per year. Tourist spending in Bermuda was increasing at a rate of approximately \$23.4 million per year during the given period.

35. (a) $\frac{1}{22} \int_5^{27} (3t^2 - 90t + 4200) \, dt =$

$\frac{1}{22} [t^3 - 45t^2 + 4200t]_5^{27} \approx 3650$ pounds

(b) $\frac{1}{2}\int_{t-2}^{t}(3x^2 - 90x + 4200)\,dx =$

$\frac{1}{2}[t^3 - 45t^2 + 4200t]_{t-2}^{t} = \frac{1}{2}\{t^3 - (t-2)^3 - 45[t^2 - (t-2)^2] + 8400\}$ **(c)** The function is quadratic; the t^3 terms cancel.

37. (a) The line through $(t, s) = (0, 240)$ and $(25, 600)$ is $s = 14.4t + 240$. **(b)** $\bar{s}(t) =$
$\frac{1}{4}\int_{t-4}^{t}(14.4x + 240)\,dx = \frac{1}{4}[7.2x^2 + 240x]_{t-4}^{t} =$
$\frac{1}{4}\{7.2[t^2 - (t-4)^2] + 960\} = \frac{1}{4}(57.6t + 844.4)$
$= 14.4t + 211.2$ **(c)** The slope of the moving average is the same as the slope of the original function (because the original is linear).

39. $\bar{f}(x) = \frac{1}{a}\int_{x-a}^{x}(mt + b)\,dt = \frac{1}{a}\left[\frac{m}{2}t^2 + bt\right]_{x-a}^{x} =$
$\frac{1}{a}\left[\frac{m}{2}[x^2 - (x-a)^2] + bx - b(x-a)\right] =$
$\frac{1}{a}\left[max - \frac{ma^2}{2} + ab\right] = mx + b - \frac{ma}{2}$

41. (a)

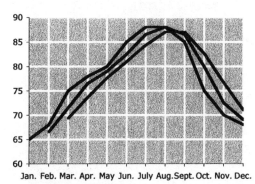

(b) The 24-month moving average is constant and equal to the year-long average of approximately

77°. **(c)** A quadratic model could not be used to predict temperatures beyond the given 12-month period, since temperature patterns are periodic, whereas parabolas are not.

43. The area above the x-axis equals the area below the x-axis. Example: $y = x$ on $[-1,1]$.

45. The moving average "blurs" the effects of short-term oscillations in the price, and shows the longer-term trend of the stock price.

47. This need not be the case; for instance, the function $f(x) = x^2$ on $[0, 1]$ has average value $1/3$, whereas the value midway between the maximum and minimum is $1/2$.

49. (c): A shorter term moving average most closely approximates the original function, since it averages the function over a shorter period, and continuous functions change by only a small amount over a small period.

Section 7.4

7.4

1. $TV = \int_0^{10} 30{,}000 \, dt = [30{,}000t]_0^{10} = \$300{,}000$; $FV = \int_0^{10} 30{,}000 e^{0.07(10-t)} \, dt = \left[-\frac{30{,}000}{0.07} e^{0.07(10-t)}\right]_0^{10} =$ $\$434{,}465.45$

3. $TV = \int_0^{10} (30{,}000 + 1000t) \, dt = [30{,}000t + 500t^2]_0^{10} = \$350{,}000$; $FV = \int_0^{10} (30{,}000 + 1000t) e^{0.07(10-t)} \, dt$
$= \left[-\frac{1}{0.07}(30{,}000 + 1000t) e^{0.07(10-t)} - \frac{1000}{0.07^2} e^{0.07(10-t)}\right]_0^{10}$ (using integration by parts) $= \$498{,}496.61$

5. $TV = \int_0^{10} 30{,}000 e^{0.05t} \, dt = [600{,}000 e^{0.05t}]_0^{10} = \$389{,}232.76$; $FV = \int_0^{10} 30{,}000 e^{0.05t} e^{0.07(10-t)} \, dt =$
$\int_0^{10} 30{,}000 e^{0.7} e^{-0.02t} \, dt = [-1{,}500{,}000 e^{0.7} e^{-0.02t}]_0^{10} = \$547{,}547.16$

7. $TV = \int_0^5 20{,}000 \, dt = [20{,}000t]_0^5 = \$100{,}000$; $PV = \int_0^5 20{,}000 e^{-0.08t} \, dt = [-250{,}000 e^{-0.08t}]_0^5 = \$82{,}419.99$

9. $TV = \int_0^5 (20{,}000 + 1000t) \, dt = [20{,}000t + 500t^2]_0^5 = \$112{,}500$; $PV = \int_0^5 (20{,}000 + 1000t) e^{-0.08t} \, dt =$
$[-(250{,}000 + 12{,}500t) e^{-0.08t} - 156{,}250 e^{-0.08t}]_0^5$ (by integration by parts) $= \$92{,}037.48$

11. $TV = \int_0^5 20{,}000 e^{0.03t} \, dt = \left[\frac{20{,}000}{0.03} e^{0.03t}\right]_0^5 = \$107{,}889.50$; $PV = \int_0^5 20{,}000 e^{0.03t} e^{-0.08t} \, dt =$
$\int_0^5 20{,}000 e^{-0.05t} \, dt = [-400{,}000 e^{-0.05t}]_0^5 = \$88{,}479.69$

13. Total revenue $= \int_5^{11} (0.31t^2 - 1.1t + 4.4) \, dt = [0.1033t^3 - 0.55t^2 + 4.4t]_5^{11} = \$98$ billion

Section 7.4

15. Total revenue $= \int_0^7 83e^{0.15t}\, dt = [553.3e^{0.15t}]_0^7 = \$1028$ billion

17. $FV = \int_5^{11} (0.31t^2 - 1.1t + 4.4)e^{0.02(11-t)}\, dt =$

$\left[-\dfrac{(0.31t^2 - 1.1t + 4.4)e^{0.02(11-t)}}{0.02} - \dfrac{(0.62t - 1.1)e^{0.02(11-t)}}{0.02^2} - \dfrac{0.62e^{0.02(11-t)}}{0.02^3} \right]_5^{11} = \$103$ billion

19. $FV = \int_0^7 83e^{0.15t}e^{-0.05(7-t)}\, dt = \int_0^7 83e^{-0.35}e^{0.2t}\, dt = [415e^{-0.35}e^{0.2t}]_0^7 = \$893$ billion

21. $R(t) = 12 \times 700 = \$8400$/year. $FV = \int_0^{45} 8400 e^{0.06(45-t)}\, dt = [-140{,}000 e^{0.06(45-t)}]_0^{45} = \$1{,}943{,}162.44$

23. $R(t) = 12 \times 700 e^{0.03t} = 8400 e^{0.03t}$. $FV = \int_0^{45} 8400 e^{0.03t} e^{0.06(45-t)}\, dt = \int_0^{45} 8400 e^{2.7} e^{-0.03t}\, dt =$

$[-280{,}000 e^{2.7} e^{-0.03t}]_0^{45} = = \$3{,}086{,}245.73$

25. $R(t) = 3375$. $PV = \int_0^{30} 3375 e^{-0.04t}\, dt = [-84{,}375 e^{-0.04t}]_0^{30} = \$58{,}961.74$

27. $R(t) = 100{,}000 + 5000t$. $PV = \int_0^{20} (100{,}000 + 5000t)e^{-0.05t}\, dt = [-20(100{,}000 + 5000t)e^{-0.05t} -$

$400(5000)e^{-0.05t}]_0^{20} = \$1{,}792{,}723.35$

29. Total

31. She is correct, provided there is a positive rate of return, in which case the future value (which includes interest) is greater than the total value (which does not).

33. $PV < TV < FV$

Section 7.5

7.5

1. $\int_{1}^{+\infty} x\, dx = \lim_{M\to+\infty} \int_{1}^{M} x\, dx = \lim_{M\to+\infty} \left[\frac{1}{2}x^2\right]_{1}^{M} = \lim_{M\to+\infty}\left(\frac{1}{2}M^2 - \frac{1}{2}\right) = +\infty$; diverges

3. $\int_{-2}^{+\infty} e^{-0.5x}\, dx = \lim_{M\to+\infty}\int_{-2}^{M} e^{-0.5x}\, dx = \lim_{M\to+\infty}[-2e^{-0.5x}]_{-2}^{M} = \lim_{M\to+\infty}(-2e^{-0.5M} + 2e) = 2e$; converges

5. $\int_{-\infty}^{2} e^x\, dx = \lim_{M\to-\infty}\int_{M}^{2} e^x\, dx = \lim_{M\to-\infty}[e^x]_{M}^{2} = \lim_{M\to-\infty}(e^2 - e^M) = e^2$; converges

7. $\int_{-\infty}^{-2}\frac{1}{x^2}\, dx = \lim_{M\to-\infty}\int_{M}^{-2}\frac{1}{x^2}\, dx = \lim_{M\to-\infty}[-x^{-1}]_{M}^{-2} = \lim_{M\to-\infty}\left(\frac{1}{2} + \frac{1}{M}\right) = \frac{1}{2}$; converges

9. $\int_{0}^{+\infty} x^2 e^{-6x}\, dx = \lim_{M\to+\infty}\int_{0}^{M} x^2 e^{-6x}\, dx = \lim_{M\to+\infty}\left[-\frac{1}{6}x^2 e^{-6x} - \frac{2}{36}x e^{-6x} - \frac{2}{216}e^{-6x}\right]_{0}^{M} =$

$\lim_{M\to+\infty}\left(-\frac{1}{6}M^2 e^{-6M} - \frac{2}{36}Me^{-6M} - \frac{2}{216}e^{-6M} + \frac{2}{216}\right) = \frac{2}{216} = \frac{1}{108}$; converges

11. $\int_{0}^{5}\frac{2}{x^{1/3}}\, dx = \lim_{r\to 0^+}\int_{r}^{5}\frac{2}{x^{1/3}}\, dx = \lim_{r\to 0^+}[3x^{2/3}]_{r}^{5} = \lim_{r\to 0^+}(3\times 5^{2/3} - 3r^{2/3}) = 3\times 5^{2/3}$; converges

13. $\int_{-1}^{2}\frac{3}{(x+1)^2}\, dx = \lim_{r\to-1^+}\int_{r}^{2}\frac{3}{(x+1)^2}\, dx = \lim_{r\to-1^+}[-3(x+1)^{-1}]_{r}^{2} = \lim_{r\to-1^+}\left(-1 + \frac{3}{r+1}\right) = +\infty$; diverges

15. $\int_{-1}^{2}\frac{3x}{x^2-1}\, dx = \lim_{r\to-1^+}\int_{-1}^{r}\frac{3x}{x^2-1}\, dx + \lim_{r\to 1^-}\int_{r}^{0}\frac{3x}{x^2-1}\, dx + \lim_{r\to 1^+}\int_{r}^{2}\frac{3x}{x^2-1}\, dx$. Now, $\int\frac{3x}{x^2-1}\, dx =$

Section 7.5

$\frac{3}{2}\ln|x^2 - 1| + C$ by substitution, so $\lim_{r \to -1^+} \int_r^0 \frac{3x}{x^2 - 1} dx = \lim_{r \to -1^+} \left[\frac{3}{2}\ln|x^2 - 1|\right]_r^0 =$

$\lim_{r \to -1^+} \left(-\frac{3}{2}\ln|r^2 - 1|\right) = +\infty$. Since this one part diverges, the whole integral diverges. (In fact, all three parts diverge.)

17. $\int_{-2}^{2} \frac{1}{(x+1)^{1/5}} dx = \lim_{r \to -1^-} \int_{-2}^{r} \frac{1}{(x+1)^{1/5}} dx + \lim_{r \to -1^+} \int_{r}^{2} \frac{1}{(x+1)^{1/5}} dx = \lim_{r \to -1^-} \left[\frac{5}{4}(x+1)^{4/5}\right]_{-2}^{r} +$

$\lim_{r \to -1^+} \left[\frac{5}{4}(x+1)^{4/5}\right]_{r}^{2} = \lim_{r \to -1^-} \left[\frac{5}{4}(r+1)^{4/5} - \frac{5}{4}\right] + \lim_{r \to -1^+} \left[\frac{5}{4}3^{4/5} - \frac{5}{4}(r+1)^{4/5}\right] = \frac{5}{4}(3^{4/5} - 1)$;

converges

19. $\int_{-1}^{1} \frac{2x}{x^2 - 1} dx = \lim_{r \to -1^+} \int_{-1}^{0} \frac{2x}{x^2 - 1} dx + \lim_{r \to 1^-} \int_{0}^{r} \frac{2x}{x^2 - 1} dx = \lim_{r \to -1^+} [\ln|x^2 - 1|]_r^0 + \lim_{r \to 1^-} [\ln|x^2 - 1|]_0^r$

(use the substitution $u = x^2 - 1$) $= \lim_{r \to -1^+} (-\ln|r^2 - 1|) + \lim_{r \to 1^-} \ln|r^2 - 1| = +\infty - \infty$; diverges (Note that the infinities don't cancel. For convergence we need each part of the integral to converge on its own.)

21. $\int_{-\infty}^{+\infty} xe^{-x^2} dx = \lim_{M \to -\infty} \int_{M}^{0} xe^{-x^2} dx + \lim_{M \to +\infty} \int_{0}^{M} xe^{-x^2} dx = \lim_{M \to -\infty}\left[-\frac{1}{2}e^{-x^2}\right]_M^0 + \lim_{M \to +\infty}\left[-\frac{1}{2}e^{-x^2}\right]_0^M$

(use the substitution $u = -x^2$) $= \lim_{M \to -\infty}\left(-\frac{1}{2} + \frac{1}{2}e^{-M^2}\right) + \lim_{M \to +\infty}\left(-\frac{1}{2}e^{-M^2} + \frac{1}{2}\right) = -\frac{1}{2} + \frac{1}{2} = 0$;

converges

23. $\int_0^{+\infty} \frac{1}{x \ln x} dx = \lim_{r \to 0^+} \int_r^{1/2} \frac{1}{x \ln x} dx + \lim_{r \to 1^-} \int_{1/2}^{r} \frac{1}{x \ln x} dx + \lim_{r \to 1^+} \int_r^{2} \frac{1}{x \ln x} dx + \lim_{M \to +\infty} \int_2^{M} \frac{1}{x \ln x} dx$.

$\lim_{r \to 0^+} \int_r^{1/2} \frac{1}{x \ln x} dx = \lim_{r \to 0^+} [\ln |\ln x|]_r^{1/2}$ (use the substitution $u = \ln x$) $= \lim_{r \to 0^+} [\ln |\ln(1/2)| - \ln |\ln r|] =$

$-\infty$. Without checking the remaining parts of the integral we can say that the whole integral diverges.

Section 7.5

25. $\int_0^{+\infty} \frac{2x}{x^2-1} dx = \lim_{r \to 1^-} \int_0^r \frac{2x}{x^2-1} dx + \lim_{r \to 1^+} \int_r^2 \frac{2x}{x^2-1} dx + \lim_{M \to +\infty} \int_2^M \frac{2x}{x^2-1} dx = \lim_{r \to 1^-} [\ln|x^2-1|]_0^r +$

$\lim_{r \to 1^+} [\ln|x^2-1|]_r^2 + \lim_{M \to +\infty} [\ln|x^2-1|]_2^M = \lim_{r \to 1^-} \ln|r^2-1| + \lim_{r \to 1^+} (\ln 3 - \ln|r^2-1|) +$

$\lim_{M \to +\infty} (\ln|M^2-1| - \ln 3) = -\infty + \infty + \infty;$ diverges

27. Total revenue = $\int_0^{+\infty} 91.7(0.90)^t dt = \lim_{M \to +\infty} \int_0^M 91.7(0.90)^t dt = \lim_{M \to +\infty} \left[\frac{91.7}{\ln 0.90}(0.90)^t\right]_0^M =$

$\lim_{M \to +\infty} \left[\frac{91.7}{\ln 0.90}(0.90)^M - \frac{91.7}{\ln 0.90}\right] = -\frac{91.7}{\ln 0.90} \approx \$870$ million

29. Annual sales = $S(t) = 485.5(0.897)^t$ billion cigarettes per year. Total sales = $\int_0^{+\infty} 485.5(0.897)^t dt =$

$\lim_{M \to +\infty} \int_0^M 485.5(0.897)^t dt = \lim_{M \to +\infty} \left[\frac{485.5}{\ln 0.897}(0.927)^t\right]_0^M = \lim_{M \to +\infty} \left[\frac{485.5}{\ln 0.897}(0.897)^M - \frac{485.5}{\ln 0.897}\right] =$

$-\frac{485.5}{\ln 0.897} \approx 4466$ billion cigarettes

31. Annual sales = $S(t) = 200(0.90)^t$. Total sales = $\int_0^{+\infty} 200(0.90)^t dt = \lim_{M \to +\infty} \int_0^M 200(0.90)^t dt =$

$\lim_{M \to +\infty} \left[\frac{200}{\ln 0.90}(0.90)^t\right]_0^M = \lim_{M \to +\infty} \left[\frac{200}{\ln 0.90}(0.90)^M - \frac{200}{\ln 0.90}\right] = -\frac{200}{\ln 0.90} \approx 1900.$ No, you will not sell more than about 2000 of them.

33. $\lim_{t \to +\infty} N(t) = 2.5 + \int_1^{+\infty} 0.214 t^{-0.91} dt = 2.5 + \lim_{M \to +\infty} \int_1^M 0.214 t^{-0.91} dt = 2.5 + \lim_{M \to +\infty} [2.378 t^{0.09}]_1^M =$

$2.5 + \lim_{M \to +\infty} [2.378 M^{0.09} - 2.378] = +\infty.$ The integral diverges, so the number of graduates each year

will rise without bound.

35. (a) The revenue per cell phone user is $P(t) = 350e^{-0.1t}$; multiplying by the number of users gives the annual revenue as $R(t) = 350e^{-0.1t}(39t + 68)$ million dollars per year. **(b)** Total revenue =

Section 7.5

$$\int_0^{+\infty} 350e^{-0.1t}(39t + 68)\, dt = \lim_{M \to +\infty} \int_0^M 350e^{-0.1t}(39t + 68)\, dt = \lim_{M \to +\infty} [-3500e^{-0.1t}(39t + 68) -$$

$1,365,000e^{-0.1t}]_0^M = \lim_{M \to +\infty} [-3500e^{-0.1M}(39M + 68) - 1,365,000e^{-0.1M} + 3500(68) + 1,365,000] =$

$3500(68) + 1,365,000 = \$1,603,000$ million.

37. The annual investment, in billions of constant dollars per year, is $Q(t) = (10 + 1.56t^2)e^{-0.05t}$. The total

investment is $\int_0^{+\infty} (10 + 1.56t^2)e^{-0.05t}\, dt = \lim_{M \to +\infty} \int_0^M (10 + 1.56t^2)e^{-0.05t}\, dt =$

$\lim_{M \to +\infty} [-20(10 + 1.56t^2)e^{-0.05t} - 1248te^{-0.05t} - 24,960e^{-0.05t}]_0^M = \lim_{M \to +\infty} [-20(10 + 1.56M^2)e^{-0.05M} -$

$1248Me^{-0.05M} - 24,960e^{-0.05M} + 200 + 24,960] = \$25,160$ billion.

39. $\int_0^{+\infty} N(t)\, dt = \int_0^{+\infty} \dfrac{82.8(7.14)^t}{21.8 + (7.14)^t}\, dt = \lim_{M \to +\infty} \int_0^M \dfrac{82.8(7.14)^t}{21.8 + (7.14)^t}\, dt = \lim_{M \to +\infty} \left[\dfrac{82.8}{\ln 7.14} \ln(21.8 + 7.14^t)\right]_0^M$

(use the substitution $u = 21.8 + 7.14^t$) $= \lim_{M \to +\infty} \left[\dfrac{82.8}{\ln 7.14} \ln(21.8 + 7.14^M) - \dfrac{82.8}{\ln 7.14} \ln(22.8)\right] = +\infty.$

$\int_0^{+\infty} N(t)\, dt$ diverges, indicating that there is no bound to the expected total future online sales of books.

$\int_{-\infty}^0 N(t)\, dt = \int_{-\infty}^0 \dfrac{82.8(7.14)^t}{21.8 + (7.14)^t}\, dt = \lim_{M \to -\infty} \int_M^0 \dfrac{82.8(7.14)^t}{21.8 + (7.14)^t}\, dt = \lim_{M \to -\infty} \left[\dfrac{82.8}{\ln 7.14} \ln(21.8 + 7.14^t)\right]_M^0 =$

$\lim_{M \to -\infty} \left[\dfrac{82.8}{\ln 7.14} \ln(22.8) - \dfrac{82.8}{\ln 7.14} \ln(21.8 + 7.14^M)\right] = \dfrac{82.8}{\ln 7.14} \ln(22.8) - \dfrac{82.8}{\ln 7.14} \ln(21.8)) \approx 1.889.$

$\int_{-\infty}^0 N(t)\, dt$ converges to approximately 1.889, indicating that total online sales of books prior to 1997

amounted to approximately 1.889 million books.

41. 1   **43.** 0.1587

45. The value per bottle is $P(t) = 85e^{0.4t}$. The annual sales rate is $Q(t) = 500e^{-t}$. The annual income is

$R(t) = P(t)Q(t) = 42,500e^{-0.6t}$. The total income is $\int_0^{+\infty} 42,500e^{-0.6t}\, dt = \lim_{M \to +\infty} \int_0^M 42,500e^{-0.6t}\, dt =$

169

Section 7.5

$$\lim_{M \to +\infty} [-70{,}833 e^{-0.6t}]_0^M = \lim_{M \to +\infty} [-70{,}833 e^{-0.6M} + 70{,}833] = \$70{,}833.$$

47. (a) $\displaystyle\int_{0.2}^{+\infty} \frac{1}{5.6997 k^{1.081}} \, dk = \lim_{M \to +\infty} \int_{0.2}^{M} \frac{1}{5.6997 k^{1.081}} \, dk = \lim_{M \to +\infty} [-2.166 k^{-0.081}]_{0.2}^{M} =$

$\displaystyle\lim_{M \to +\infty} [-2.166 M^{-0.081} + 2.166(0.2)^{-0.081}] \approx 2.468$ meteors on average. **(b)** $\displaystyle\int_{0}^{1} \frac{1}{5.6997 k^{1.081}} \, dk =$

$\displaystyle\lim_{r \to 0} \int_{r}^{1} \frac{1}{5.6997 k^{1.081}} \, dk = \lim_{r \to 0} [-2.166 k^{-0.081}]_{r}^{1} = \lim_{r \to 0} [-2.166 + 2.166 r^{-0.081}] = +\infty$; the integral diverges.

We can interpret this as saying that the number of impacts by meteors smaller than 1 megaton is very large. (This makes sense because, for example, this number includes meteors no larger than a grain of dust.)

49. (a) $\Gamma(1) = \displaystyle\int_{0}^{+\infty} e^{-t} \, dt = \lim_{M \to +\infty} \int_{0}^{M} e^{-t} \, dt = \lim_{M \to +\infty} [-e^{-t}]_{0}^{M} = \lim_{M \to +\infty} [-e^{-M} + 1] = 1.$ $\Gamma(2) =$

$\displaystyle\int_{0}^{+\infty} t e^{-t} \, dt = \lim_{M \to +\infty} \int_{0}^{M} t e^{-t} \, dt = \lim_{M \to +\infty} [-t e^{-t} - e^{-t}]_{0}^{M} = \lim_{M \to +\infty} [-M e^{-M} - e^{-M} + 1] = 1.$ **(b)** $\Gamma(n+1) =$

$\displaystyle\int_{0}^{+\infty} t^n e^{-t} \, dt = \lim_{M \to +\infty} \int_{0}^{M} t^n e^{-t} \, dt = \lim_{M \to +\infty} \left([-t^n e^{-t}]_{0}^{M} + \int_{0}^{M} n t^{n-1} e^{-t} \, dt \right) = \lim_{M \to +\infty} [-M^n e^{-M} + 0] +$

$n \displaystyle\int_{0}^{+\infty} t^{n-1} e^{-t} \, dt = n \Gamma(n)$. **(c)** If n is a positive integer, then applying part (b) several times we get $\Gamma(n) =$

$(n-1)\Gamma(n-1) = (n-1)(n-2)\Gamma(n-2) = \ldots = (n-1)(n-2)\cdots 1 \Gamma(1) = (n-1)!$ by part (a).

51. The integral does not converge, so the number given by the FTC is meaningless.

53. Yes; the integrals converge to 0, and the FTC also gives 0.

55. In all cases, you need to rewrite the improper integral as a limit and use technology to evaluate the integral of which you are taking the limit.

Evaluate for several values of the endpoint approaching the limit. In the case of an integral in which one of the limits of integration is infinite, you may have to instruct the calculator or computer to use more subdivisions as you approach $+\infty$.

57. Answers will vary.

7.6

1. $y = \int (x^2 + \sqrt{x})\, dx = \dfrac{x^3}{3} + \dfrac{2x^{3/2}}{3} + C$

3. $y\, dy = x\, dx$; $\int y\, dy = \int x\, dx$; $\dfrac{y^2}{2} = \dfrac{x^2}{2} + C$

5. $\dfrac{1}{y}\, dy = x\, dx$; $\int \dfrac{1}{y}\, dy = \int x\, dx$; $\ln|y| = \dfrac{x^2}{2} + C$;

$|y| = e^{x^2/2 + C} = e^C e^{x^2/2}$; $y = Ae^{x^2/2}$ (where $A = \pm e^C$)

7. $\dfrac{1}{y^2}\, dy = (x+1)\, dx$; $\int \dfrac{1}{y^2}\, dy = \int (x+1)\, dx$;

$-\dfrac{1}{y} = \dfrac{1}{2}(x+1)^2 + K = \dfrac{(x+1)^2 + C}{2}$;

$y = -\dfrac{2}{(x+1)^2 + C}$

9. $y\, dy = \dfrac{\ln x}{x}\, dx$; $\int y\, dy = \int \dfrac{\ln x}{x}\, dx$; $\dfrac{y^2}{2} = $

$(\ln x)^2 + C$ (substitute $u = \ln x$);
$y = \pm\sqrt{(\ln x)^2 + C}$

11. $y = \int (x^3 - 2x)\, dx = \dfrac{x^4}{4} - x^2 + C$; $1 = 0 -$

$0 + C$; $C = 1$; $y = \dfrac{x^4}{4} - x^2 + 1$

13. $y^2\, dy = x^2\, dx$; $\int y^2\, dy = \int x^2\, dx$; $\dfrac{y^3}{3} = \dfrac{x^3}{3} +$

C; $y^3 = x^3 + K$; $8 = 0 + K$; $K = 8$; $y^3 = x^3 + 8$;
$y = (x^3 + 8)^{1/3}$

15. $\dfrac{1}{y}\, dy = \dfrac{1}{x}\, dx$; $\int \dfrac{1}{y}\, dy = \int \dfrac{1}{x}\, dx$; $\ln|y| =$

$\ln|x| + C$; $|y| = e^{\ln|x|+C} = e^C e^{\ln|x|} = e^C |x|$; $y = Ax$; $2 = A\times 1$; $A = 2$; $y = 2x$

17. $\dfrac{1}{y+1}\, dy = x\, dx$; $\int \dfrac{1}{y+1}\, dy = \int x\, dx$;

$\ln|y+1| = \dfrac{x^2}{2} + C$; $\ln 1 = 0 + C$; $C = 0$;

$\ln|y+1| = \dfrac{x^2}{2}$; $|y+1| = e^{x^2/2}$; $y + 1 = e^{x^2/2}$
(note that $y(0) + 1 = 1 > 0$); $y = e^{x^2/2} - 1$

19. $\dfrac{1}{y^2}\, dy = \dfrac{x}{x^2+1}\, dx$; $\int \dfrac{1}{y^2}\, dy = \int \dfrac{x}{x^2+1}\, dx$;

$-\dfrac{1}{y} = \dfrac{1}{2}\ln(x^2+1) + C$; $1 = 0 + C$; $C = 1$;

$-\dfrac{1}{y} = \dfrac{1}{2}\ln(x^2+1) + 1$; $y = -\dfrac{2}{\ln(x^2+1) + 2}$

21. With $s(t) =$ monthly sales after t months, $\dfrac{ds}{dt} = -0.05s$; $s = 1000$ when $t = 0$. $\dfrac{1}{s}\, ds = $

$-0.05\, dt$; $\int \dfrac{1}{s}\, ds = \int (-0.05)\, dt$; $\ln|s| =$

$-0.05t + C$; $|s| = e^{-0.05t+C} = e^C e^{-0.05t}$; $s = Ae^{-0.05t}$;
$1000 = A\times 1$; $A = 1000$; $s = 1000e^{-0.05t}$ quarts per month.

23. $\dfrac{dH}{dt} = -k(H-75)$; $\dfrac{1}{H-75}\, dH = -k\, dt$;

$\int \dfrac{1}{H-75}\, dH = \int (-k)\, dt$; $\ln(H-75) = -kt +$

C; $H(t) = 75 + Ae^{-kt}$. $H(0) = 190$, so $190 = 75 + A$; $A = 115$. $H(10) = 150$, so $150 = 75 + 115e^{-10k}$; $k = -\dfrac{1}{10}\ln\left(\dfrac{150-75}{115}\right) \approx 0.04274$.

$H(t) = 75 + 115e^{-0.04274t}$ degrees Fahrenheit after t minutes.

25. With $S(t)$ = total sales after t months, $\frac{dS}{dt} =$ 0.1(100,000 − S); $S(0) = 0$. $\frac{1}{100,000 - S} dS =$ 0.1 dt; $\int \frac{1}{100,000 - S} dS = \int 0.1\, dt$;

$-\ln(100,000 - S) = 0.1t + C$; $S(t) = 100,000 - Ae^{-0.1t}$. $0 = 100,000 - A$; $A = 100,000$. $S(t) = 100,000 - 100,000e^{-0.1t} = 100,000(1 - e^{-0.1t})$ monitors after t months.

27. (a) $\frac{dp}{dt} = k[D(p) - S(p)] = k(20,000 - 1000p)$ **(b)** $\frac{1}{20,000 - 1000p} dp = k\, dt$;

$\int \frac{1}{20,000 - 1000p} dp = \int k\, dt$;

$-\ln(20,000 - 1000p) = kt + C$; $p(t) = 20 - Ae^{-kt}$. **(c)** $p(0) = 10$ and $p(1) = 12$, so $10 = 20 - A$; $A = 10$; $12 = 20 - 10e^{-k}$; $k = -\ln\left(\frac{20-12}{10}\right) \approx 0.2231$; $p(t) = 20 - 10e^{-0.2231t}$ dollars after t months.

29. $-\frac{p}{q}\frac{dq}{dp} = 0.05p - 1.5$; $\frac{1}{q} dq = \left(-0.05 + \frac{1.5}{p}\right) dp$; $\int \frac{1}{q} dq = \int \left(-0.05 + \frac{1.5}{p}\right) dp$;

$\ln q = -0.05p + 1.5 \ln p + C$; $q = Ae^{-0.05p} p^{1.5}$. $q(20) = 20$, so $20 = Ae^{-1}(20)^{1.5}$; $A = e(20)^{-0.5} \approx 0.6078$. $q = 0.6078 e^{-0.05p} p^{1.5}$.

31. If $y = \frac{CL}{e^{-aLt} + C}$, then $\frac{dy}{dt} = \frac{-aCL^2 e^{-aLt}}{(e^{-aLt} + C)^2}$ and

$ay(L - y) = a\left(\frac{CL}{e^{-aLt} + C}\right)\left(\frac{-Le^{-aLt}}{e^{-aLt} + C}\right) = \frac{-aCL^2 e^{-aLt}}{(e^{-aLt} + C)^2}$ also, so this y satisfies the differential equation.

33. $a = 1/4$ and $L = 2$, so $S = \frac{2C}{e^{-0.5t} + C}$ for some C. $S = 0.001$ when $t = 0$, so $0.001 = \frac{2C}{1 + C}$; $C = \frac{0.001}{2 - 0.001} = \frac{1}{1999}$. $S = \frac{2/1999}{e^{-0.5t} + 1/1999}$. Graph:

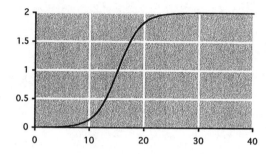

It will take about 27 months to saturate the market.

35. (a) $\frac{1}{y \ln(y/b)} dy = -a\, dt$; $\int \frac{1}{y \ln(y/b)} dy = \int (-a)\, dt$; $\ln[\ln(y/b)] = -at + C$ [use the substitution $u = \ln(y/b)$]; $y = be^{Ae^{-at}}$, for some constant A. **(b)** $5 = 10e^A$, $A = \ln 0.5 \approx -0.69315$; $y = 10e^{-0.69315e^{-t}}$. Graph:

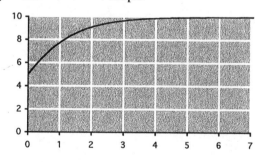

37. A general solution gives all possible solutions to the equation, using at least one arbitrary constant. A particular solution is one specific function that satisfies the equation. We obtain a

Section 7.6

particular solution by substituting specific values for any arbitrary constants in the general solution.

39. Example: $\frac{d^2y}{dx^2} = 1$ has general solution $y = \frac{1}{2}x^2 + Cx + D$ (integrate twice).

41. Differentiate to get the differential equation $\frac{dy}{dx} = -4e^{-x} + 3$.

Chapter 7 Review Test

1. (a)

D	I
+ $x^2 + 2$	e^x
− $2x$	e^x
+ 2	e^x
−∫ 0 →	e^x

$\int (x^2 + 2)e^x \, dx = (x^2 + 2)e^x - 2xe^x + 2e^x + C =$

$(x^2 - 2x + 4)e^x + C$

(b)

D	I
+ $\ln(2x)$	x^2
−∫ $1/x$ →	$x^3/3$

$\int x^2 \ln(2x) \, dx = \frac{1}{3} x^3 \ln(2x) - \int \frac{1}{3} x^2 \, dx =$

$\frac{1}{3} x^3 \ln(2x) - \frac{1}{9} x^3 + C$

(c)

D	I
+ $x^3 + 1$	e^{-x}
− $3x^2$	$-e^{-x}$
+ $6x$	e^{-x}
− 6	$-e^{-x}$
+∫ 0 →	e^{-x}

$\int_{-2}^{2} (x^3 + 1)e^{-x} \, dx = [-(x^3 + 1)e^{-x} - 3x^2 e^{-x} -$

$6xe^{-x} - 6e^{-x}]_{-2}^{2} = [-(x^3 + 3x^2 + 6x + 7)e^{-x}]_{-2}^{2} =$

$-39e^{-2} - e^2 \approx -12.67$

(d)

D	I
+ $\ln x$	x^2
−∫ $1/x$ →	$x^3/3$

$\int_1^e x^2 \ln x \, dx = \left[\frac{1}{3} x^3 \ln x\right]_1^e - \int_1^e \frac{1}{3} x^2 \, dx =$

$\left[\frac{1}{3} x^3 \ln x - \frac{1}{9} x^3\right]_1^e = \frac{1}{3} e^3 - \frac{1}{9} e^3 + \frac{1}{9} = \frac{2e^3 + 1}{9}$

(e) $\int_1^{+\infty} \frac{1}{x^5} \, dx = \lim_{M \to +\infty} \int_1^M \frac{1}{x^5} \, dx =$

$\lim_{M \to +\infty} \left[-\frac{1}{4} x^{-4}\right]_1^M =$

$\lim_{M \to +\infty} \left[-\frac{1}{4} M^{-4} + \frac{1}{4}\right] = \frac{1}{4}$

(f) $\int_0^1 \frac{1}{\sqrt{1-x}} \, dx = \lim_{r \to 1^-} \int_0^r \frac{1}{\sqrt{1-x}} \, dx =$

$\lim_{r \to 1^-} [-2\sqrt{1-x}]_0^r = \lim_{r \to 1^-} (-2\sqrt{1-r} + 2) = 2$

2. (a) $x^3 = 1 - x^3$ when $x^3 = 1/2$, $x = 1/2^{1/3}$.

Area $= \int_0^{1/2^{1/3}} [(1 - x^3) - x^3] \, dx +$

$\int_{1/2^{1/3}}^{1} [x^3 - (1 - x^3)] \, dx = \int_0^{1/2^{1/3}} (1 - 2x^3) \, dx +$

$\int_{1/2^{1/3}}^{1} (2x^3 - 1) \, dx = \left[x - \frac{1}{2} x^4\right]_0^{1/2^{1/3}} +$

174

Chapter 7 Review Test

$\left[\frac{1}{2}x^4 - x\right]_{1/2^{1/3}}^{1} = \frac{1}{2^{1/3}} - \frac{1}{4 \cdot 2^{1/3}} + \frac{1}{2} - 1 - \frac{1}{4 \cdot 2^{1/3}}$

$+ \frac{1}{2^{1/3}} = \frac{3}{2 \cdot 2^{1/3}} - \frac{1}{2} \approx 0.6906$

(b) $e^x \geq 1 \geq e^{-x}$ for x in [0, 2], so the area =

$\int_0^2 (e^x - e^{-x}) \, dx = [e^x + e^{-x}]_0^2 = e^2 + e^{-2} - 2$

(c) $1 - x^2 = x^2$ when $x = \pm 1/\sqrt{2}$, so the area =

$\int_{-1/\sqrt{2}}^{1/\sqrt{2}} [(1 - x^2) - x^2] \, dx = \int_{-1/\sqrt{2}}^{1/\sqrt{2}} (1 - 2x^2) \, dx =$

$\left[x - \frac{2}{3}x^3\right]_{-1/\sqrt{2}}^{1/\sqrt{2}} = \frac{1}{\sqrt{2}} - \frac{2}{6\sqrt{2}} + \frac{1}{\sqrt{2}} - \frac{2}{6\sqrt{2}} = \frac{4}{3\sqrt{2}}$

$= \frac{2\sqrt{2}}{3}$

(d) $x \geq xe^{-x}$ for x in [0, 2] because $e^{-x} \leq 1$ in that range. Area = $\int_0^2 (x - xe^{-x}) \, dx =$

$\left[\frac{1}{2}x^2 + xe^{-x} + e^{-x}\right]_0^2$ (using integration by parts) =

$2 + 2e^{-2} + e^{-2} - 1 = 1 + 3e^{-2}$.

3. (a) Average = $\frac{1}{1 - 0}\int_0^1 x^2 e^x \, dx = [x^2 e^x -$

$2xe^x + 2e^x]_0^1$ (using integration by parts) = $e - 2$

(b) Average = $\frac{1}{2e - 1}\int_1^{2e} (x + 1) \ln x \, dx =$

$\frac{1}{2e - 1}\left[\left(\frac{1}{2}x^2 + x\right) \ln x\right]_1^{2e} -$

$\frac{1}{2e - 1}\int_1^{2e} \left(\frac{1}{2}x + 1\right) dx$ (using integration by parts) = $\frac{1}{2e - 1}\left[\left(\frac{1}{2}x^2 + x\right) \ln x - \frac{1}{4}x^2 - x\right]_1^{2e} =$

$\frac{1}{2e - 1}\left[(2e^2 + 2e) \ln(2e) - e^2 - 2e + \frac{1}{4} + 1\right] =$

$\frac{2(e^2 + e) \ln 2 + e^2 + 5/4}{2e - 1} \approx 5.10548$

4. (a) $\frac{1}{2}\int_{x-2}^{x} t^{4/3} \, dt = \left[\frac{3}{14} t^{7/3}\right]_{x-2}^{x} =$

$\frac{3}{14} [x^{7/3} - (x - 2)^{7/3}]$

(b) $\frac{1}{2}\int_{x-2}^{x} \ln t \, dt = \frac{1}{2}[t \ln t - t]_{x-2}^{x} =$

$\frac{1}{2}[x \ln x - x - (x - 2) \ln(x - 2) + x - 2] =$

$\frac{1}{2}[x \ln x - (x - 2) \ln(x - 2) - 2]$

5. (a) $\frac{1}{y^2} \, dy = x^2 \, dx$; $\int \frac{1}{y^2} \, dy = \int x^2 \, dx$;

$-\frac{1}{y} = \frac{x^3}{3} + K = \frac{x^3 + C}{3}$; $y = -\frac{3}{x^3 + C}$

(b) $\frac{1}{y + 2} \, dy = x \, dx$; $\int \frac{1}{y + 2} \, dy = \int x \, dx$;

$\ln |y + 2| = \frac{x^2}{2} + C$; $y + 2 = Ae^{x^2/2}$;

$y = Ae^{x^2/2} - 2$

(c) $y \, dy = \frac{1}{x} \, dx$; $\int y \, dy = \int \frac{1}{x} \, dx$; $\frac{y^2}{2} = \ln |x| +$

C; $\frac{1}{2} = 0 + C$; $\frac{y^2}{2} = \ln |x| + \frac{1}{2}$; $y = \sqrt{2 \ln |x| + 1}$

(note that $y(1) > 0$)

(d) $\frac{1}{y} \, dy = \frac{x}{x^2 + 1} \, dx$; $\int \frac{1}{y} \, dy = \int \frac{x}{x^2 + 1} \, dx$;

$\ln |y| = \frac{1}{2} \ln(x^2 + 1) + C$ (substitute $u = x^2 + 1$);

$\ln 2 = 0 + C$; $\ln y = \frac{1}{2} \ln(x^2 + 1) + \ln 2 =$

$\ln (2\sqrt{x^2 + 1})$; $y = 2\sqrt{x^2 + 1}$

Chapter 7 Review Test

6. The price follows the function $p(t) = 40 - 2t$ while the quantity sold per week follows $q(t) = 5000e^{-0.1t}$, where t is measured in weeks. The revenue per week is therefore $R(t) = p(t)q(t)$ and the total revenue over the next 8 weeks is

$$\int_0^8 5000(40 - 2t)e^{-0.1t}\, dt =$$

$5000[-(400 - 20t)e^{-0.1t} + 200e^{-0.1t}]_0^8$ (using integration by parts) $= 5000[(20t - 200)e^{-0.1t}]_0^8 = 5000[-40e^{-0.8} + 200] \approx \$910{,}000$.

7. (a) $1000\sqrt{200 - 2p} = 1000\sqrt{10p - 400}$; $200 - 2p = 10p - 400$; $\bar{p} = 50$; $\bar{q} = 1000\sqrt{200 - 2(50)} = 10{,}000$ **(b)** Solve the demand equation for $p = 100 - q^2/2{,}000{,}000$.

$$CS = \int_0^{10{,}000} \left(100 - \frac{q^2}{2{,}000{,}000} - 50\right) dq =$$

$$\int_0^{10{,}000} \left(50 - \frac{q^2}{2{,}000{,}000}\right) dq =$$

$$\left[50q - \frac{q^3}{6{,}000{,}000}\right]_0^{10{,}000} \approx \$333{,}000. \text{ Solve the}$$

supply equation for $p = 40 + q^2/10{,}000{,}000$.

$$PS = \int_0^{10{,}000} \left(50 - 40 - \frac{q^2}{10{,}000{,}000}\right) dq =$$

$$\int_0^{10{,}000} \left(10 - \frac{q^2}{10{,}000{,}000}\right) dq =$$

$$\left[10q - \frac{q^3}{30{,}000{,}000}\right]_0^{10{,}000} \approx \$66{,}700.$$

8. (a) The amount in the account at any given time is $1{,}000{,}000e^{0.06t}$ dollars after t years, so the average amount over two years is $\dfrac{1{,}000{,}000}{2 - 0}$

$$\int_0^2 e^{0.06t}\, dt = 500{,}000\left[\frac{e^{0.06t}}{0.06}\right]_0^2 = 500{,}000\,\frac{e^{0.12} - 1}{0.06}$$

$\approx \$1{,}062{,}500$ **(b)** $\dfrac{1{,}000{,}000}{1/12 - 0} \displaystyle\int_{t-1/12}^{t} e^{0.06s}\, ds =$

$200{,}000{,}000[e^{0.06s}]_{t-1/12}^{t} =$
$200{,}000{,}000(e^{0.06t} - e^{0.06(t-1/12)}) =$
$200{,}000{,}000e^{0.06t}(1 - e^{-0.06/12}) \approx 997{,}500 e^{0.06t}$

(c) The rate at which money is deposited is $100{,}000 + 10{,}000(12t)$ dollars per month after t years; converting to dollars per year gives $R(t) = 1{,}200{,}000 + 1{,}440{,}000t$ dollars per year after t years. The total deposited over two years is

$$\int_0^2 (1{,}200{,}000 + 1{,}440{,}000t)e^{0.06(2-t)}\, dt =$$

$$\left[-\frac{1{,}200{,}000 + 1{,}440{,}000t}{0.06}e^{0.06(2-t)} - \frac{1{,}440{,}000}{0.06^2}e^{0.06(2-t)}\right]_0^2 \approx \$5{,}549{,}000$$ **(d)** The principal is given by

$$\int_0^2 (1{,}200{,}000 + 1{,}440{,}000t)\, dt = [1{,}200{,}000t + 720{,}000t^2]_0^2 = \$5{,}280{,}000,$$ so the interest is the remaining $\$269{,}000$.

9. The revenue stream is $R(t) = 50e^{0.1t}$ million dollars per year. The present value if the next year's revenue is $\displaystyle\int_0^1 50e^{0.1t}e^{-0.06t}\, dt =$

$$\int_0^1 50e^{0.04t}\, dt = [1250e^{0.04t}]_0^1 \approx \$51 \text{ million}$$

176

10. The money $y(t)$ in the account satisfies the differential equation $\frac{dy}{dt} = 0.0001y^2$. We solve this equation: $\frac{1}{y^2} dy = 0.0001\ dt$; $\int \frac{1}{y^2} dy = \int 0.0001\ dt$; $-\frac{1}{y} = 0.0001t + C$; $y(0) = 10{,}000$, so $C = -1/10{,}000 = -0.0001$; $y = \frac{1}{0.0001 - 0.0001t} = \frac{10{,}000}{1 - t}$. The amount in the account would approach infinity one year after the deposit.

Chapter 7 Review Test

Chapter 8
8.1

1. (a) $f(0, 0) = 0^2 + 0^2 - 0 + 1 = 1$ (b) $f(1, 0) = 1^2 + 0^2 - 1 + 1 = 1$ (c) $f(0, -1) = 0^2 + (-1)^2 - 0 + 1 = 2$ (d) $f(a, 2) = a^2 + 2^2 - a + 1 = a^2 - a + 5$ (e) $f(y, x) = y^2 + x^2 - y + 1$ (f) $f(x + h, y + k) = (x + h)^2 + (y + k)^2 - (x + h) + 1$

3. (a) $f(0, 0) = 0 + 0 - 0 = 0$ (b) $f(1, 0) = 0.2 + 0 - 0 = 0.2$ (c) $f(0, -1) = 0 + 0.1(-1) - 0 = -0.1$ (d) $f(a, 2) = 0.2a + 0.1(2) - 0.01a(2) = 0.18a + 0.2$ (e) $f(y, x) = 0.2y + 0.1x - 0.01xy = 0.1x + 0.2y - 0.01xy$ (f) $f(x + h, y + k) = 0.2(x + h) + 0.1(y + k) - 0.01(x + h)(y + k)$

5. (a) $g(0, 0, 0) = e^{0+0+0} = 1$ (b) $g(1, 0, 0) = e^{1+0+0} = e$ (c) $g(0, 1, 0) = e^{0+1+0} = e$ (d) $g(z, x, y) = e^{z+x+y} = e^{x+y+z}$ (e) $g(x + h, y + k, z + l) = e^{x+h+y+k+z+l}$

7. (a) $g(0, 0, 0) = \dfrac{0}{0 + 0 + 0}$ does not exist (b) $g(1, 0, 0) = \dfrac{0}{1 + 0 + 0} = 0$ (c) $g(0, 1, 0) = \dfrac{0}{0 + 1 + 0} = 0$ (d) $g(z, x, y) = \dfrac{zxy}{z^2 + x^2 + y^2} = \dfrac{xyz}{x^2 + y^2 + z^2}$ (e) $g(x + h, y + k, z + l) = \dfrac{(x + h)(y + k)(z + l)}{(x + h)^2 + (y + k)^2 + (z + l)^2}$

9. (a) f increases by 2.3 units for every 1 unit of increase in x. (b) f decreases by 1.4 units for every 1 unit of increase in y. (c) f decreases by 2.5 units for every 1 unit increase in z.

11. Neither, because of the y^2 term

13. Linear

15. Linear

17. Interaction, because $g(x, y, z) = \frac{1}{4}x - \frac{3}{4}y + \frac{1}{4}yz$

19. (a) $f(20, 10) = 107$ (b) $f(40, 20) = -14$ (c) $f(10, 20) - f(20, 10) = -6 - 107 = -113$

21.

	$x \to$			
	10	20	30	40
y 10	52	107	162	217
↓ 20	94	194	294	394
30	136	281	426	571
40	178	368	558	748

25.

18
4
0.0965
47,040

27.

6.9078
1.5193
5.4366
0

29. Let z = annual sales of Z (in millions of dollars), x = annual sales of X, and y = annual sales of Y. A linear model has the form $z = ax + by + c$. We are told that $a = -2.1$ and $b = 0.4$, and that $z(6, 6) = 6$. Thus, $6 = -2.1(6) + 0.4(6) + c$, so $c = 16.2$. The model is $z = -2.1x + 0.4y + 16.2$.

31. $\sqrt{(2 - 1)^2 + (-2 + 1)^2} = \sqrt{2}$

33. $\sqrt{(0 - a)^2 + (b - 0)^2} = \sqrt{a^2 + b^2}$

Section 8.1

35. Set the two distances equal and solve:
$\sqrt{(1-0)^2 + (k-0)^2} = \sqrt{(1-2)^2 + (k-1)^2}$;
$\sqrt{1+k^2} = \sqrt{2-2k+k^2}$; $1+k^2 = 2-2k+k^2$; $2k = 1$; $k = \frac{1}{2}$

37. Rewrite the equation as $\sqrt{(x-2)^2 + (y+1)^2} = 3$. The set is the set of points at a distance of 3 from $(2, -1)$, that is, the circle with center $(2, -1)$ and radius 3.

39. The marginal cost of cars is $6000 per car, the coefficient of x. The marginal cost of trucks is $4000 per truck, the coefficient of y.

41. $C(x, y) = 10 + 0.03x + 0.04y$ where C is the cost in dollars, x is the number of video clips sold per month, and y is the number of audio clips sold per month.

43. (a) $c(18, 12, 12) = 48.4 - 0.06(18) - 0.40(12) - 1.3(12) \approx 29\%$ **(b)** $32 = c(x, 16, 8) = 48.4 + 0.06x - 0.40(16) - 1.3(8)$; $x \approx 7\%$ **(c)** 0.4-point drop (the coefficient of y)

45. (a) CBS: If CBS's rating goes up, Fox's goes down, but if either ABC's or NBC's rating goes up, Fox's also goes up. **(b)** ABC: Fox's rating goes up the quickest when ABC's rating goes up. **(c)** $F(12.2, 11.3, 11.3) = 3.1(12.2) - 0.27(11.3) + 0.87(11.3) - 36.7 \approx 7.9$ **(d)** Writing f for Fox's rating, we have $f = F(a, c, n) = 3.1a - 0.27c + 0.87n - 36.7$. Solving for a as a function of the others gives
$A(f, n, c) = \dfrac{f + 0.27c - 0.87n + 36.7}{3.1}$.

47. (a) $R(12{,}000, 5000, 5000) = \$9980$

(b) $R(z) = R(5000, 5000, z) = 10{,}000 - 0.01(5000) - 0.02(5000) - 0.01z + 0.00001(5000)z = 9850 + 0.04z$

49. We have three data points: $(c, t, s) = (8, 0, 10)$, $(48, 5, 50)$, and $(120, 10, 130)$. We want $s(c, t) = Ac + Bt + C$, so we must have $8A + C = 10$, $48A + 5B + C = 50$, and $120A + 10B + C = 130$. Solving this system of equations for A, B, and C gives $A = 1.25$, $B = -2$, and $C = 0$, so $s(c, t) = 1.25c - 2t$.

51. $U(11, 10) - U(10, 10) \approx 5.75$. This means that, if your company now has 10 copies of Macro Publish and 10 copies of Turbo Publish, then the purchase of one additional copy of Macro Publish will result in a productivity increase of approximately 5.75 pages per day.

53. (a) $(a, b, c) = (3, 1/4, 1/\pi)$ and $(a, b, c) = (1/\pi, 3, 1/4)$ both work. In fact, if we take any positive values for a and b we can take $c = \dfrac{3}{4\pi ab}$. **(b)** $V(a, a, a) = \frac{4}{3}\pi a^3 = 1$ gives $a = \left(\dfrac{3}{4\pi}\right)^{1/3}$. The resulting ellipsoid is a sphere with radius a.

55. $P(100, 500{,}000) = 1000(100^{0.5})(500{,}000^{0.5}) \approx 7{,}000{,}000$

57. (a) $100 = K(1000)^a(1{,}000{,}000)^{1-a}$, $10 = K(1000)^a(10{,}000)^{1-a}$ **(b)** Taking logs of both sides of the first equation we get $\log 100 = \log K + a \log 1000 + (1 - a) \log 1{,}000{,}000$; $2 = \log K + 3a + 6(1 - a)$; $\log K - 3a = -4$. From the second equation we get $\log K - a = -3$ similarly. **(c)** Solving we get $\log K = -2.5$ and $a = 0.5$, so $K = 10^{-2.5} \approx 0.003162$

Section 8.1

(d) $P(500, 1{,}000{,}000) =$
$0.003162(500^{0.5})(1{,}000{,}000^{0.5}) = 71$ pianos (to the nearest piano)

59. (a) We first need to convert n into years: 5 days = 5/365 years. So, $B(1.5 \times 10^{14}, 5/365) = \dfrac{1.5 \times 10^{14} \times 5}{5.1 \times 10^{14} \times 365} \approx 4 \times 10^{-3}$ grams per square meter. **(b)** The total weight of sulfates in the earth's atmosphere.

61. (a) The value of N would be doubled.
(b) $N(R, f_p, n_e, f_l, f_i, L) = R f_p n_e f_l f_i L$, where here L is the average lifetime of an intelligent civilization. **(c)** Take the logarithm of both sides, since this would yield the linear function $\ln N = \ln R + \ln f_p + \ln n_e + \ln f_l + \ln f_i + \ln f_c + \ln L$.

63. (a)

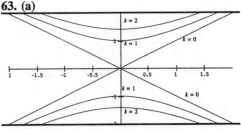

(b) The level curve $f(x, y) = 3$ is similar to the curves for 1 and 2, just further away from the origin.

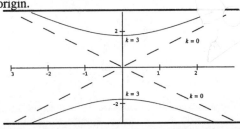

(c) These level curves would be similar to those in (a), but with the roles of x and y reversed.

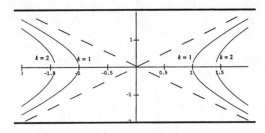

65. They are reciprocals of each other.

67. For example, $f(x, y) = x^2 + y^2$.

69. For example, $f(x, y, z) = xyz$

71. For example, take $f(x, y) = x + y$. Then setting $y = 3$ gives $f(x, 3) = x + 3$. This can be viewed as a function of the single variable x. Choosing other values for y gives other functions of x.

73. If $f = ax + by + c$, then fixing $y = k$ gives $f = ax + (bk + c)$, a linear function with slope a and intercept $bk + c$. The slope is independent of the choice of k.

75. That CDs cost more than cassettes.

Section 8.2

8.2

1.

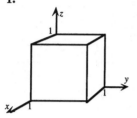

3.

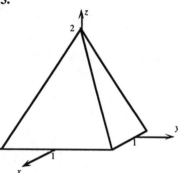

5.

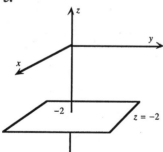

7.

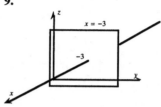

9.

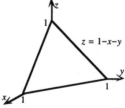

11. (H): The plane with z intercept 1, x intercept 1/3, and y intercept $-1/2$.

13. (B): An inverted paraboloid with apex at 1 on the z axis.

15. (F): A hemisphere lying below the xy plane.

17. (C): A surface of revolution having the z axis as a vertical asymptote.

19. The plane with x, y, and z intercepts all 1.

21. The plane with x intercept 1, y intercept 2, and z intercept -2.

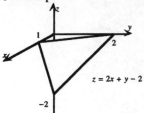

$z = 2x + y - 2$

23. The plane with x intercept -2 and z intercept 2, parallel to the y axis.

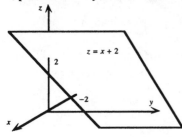

$z = x + 2$

25. The plane passing through the origin, the line $z = x$ in the xz plane, and the line $z = y$ in the yz plane.

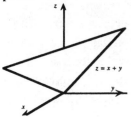

$z = x + y$

27. The cross section at $z = 1$ is the circle $x^2 + y^2 = 1/2$ and the cross section at $z = 2$ is the circle $x^2 + y^2 = 1$.

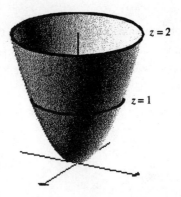

29. The cross section at $x = 0$ is the parabola $z = 2y^2$ and the cross section at $z = 1$ is the ellipse $x^2 + 2y^2 = 1$.

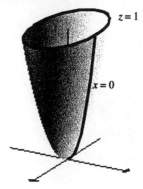

31. The graph is a cone with vertex at 2 on the z axis. The cross section at $y = 0$ is $z = 2 + |x|$ and the cross section at $z = 3$ is the circle $x^2 + y^2 = 1$.

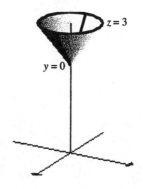

33. The graph is an inverted cone with apex at the origin. The cross section at $z = -4$ is the circle $x^2 + y^2 = 4$ and the cross section at $y = 1$ is a part of the hyperbola $z^2 - 4x^2 = 4$.

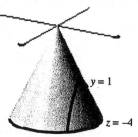

35. The cross sections $x = c$ are all copies of the parabola $z = y^2$.

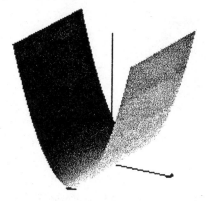

37. The cross sections $x = c$ are all copies of the hyperbola $z = 1/y$.

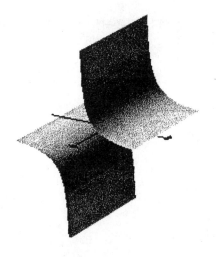

39. The cross sections $z = c$ are circles. The cross section $y = 0$ is the bell-shaped curve $z = e^{-x^2}$.

41. (a) The graph is a plane with x intercept -40, y intercept -60, and z intercept $240{,}000$. **(b)** The slice $x = 10$ is the straight line with equation $z = 300{,}000 + 4000y$. It describes the cost function for the manufacture of trucks if car production is held fixed at 10 cars per week. **(c)** The level curve $z = 480{,}000$ is the straight line $6000x + 4000y = 240{,}000$. It describes the number of cars and trucks you can manufacture to maintain weekly costs at $\$480{,}000$.

43. The graph is a plane with x_1 intercept 0.3, x_2 intercept 33, and x_3 intercept 0.66. The slices by $x_1 = constant$ are straight lines that are parallel to each other. Thus, the rate of change of General Motors' share as a function of Ford's share does not depend on Chrysler's share. Specifically, GM's share decreases by 0.02 percentage points per one percentage-point increase in Ford's market share, regardless of Chrysler's share.

45. (a) The slices $x = $ *constant* and $y = $ *constant* are straight lines. **(b)** No. Even though the slices $x = $ *constant* and $y = $ *constant* are straight lines, the level curves are not, and so the surface is not a plane. **(c)** The slice $x = 10$ has a slope of 3800. The slice $x = 20$ has a slope of 3600. Manufacturing more cars lowers the marginal cost of manufacturing trucks.

47. Both level curves are quarter-circles. (We see only the portion in the first quadrant because $e \geq 0$ and $k \geq 0$.) The level curve $C = 30,000$ represents the relationship between the number of electricians and the number of carpenters used in building a home that costs \$30,000. Similarly for the level curve $C = 40,000$.

49. The following figure shows several level curves together with several lines of the form $h + w = c$. (The horizontal axis is the h axis and the vertical axis is the w axis.)

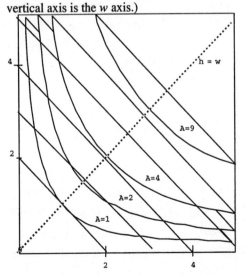

From the figure, thinking of the curves as contours on a map, we see that the largest value of A anywhere along any of the lines $h + w = c$ occurs midway along the line, when $h = w$. Thus, the largest area rectangle with a fixed perimeter occurs when $h = w$ (that is, when the rectangle is a square).

51. The level curve at $z = 3$ has the form $3 = x^{0.5}y^{0.5}$, or $y = 9/x$, and shows the relationship between the number of workers and the operating budget at a production level of 3 units

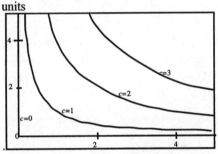

53. The level curve at $z = 0$ consists of the nonnegative y axis ($x = 0$) and tells us that zero utility corresponds to zero copies of Macro Publish, regardless of the number of copies of Turbo Publish. (Zero copies of Turbo Publish does not necessarily result in zero utility, according to the formula.)

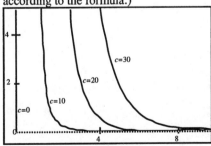

55. plane

Section 8.2

57. The graph of a function of three or more variables lives in four-dimensional (or higher) space, which makes it difficult to draw and visualize.

59. Disagree: For example, the function $f(x, y) = xy$ has such slices, but its graph is not a plane.

61. Referring to the diagram, we first compute the distance s. Since (a, b, z) and (x, y, z) lie in the same plane, the distance between them is the same as the distance between (a, b) and (x, y) in two dimensions, so $s = \sqrt{(x - a)^2 + (y - b)^2}$. The distance t is just the difference in height between (a, b, c) and (a, b, z), so $t = z - c$. Now d is the length of the hypotenuse of a right triangle in which the other two sides are s and t, so we have $d = \sqrt{s^2 + t^2} = \sqrt{(x - a)^2 + (y - b)^2 + (z - c)^2}$ as claimed.

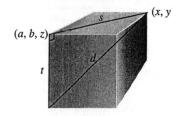

63. We need one dimension for each of the variables plus one dimension for the value of the function.

8.3

1. $f_x(x, y) = -40$; $f_y(x, y) = 20$; $f_x(1, -1) = -40$; $f_y(1, -1) = 20$

3. $f_x(x, y) = 6x + 1$; $f_y(x, y) = -3y^2$; $f_x(1, -1) = 7$; $f_y(1, -1) = -3$

5. $f_x(x, y) = -40 + 10y$; $f_y(x, y) = 20 + 10x$; $f_x(1, -1) = -50$; $f_y(1, -1) = 30$

7. $f_x(x, y) = 6xy$; $f_y(x, y) = 3x^2$; $f_x(1, -1) = -6$; $f_y(1, -1) = 3$

9. $f_x(x, y) = 2xy^3 - 3x^2y^2 - y$; $f_y(x, y) = 3x^2y^2 - 2x^3y - x$; $f_x(1, -1) = -4$; $f_y(1, -1) = 4$

11. $f_x(x, y) = 6y(2xy + 1)^2$; $f_y(x, y) = 6x(2xy + 1)^2$; $f_x(1, -1) = -6$; $f_y(1, -1) = 6$

13. $f_x(x, y) = e^{x+y}$; $f_y(x, y) = e^{x+y}$; $f_x(1, -1) = 1$; $f_y(1, -1) = 1$

15. $f_x(x, y) = 3x^{-0.4}y^{0.4}$; $f_y(x, y) = 2x^{0.6}y^{-0.6}$; $f_x(1, -1)$ is undefined; $f_y(1, -1)$ is undefined

17. $f_x(x, y) = 0.2ye^{0.2xy}$; $f_y(x, y) = 0.2xe^{0.2xy}$; $f_x(1, -1) = -0.2e^{-0.2}$; $f_y(1, -1) = 0.2e^{-0.2}$

19. $f_{xx}(x, y) = 0$; $f_{yy}(x, y) = 0$; $f_{xy}(x, y) = f_{yx}(x, y) = 0$; $f_{xx}(1, -1) = 0$; $f_{yy}(1, -1) = 0$; $f_{xy}(1, -1) = f_{yx}(1, -1) = 0$

21. $f_{xx}(x, y) = 0$; $f_{yy}(x, y) = 0$; $f_{xy}(x, y) = f_{yx}(x, y) = 10$; $f_{xx}(1, -1) = 0$; $f_{yy}(1, -1) = 0$; $f_{xy}(1, -1) = f_{yx}(1, -1) = 10$

23. $f_{xx}(x, y) = 6y$; $f_{yy}(x, y) = 0$; $f_{xy}(x, y) = f_{yx}(x, y) = 6x$; $f_{xx}(1, -1) = -6$; $f_{yy}(1, -1) = 0$; $f_{xy}(1, -1) = f_{yx}(1, -1) = 6$

25. $f_{xx}(x, y) = e^{x+y}$; $f_{yy}(x, y) = e^{x+y}$; $f_{xy}(x, y) = f_{yx}(x, y) = e^{x+y}$; $f_{xx}(1, -1) = 1$; $f_{yy}(1, -1) = 1$; $f_{xy}(1, -1) = f_{yx}(1, -1) = 1$

27. $f_{xx}(x, y) = -1.2x^{-1.4}y^{0.4}$; $f_{yy}(x, y) = -1.2x^{0.6}y^{-1.6}$; $f_{xy}(x, y) = f_{yx}(x, y) = 1.2x^{-0.4}y^{-0.6}$; $f_{xx}(1, -1)$ is undefined; $f_{yy}(1, -1)$ is undefined; $f_{xy}(1, -1)$ and $f_{yx}(1, -1)$ are undefined

29. $f_x(x, y, z) = yz$; $f_y(x, y, z) = xz$; $f_z(x, y, z) = xy$; $f_x(0, -1, 1) = -1$; $f_y(0, -1, 1) = 0$; $f_z(0, -1, 1) = 0$

31. $f_x(x, y, z) = \dfrac{4}{(x + y + z^2)^2}$; $f_y(x, y, z) = \dfrac{4}{(x + y + z^2)^2}$; $f_z(x, y, z) = \dfrac{8z}{(x + y + z^2)^2}$; $f_x(0, -1, 1)$ is undefined; $f_y(0, -1, 1)$ is undefined; $f_z(0, -1, 1)$ is undefined

33. $f_x(x, y, z) = e^{yz} + yze^{xz}$; $f_y(x, y, z) = xze^{yz} + e^{xz}$; $f_z(x, y, z) = xy(e^{yz} + e^{xz})$; $f_x(0, -1, 1) = e^{-1} - 1$; $f_y(0, -1, 1) = 1$; $f_z(0, -1, 1) = 0$

35. $f_x(x, y, z) = 0.1x^{-0.9}y^{0.4}z^{0.5}$; $f_y(x, y, z) = 0.4x^{0.1}y^{-0.6}z^{0.5}$; $f_z(x, y, z) = 0.5x^{0.1}y^{0.4}z^{-0.5}$; $f_x(0, -1, 1)$ is undefined; $f_y(0, -1, 1)$ is undefined, $f_z(0, -1, 1)$ is undefined

37. $f_x(x, y, z) = yze^{xyz}$, $f_y(x, y, z) = xze^{xyz}$, $f_z(x, y, z) = xye^{xyz}$; $f_x(0, -1, 1) = -1$; $f_y(0, -1, 1) = f_z(0, -1, 1) = 0$

39. $f_x(x, y, z) = 0$; $f_y(x, y, z) = -\dfrac{600z}{y^{0.7}(1 + y^{0.3})^2}$; $f_z(x, y, z) = \dfrac{2000}{1 + y^{0.3}}$; $f_x(0, -1, 1)$ is undefined (because $f(0, -1, 1)$ is); $f_y(0, -1, 1)$ is undefined; $f_z(0, -1, 1)$ is undefined

41. $\partial C/\partial x = 6000$: The marginal cost to manufacture each car is $6000. $\partial C/\partial y = 4000$: The marginal cost to manufacture each truck is $4000.

43. $\partial x_3/\partial x_1 = -2.2$: General Motors' market share decreases by 2.2 percentage points per one percentage-point increase in Chrysler's market share if Ford's share is unchanged. $\partial x_1/\partial x_3 = -1/2.2$ (solve for x_1 first): Chrysler's market share decreases by 1 percentage point per 2.2 percentage-point increase in General Motors' market share if Ford's share is unchanged. The two partial derivatives are reciprocals of each other.

45. $C_x(x, y) = 6000 - 20y$; $C_x(10, 20) = \$5600$ per car

47. (a) $\partial M/\partial c = -3.8$, $\partial M/\partial f = 2.2$. For every 1 point increase in the percentage of Chrysler owners who remain loyal, the percentage of Mazda owners who remain loyal decreases by 3.8 points. For every 1 point increase in the percentage of Ford owners who remain loyal, the percentage of Mazda owners who remain loyal increases by 2.2 points.
(b) $M(0.56, 0.56, 0.72, 0.50, 0.43) \approx 16\%$

49. (a) Writing $z(t, x)$, $z(10, 0) = \$16,500$
(b) $z(10, 1) = \$28,600$ **(c)** $z_t(t, x) = 350 + 220x$; $z_t(10, 0) = \$350$ per year **(d)** $z_t(10, 1) = \$570$ per year **(e)** The gap was widening: The median income of a white family was rising faster.

51. The marginal cost of a car is $C_x(x, y) = \$6000 + 1000e^{-0.01(x+y)}$ per car. The marginal cost of a truck is $C_y(x, y) = \$4000 + 1000e^{-0.01(x+y)}$ per truck. Both marginal costs decrease as production rises

53. $\bar{C}(x, y) = \dfrac{200{,}000 + 6000x + 4000y - 100{,}000e^{-0.01(x+y)}}{x + y}$;

$\bar{C}_x(x, y) = \dfrac{(6000 + 1000e^{-0.1(x+y)})(x + y) - (200{,}000 + 6000x + 4000y - 100{,}000e^{-0.01(x+y)})}{(x + y)^2} =$

$\dfrac{-200{,}000 + 2000y + (1000x + 1000y + 100{,}000)e^{-0.01(x+y)}}{(x + y)^2}$; $\bar{C}_x(50, 50) = -\$2.64$ per car. This means that at a production level of 50 cars and 50 trucks per week, the average cost per vehicle is decreasing by \$2.64 for each additional car manufactured.

$\bar{C}_y(x, y) = \dfrac{(4000 + 1000e^{-0.01(x+y)})(x + y) - (200{,}000 + 6000x + 4000y - 100{,}000e^{-0.01(x+y)})}{(x + y)^2} =$

$\dfrac{-200{,}000 - 2000x + (1000x + 1000y + 100{,}000)e^{-0.01(x+y)}}{(x + y)^2}$; $\bar{C}_y(50, 50) = -\$22.64$ per truck. This means that at a production level of 50 cars and 50 trucks per week, the average cost per vehicle is decreasing by \$22.64 for each additional truck manufactured.

Section 8.3

55. No. Your revenue function is $R(x, y) = 15{,}000x + 10{,}000y - 5000\sqrt{x+y}$, so your marginal revenue from the sale of cars is $R_x(x, y) = \$15{,}000 - \dfrac{2500}{\sqrt{x+y}}$ per car and your marginal revenue from the sale of trucks is $R_y(x, y) = \$10{,}000 - \dfrac{2500}{\sqrt{x+y}}$ per truck. These increase with increasing x and y. In other words, you will earn more revenue per vehicle with increasing sales, and so the rental company will pay more for each additional vehicle it buys.

57. $P_z(x, y, z) = 0.016x^{0.4}y^{0.2}z^{-0.6}$; $P_z(10, 100{,}000, 1{,}000{,}000) \approx 0.0001010$ papers/\$, or approximately 1 paper per \$10,000 increase in the subsidy.

59. (a) $U_x(x, y) = 4.8x^{-0.2}y^{0.2} + 1$; $U_y(x, y) = 1.2x^{0.8}y^{-0.8}$; $U_x(10, 5) = 5.18$, $U_y(10, 5) = 2.09$. This means that, if 10 copies of Macro Publish and 5 copies of Turbo Publish are purchased, the company's daily productivity is increasing at a rate of 5.18 pages per day for each additional copy of Macro purchased and by 2.09 pages per day for each additional copy of Turbo purchased.
(b) $\dfrac{U_x(10, 5)}{U_y(10, 5)} \approx 2.48$ is the ratio of the usefulness of one additional copy of Macro to one of Turbo. Thus, with 10 copies of Macro and 5 copies of Turbo, the company can expect approximately 2.48 times the productivity per additional copy of Macro compared to Turbo.

61. $F_y(x, y, z) = -\dfrac{2KQq(y - b)}{[(x - a)^2 + (y - b)^2 + (z - c)^2]^2}$. With $(a, b, c) = (0, 0, 0)$, $K = 9\times 10^9$, $Q = 10$, and $q = 5$, $F_y(2, 3, 3) \approx -6\times 10^9$ N/sec.

63. (a) $A_P(P, r, t) = (1 + r)^t$; $A_r(P, r, t) = tP(1 + r)^{t-1}$; $A_t(P, r, t) = P(1 + r)^t \ln(1 + r)$; $A_P(100, 0.1, 10) = 2.59$; $A_r(100, 0.1, 10) = 2{,}357.95$; $A_t(100, 0.1, 10) = 24.72$. Thus, for a \$100 investment at 10% interest, after 10 years the accumulated amount is increasing at a rate of \$2.59 per \$1 of principal, at a rate of \$2,357.95 per increase of 1 in r (note that this would correspond to an increase in the interest rate of 100%), and at a rate of \$24.72 per year.
(b) $A_P(100, 0.1, t)$ tells you the rate at which the accumulated amount in an account bearing 10% interest with a principal of \$100 is growing per \$1 increase in the principal, t years after the investment.

65. (a) $P_x = Ka\left(\dfrac{y}{x}\right)^b$ and $P_y = Kb\left(\dfrac{x}{y}\right)^a$. They are equal precisely when $\dfrac{a}{b} = \left(\dfrac{x}{y}\right)^b\left(\dfrac{x}{y}\right)^a$. Substituting $b = 1-a$ now gives $\dfrac{a}{b} = \dfrac{x}{y}$. **(b)** The given information implies that $P_x(100, 200) = P_y(100, 200)$. By part (a), this occurs precisely when $a/b = x/y = 100/200 = 1/2$. But $b = 1 - a$, so $a/(1 - a) = 1/2$, giving $a = 1/3$ and $b = 2/3$.

67. $u_t(r, t) = -\dfrac{1}{4\pi Dt^2}e^{-r^2/(4Dt)} + \dfrac{r^2}{16\pi D^2t^3}e^{-r^2/(4Dt)}$. Taking $D = 1$, $u_t(1, 3) \approx -0.0075$, so the concentration is decreasing at 0.0075 parts of nutrient per part of water/sec.

69. f is increasing at a rate of s units per unit of x, f is increasing at a rate of t units per unit of y, and the value of f is r when $x = a$ and $y = b$.

71. The marginal cost of building an additional orbicus; zonars per unit.

73. One example is $f(x, y) = -2x + 3y + 9$. Another is $f(x, y) = xy - 3x + 2y + 10$.

75. (a) b is the z-intercept of the plane, m is the slope of the intersection of the plane with the xz-plane, n is the slope of the intersection of the plane with the yz-plane. **(b)** Write $z = b + rx + sy$. We are told that $\partial z/\partial x = m$, so $r = m$. Similarly, $s = n$. Thus, $z = b + mx + ny$. We are also told that the plane passes through (h, k, l). Substituting gives $l = b + mh + nk$. This gives b as $l - mh - nk$. Substituting in the equation for z therefore gives $z = l - mh - nk + mx + ny = l + m(x - h) + n(y - k)$, as required.

8.4

1. P: relative minimum; Q: none of the above; R: relative maximum

3. P: saddle point; Q: relative maximum; R: none of the above

5. relative minimum

7. Neither

9. Saddle point

11. $f_x = 2x$; $f_y = 2y$; $f_{xx} = 2$; $f_{yy} = 2$; $f_{xy} = 0$. $f_x = 0$ when $x = 0$; $f_y = 0$ when $y = 0$, so $(0, 0)$ is the only critical point. $H = 4$ and $f_{xx} > 0$, so f has a relative minimum at $(0, 0, 1)$.

13. $g_x = -2x - 1$; $g_y = -2y + 1$; $g_{xx} = -2$; $g_{yy} = -2$; $g_{xy} = 0$. $g_x = 0$ when $x = -1/2$; $g_y = 0$ when $y = 1/2$, so $(-1/2, 1/2)$ is the only critical point. $H = 4$ and $g_{xx} < 0$, so g has a relative maximum at $(-1/2, 1/2, 3/2)$.

15. $h_x = 2xy - 4x$; $h_y = x^2 - 8y$; $h_{xx} = 2y - 4$; $h_{yy} = -8$; $h_{xy} = 2x$. $h_x = 0$ when $x = 0$ or $y = 2$; $h_y = 0$ when $x^2 = 8y$. The two possibilities are $x = 0$, so $y = 0$, or $y = 2$, so $x^2 = 16$ or $x = \pm 4$. This gives three critical points: $(0, 0)$, $(-4, 2)$, and $(4, 2)$. $H(x, y) = -8(2y - 4) - 4x^2 = 32 - 16y - 4x^2$; $H(0, 0) = 32$ and $h_{xx}(0, 0) = -4 < 0$; $H(-4, 2) = -64 = H(4, 2)$. Hence h has a relative maximum at $(0, 0, 0)$ and saddle points at $(\pm 4, 2, -16)$.

17. $s_x = 2xe^{x^2+y^2}$; $s_y = 2ye^{x^2+y^2}$; $s_{xx} = (2 + 4x^2)e^{x^2+y^2}$; $s_{yy} = (2 + 4y^2)e^{x^2+y^2}$; $s_{xy} = 4xye^{x^2+y^2}$. $s_x = 0$ when $x = 0$; $s_y = 0$ when $y = 0$; so the only critical point is $(0, 0)$.

$H(0, 0) = 4 - 0 = 4$ and $s_{xx}(0, 0) = 2 > 0$, so s has a relative minimum at $(0, 0, 1)$.

19. $t_x = 4x^3 + 8y^2$; $t_y = 16xy + 8y^3$; $t_{xx} = 12x^2$; $t_{yy} = 16x + 24y^2$; $t_{xy} = 16y$. $t_x = 0$ when $x^3 = -2y^2$ (notice that $x \le 0$ in this case); $t_y = 0$ when $y = 0$ or $y^2 = -2x$; if $y = 0$ then $x = 0$; if $y^2 = -2x$ then $x^3 = 4x$, so $x = 0$ (and $y = 0$) or $x = -2$ and $y = \pm 2$. This gives three critical points: $(0, 0)$ and $(-2, \pm 2)$. $H(-2, \pm 2) = 3072 \pm 32 > 0$ and $t_{xx}(-2, \pm 2) = 48 > 0$. $H(0, 0) = 0$ so the second derivative test is inconclusive. We can see from the graph of t that the origin is not a max, min, or saddle point. Or we can look at the slice along $x = y$ (suggested by the graph) where $t = 3x^4 + 8x^3$; this function increases as x approaches 0 and then increases again as x becomes larger than 0. So, t has two relative minima at $(-2, \pm 2, -16)$ and $(0, 0)$ is a critical point that is not a relative extremum or saddle point.

21. $f_x = 2x$; $f_y = 1 - e^y$; $f_{xx} = 2$; $f_{yy} = -e^y$; $f_{xy} = 0$. $f_x = 0$ when $x = 0$; $f_y = 0$ when $y = 0$. This gives one critical point, $(0, 0)$. $H(0, 0) = -2$, so f has a saddle point at $(0, 0, -1)$ and no other critical points.

23. $f_x = -(2x + 2)e^{-(x^2+y^2+2x)}$; $f_y = -2ye^{-(x^2+y^2+2x)}$; $f_{xx} = (4x^2 + 8x + 2)e^{-(x^2+y^2+2x)}$; $f_{yy} = (4y^2 - 2)e^{-(x^2+y^2+2x)}$; $f_{xy} = 2y(2x + 2)e^{-(x^2+y^2+2x)}$. $f_x = 0$ when $x = -1$; $f_y = 0$ when $y = 0$. This gives one critical point, $(-1, 0)$. $H(-1, 0) = 4e^2 > 0$ and $f_{xx}(-1, 0) = -2e < 0$, so f has a relative maximum at $(-1, 0, e)$.

25. $f_x = y - \dfrac{2}{x^2}$; $f_y = x - \dfrac{2}{y^2}$; $f_{xx} = \dfrac{4}{x^3}$; $f_{yy} = \dfrac{4}{y^3}$; $f_{xy} = 1$. $f_x = 0$ when $y = \dfrac{2}{x^2}$; $f_y = 0$ when $x = $

Section 8.4

$\frac{2}{y^2} = \frac{1}{2}x^4$, $x = 2^{1/3}$ ($x = 0$ is excluded because it is not in the domain of f). The corresponding value of y is $y = 2^{1/3}$. This gives one critical point, $(2^{1/3}, 2^{1/3})$. $H(2^{1/3}, 2^{1/3}) = 3$ and $f_{xx}(2^{1/3}, 2^{1/3}) = 2 > 0$, so f has a relative minimum at $(2^{1/3}, 2^{1/3}, 3(2^{2/3}))$.

27. $g_x = 2x - \frac{2}{x^2 y}$; $g_y = 2y - \frac{2}{xy^2}$; $g_{xx} = 2 + \frac{4}{x^3 y}$; $g_{yy} = 2 + \frac{4}{xy^3}$; $g_{xy} = \frac{2}{x^2 y^2}$. $g_x = 0$ when $y = \frac{1}{x^3}$; $g_y = 0$ when $x = \frac{1}{y^3} = x^9$, $x = \pm 1$ ($x = 0$ is excluded because it is not in the domain of f). This gives two critical points, $(1, 1)$ and $(-1, -1)$. $H(1, 1) = 32$ and $g_{xx}(1, 1) = 6 > 0$; $H(-1, -1) = 32$ and $g_{xx}(-1, -1) = 6 > 0$. So, g has relative minima at $(1, 1, 4)$ and $(-1, -1, 4)$.

29. f has an absolute minimum at $(0, 0, 1)$: $x^2 + y^2 + 1 \geq 1$ for all x and y because $x^2 + y^2 \geq 0$.

31. The relative maximum at $(0, 0, 0)$ is not absolute. For example, $h(10, 10) = 400 > h(0, 0)$.

33. $M_c = 8 + 8c - 20f$; $M_f = -40 - 20c + 80f$; $M_{cc} = 8$; $M_{ff} = 80$; $M_{cf} = -20$. $M_c = 0$ and $M_f = 0$ when $(c, f) = (2/3, 2/3)$; $H(2/3, 2/3) = 240$ and $M_{cc}(2/3, 2/3) = 8 > 0$, so M has a minimum at $(2/3, 2/3, 1/3)$. Thus, at least 1/3 of all Mazda owners would choose another new Mazda, and this lowest loyalty occurs when 2/3 of Chrysler and Ford owners remain loyal to their brands.

35. The subsidy is $S(x, y) = 500x + 100y$, so the net cost is $N(x, y) = C(x, y) - S(x, y) = 4000 + 100x^2 + 50y^2 - 500x - 100y$. $N_x = 200x - 500$; $N_y = 100y - 100$; $N_{xx} = 200$; $N_{yy} = 100$; $N_{xy} = 0$. $N_x = 0$ when $x = 2.5$; $N_y = 0$ when $y = 1$. $H = 192$

20,000 and $N_{xx} = 100 > 0$, so N has a maximum at $(2.5, 1)$. The firm should remove 2.5 pounds of sulfur and 1 pound of lead per day.

37. The total revenue is $R = p_1 q_1 + p_2 q_2 = 100,000p_1 - 100p_1^2 + 10p_1 p_2 + 150,000p_2 + 10p_1 p_2 - 100p_2^2 = 100,000p_1 + 150,000p_2 - 100p_1^2 + 20p_1 p_2 - 100p_2^2$. For convenience, write R_1 for $\partial R/\partial p_1$ and R_2 for $\partial R/\partial p_2$. Then $R_1 = 100,000 - 200p_1 + 20p_2$; $R_2 = 150,000 + 20p_1 - 200p_2$; $R_{11} = -200$; $R_{22} = -200$; $R_{12} = 20$. $R_1 = 0$ and $R_2 = 0$ when $(p_1, p_2) = (580.81, 808.08)$. $H = 39,600$ and $R_{11} = -200 < 0$, so R has a maximum at this critical point. You should charge \$580.81 for the Ultra Mini and \$808.08 for the Big Stack.

39. Let l = length, w = width, and h = height. We are told that $l + w + h \leq 62$; for the largest possible volume we will want $l + w + h = 62$, so $l = 62 - w - h$. The volume is $V = lwh = (62 - w - h)wh = 62wh - w^2 h - wh^2$. $V_w = 62h - 2wh - h^2$; $V_h = 62w - w^2 - 2wh$; $V_{ww} = -2h$; $V_{hh} = -2w$; $V_{wh} = 62 - 2w - 2h$. $V_w = 0$ when $62 - 2w - h = 0$ ($h = 0$ is not in the domain); $V_h = 0$ when $62 - w - 2h = 0$; this occurs when $w = h = 62/3 \approx 20.67$. $H \approx 1280$ and $V_{ww} < 0$, so V has a maximum at this critical point. The largest volume bag has dimensions $l = w = h \approx 20.67$ in and volume ≈ 8827 cubic inches.

41. Let l = length, w = width, and h = height. We are told that $l + 2(w + h) \leq 108$; for the largest possible volume we will want $l + 2(w + h) = 108$, so $l = 108 - 2(w + h)$. The volume is $V = lwh = (108 - 2w - 2h)wh = 108wh - 2w^2 h - 2wh^2$. $V_w = 108h - 4wh - 2h^2$; $V_h = 108w - 2w^2 - 4wh$; $V_{ww} = -4h$; $V_{hh} = $

$-4w$; $V_{wh} = 108 - 4w - 4h$. $V_w = 0$ when $108 - 4w - 2h = 0$; $V_h = 0$ when $108 - 2w - 4h = 0$; this occurs when $w = h = 18$. $H = 3888$ and $V_{ww} < 0$, so V has a maximum at this critical point. The corresponding length is $l = 108 - 2(w + h) = 36$. So, the largest volume package has dimensions 18 in × 18 in × 36 in and volume = 11,664 cubic inches.

43.

45. The function below has a relative max at $(0, 0, -1)$, has the unit circle $x^2 + y^2 = 1$ as a vertical asymptote, and takes on values larger than -1 as $x^2 + y^2$ gets large. The particular function graphed is $f(x, y) = -\dfrac{1}{|x^2 + y^2 - 1|}$.

47. H must be positive.

49. No. In order for there to be a relative maximum at (a, b), *all* vertical planes through (a, b) should yield a curve with a relative maximum at (a, b). It could happen that a slice by another vertical plane through (a, b) (such as $x - a = y - b$) does not yield a curve with a relative maximum at (a, b). An example is $f(x, y) = x^2 + y^2 - \sqrt{xy}$, at the point $(0, 0)$. Look at the slices through $x = 0$, $y = 0$ and $y = x$.

51. $\overline{C}_x = \dfrac{\partial}{\partial x}\left(\dfrac{C}{x+y}\right) = \dfrac{(x+y)C_x - C}{(x+y)^2}$. If this is zero, then $(x+y)C_x = C$, or $C_x = \dfrac{C}{x+y} = \overline{C}$. Similarly, if $\overline{C}_y = 0$ then $C_y = \overline{C}$. This is reasonable because if the average cost is decreasing with increasing x, then the average cost is greater than the marginal cost C_x. Similarly, if the average cost is increasing with increasing x, then the average cost is less than the marginal cost C_x. Thus, if the average cost is stationary with increasing x, then the average cost equals the marginal cost C_x. (The situation is similar for the case of increasing y.)

53. The equation of the tangent plane at the point (a, b) is $z = f(a, b) + f_x(a, b)(x - a) + f_y(a, b)(y - b)$. If f has a relative extremum at (a, b), then $f_x(a, b) = 0 = f_y(a, b)$. Substituting these into the equation of the tangent plane gives $z = f(a, b)$, a constant. But the graph of $z = $ constant is a plane parallel to the xy-plane.

8.5

1. *Interior:* $f_x = 2x = 0$ when $x = 0$; $f_y = 2y = 0$ when $y = 0$. The point $(0, 0)$ is on the boundary, not in the interior. *Boundary segment* $y = 0$, $0 \le x \le 2$: $f = x^2$, $f' = 2x = 0$ when $x = 0$. We get $(0, 0)$ and the other endpoint, $(2, 0)$. *Boundary segment* $x = 0$, $0 \le y \le 2$: $f = y^2$, $f' = 2y = 0$ when $y = 0$. We get $(0, 0)$ and the other endpoint, $(0, 2)$. *Boundary segment* $y = 2$, $0 \le x \le 2$: $f = x^2 + 4$, $f' = 2x = 0$ when $x = 0$. We get $(0, 2)$ and the other endpoint, $(2, 2)$. *Boundary segment* $x = 2$, $0 \le y \le 2$: $f = 4 + y^2$, $f' = 2y = 0$ when $y = 0$. We get $(2, 0)$ and the other endpoint, $(2, 2)$. Testing the four points we found, we find that f has a maximum value of 8 at $(2, 2)$ and a minimum value of 0 at $(0, 0)$.

3. *Interior:* $h_x = 2(x - 1) = 0$ when $x = 1$; $h_y = 2y = 0$ when $y = 0$; the point $(1, 0)$ is in the interior of the domain. (We do not need to test yet whether it is a max or min.) *Boundary* $x^2 + y^2 = 4$ (the circle centered at the origin of radius 2): $y^2 = 4 - x^2$ so $h = (x - 1)^2 + 4 - x^2 = -2x + 5$ and the possible values of x are $-2 \le x \le 2$. $h' = -2 \ne 0$, so we look at the endpoints $x = \pm 2$, where $y = \pm\sqrt{4 - x^2} = 0$, getting two points, $(-2, 0)$ and $(2, 0)$. Testing the three points we've found shows that h has a maximum value of 9 at $(-2, 0)$ and a minimum value of 0 at $(1, 0)$.

5. *Interior:* $f_x = 2xe^{x^2+y^2} = 0$ when $x = 0$; $f_y = 2ye^{x^2+y^2} = 0$ when $y = 0$; $(0, 0)$ is in the interior. *Boundary* $4x^2 + y^2 = 4$ (an ellipse): $y^2 = 4 - 4x^2$, $f = e^{4-3x^2}$, $-1 \le x \le 1$; $f' = -6xe^{4-3x^2} = 0$ when $x = 0$, $y = \pm 2$. So we get the points $(0, \pm 2)$ and the endpoints $(\pm 1, 0)$. Testing the five points we've found shows that f has a maximum value of e^4 at $(0, \pm 2)$ and a minimum value of 1 at $(0, 0)$.

7. *Interior:* $h_x = 8xe^{4x^2+y^2} = 0$ when $x = 0$; $h_y = 2ye^{4x^2+y^2} = 0$ when $y = 0$; $(0, 0)$ is in the interior. *Boundary* $x^2 + y^2 = 1$: $y^2 = 1 - x^2$, $h = e^{3x^2 + 1}$, $-1 \le x \le 1$; $h' = 6xe^{3x^2 + 1} = 0$ when $x = 0$, $y = \pm 1$. So we get the points $(0, \pm 1)$ and the endpoints $(\pm 1, 0)$. Testing the five points we've found shows that h has a maximum value of e^4 at $(\pm 1, 0)$ and a minimum value of 1 at $(0, 0)$.

9. *Interior:* $f_x = 1 - \dfrac{1}{x^2 y} = 0$ when $y = \dfrac{1}{x^2}$; $f_y = 1 - \dfrac{1}{xy^2} = 0$ when $x = \dfrac{1}{y^2} = x^4$, so $x = 0$ (outside the domain) or $x = 1$, $y = 1$, giving the point $(1, 1)$ in the interior of the domain. *Boundary segment* $y = 1/2$, $1/2 \le x \le 5/2$: $f = x + \dfrac{2}{x} + \dfrac{1}{2}$, $f' = 1 - \dfrac{2}{x^2} = 0$ when $x = \sqrt{2}$. With the endpoints, this gives us three points: $(1/2, 1/2)$, $(\sqrt{2}, 1/2)$, and $(5/2, 1/2)$. *Boundary segment* $x = 1/2$, $1/2 \le y \le 5/2$: This is similar and gives two additional points, $(1/2, \sqrt{2})$ and $(1/2, 5/2)$. *Boundary segment* $y = 3 - x$, $1/2 \le x \le 5/2$: $f = 3 + \dfrac{1}{x(3 - x)}$, $f' = -\dfrac{3 - 2x}{x^2(3 - x)^2} = 0$ when $x = 3/2$, $y = 3/2$, giving one more point to check, $(3/2, 3/2)$. Checking all of the points we've found, $(1, 1)$, $(1/2, 1/2)$, $(5/2, 1/2)$, $(1/2, 5/2)$, $(3/2, 3/2)$, $(\sqrt{2}, 1/2)$, and $(1/2, \sqrt{2})$, we find that f has a maximum value of 5 at $(1/2, 1/2)$ and a minimum value of 3 at $(1, 1)$.

11. *Interior:* $h_x = y - \dfrac{8}{x^2} = 0$ if $y = \dfrac{8}{x^2}$; $h_y = x - \dfrac{8}{y^2} = 0$ if $x = \dfrac{8}{y^2} = \dfrac{x^4}{8}$, so $x = 0$ (outside of the domain) or $x = 2$, $y = 2$, giving the point $(2, 2)$ in the interior of the domain. *Boundary segment* $y = $

Section 8.5

1, $1 \leq x \leq 9$: $h = x + \frac{8}{x} + 8$, $h' = 1 - \frac{8}{x^2} = 0$ when $x = \sqrt{8} = 2\sqrt{2}$. With the endpoints, this gives us three points: (1, 1), $(2\sqrt{2}, 1)$, and (9, 1). *Boundary segment* $x = 1$, $1 \leq y \leq 9$: This is similar and gives us two additional points: $(1, 2\sqrt{2})$ and (1, 9). *Boundary segment* $y = \frac{9}{x}$, $1 \leq x \leq 9$: $h = 9 + \frac{8}{x} + \frac{8}{9}x$, $h' = -\frac{8}{x^2} + \frac{8}{9} = 0$ when $x = 3$, $y = 3$. This gives one more point: (3, 3). Checking all of the points we've found, (1, 1), $(2\sqrt{2}, 1)$, (9, 1), $(1, 2\sqrt{2})$, (1, 9), (2, 2), and (3, 3), we find that h has a maximum value of 161/9 at (1, 9) and (9, 1) and a minimum value of 12 at (2, 2).

13. The region is described by the inequalities $-2 \leq x + y \leq 2$ and $-2 \leq x - y \leq 2$. *Interior:* $f_x = 2x + 2 = 0$ when $x = -1$; $f_y = 2y = 0$ when $y = 0$. This gives the point $(-1, 0)$. *Boundary segment* $x + y = 2$, $0 \leq x \leq 2$: $y = 2 - x$ so $f = x^2 + 2x + (2 - x)^2 = 2x^2 - 2x + 4$, $f' = 4x - 2 = 0$ when $x = 1/2$, $y = 3/2$. With the endpoints this gives (0, 2), (1/2, 3/2), and (2, 0). *Boundary segment* $x + y = -2$, $-2 \leq x \leq 0$: $y = -x - 2$ so $f = x^2 + 2x + (-x - 2)^2 = 2x^2 + 6x + 4$, $f' = 4x + 6 = 0$ when $x = -3/2$, $y = -1/2$. With the endpoints this gives $(-2, 0)$, $(-3/2, -1/2)$, and $(0, -2)$. *Boundary segment* $x - y = 2$, $0 \leq x \leq 2$: $y = x - 2$ so $f = x^2 + 2x + (x - 2)^2 = 2x^2 - 2x + 4$, $f' = 4x - 2 = 0$ when $x = 1/2$, $y = -3/2$. The endpoints have been listed already, so we get one more point: $(1/2, -3/2)$. *Boundary segment* $x - y = -2$, $-2 \leq x \leq 0$: $y = x + 2$, so $f = x^2 + 2x + (x + 2)^2 = 2x^2 + 6x + 4$, $f' = 4x + 6 = 0$ when $x = -3/2$, $y = 1/2$. The endpoints have been listed already, so we get one more point: $(-3/2, 1/2)$. Checking all of the points we've found, $(-1, 0)$, $(2, 0)$, $(0, 2)$, $(-2, 0)$, $(0, -2)$, $(1/2, 3/2)$, $(1/2, -3/2)$, $(-3/2, -1/2)$, and $(-3/2, 1/2)$, we find that f has a maximum value of 8 at (2, 0), and a minimum value of -1 at $(-1, 0)$.

15. The region is described by the inequalities $0 \leq y \leq x$ and $x + y \leq 2$. *Interior:* $h_x = 3x^2 = 0$ when $x = 0$; $h_y = 3y^2 = 0$ when $y = 0$. The point (0, 0) is not in the interior of the domain (but is a corner point, so will show up again). *Boundary segment* $y = 0$, $0 \leq x \leq 2$: $h = x^3$, $h' = 3x^2 = 0$ when $x = 0$. With the other endpoint we get the two points (0, 0) and (2, 0). *Boundary segment* $y = x$, $0 \leq x \leq 1$: $h = x^3 + x^3 = 2x^3$, $h' = 6x^2 = 0$ when $x = 0$. We already have (0, 0), but we also get the other endpoint, (1, 1). *Boundary segment* $x + y = 2$, $1 \leq x \leq 2$: $h = x^3 + (2 - x)^3$, $h' = 3x^2 - 3(2 - x)^2 = 3(4 - 4x) = 0$ when $x = 1$. We already have the two endpoints, so we get no more interesting points. Checking all of the points we've found, (0, 0), (2, 0), and (1, 1), we find that h has a maximum value of 8 at (2, 0) and a minimum value of 0 at (0, 0).

17. We want to maximize $f(x, y, z) = xyz$ subject to $x^2 + y^2 + z^2 - 1 = 0$. Using Lagrange multipliers, we need to solve the system $yz = 2\lambda x$, $xz = 2\lambda y$, $xy = 2\lambda z$, and $x^2 + y^2 + z^2 - 1 = 0$. We solve the first equation for $\lambda = \frac{yz}{2x}$ and substitute in the second to find $xz = \frac{y^2 z}{x}$ or $x^2 z = y^2 z$, giving $x^2 = y^2$ (assuming that $z \neq 0$, but $z = 0$ makes $xyz = 0$ and we can easily find larger values). Substituting the expression for λ into the third equation gives $x^2 = z^2$ similarly. From the last equation we then get $3x^2 - 1 = 0$, or $x = \pm 1/\sqrt{3}$. The corresponding values of y and z are also $\pm 1/\sqrt{3}$, so we get eight points to check: $(\pm 1/\sqrt{3}, \pm 1/\sqrt{3}, \pm 1/\sqrt{3})$ with all eight choices

195

of signs. Checking, we see that the largest value of xyz occurs when all of the signs or exactly two of them are positive, that is, at the points $(1/\sqrt{3}, 1/\sqrt{3}, 1/\sqrt{3})$, $(-1/\sqrt{3}, -1/\sqrt{3}, 1/\sqrt{3})$, $(1/\sqrt{3}, -1/\sqrt{3}, -1/\sqrt{3})$, and $(-1/\sqrt{3}, 1/\sqrt{3}, -1/\sqrt{3})$.

19. We wish to minimize and maximize $C(x, y)$ subject to $100 \leq x \leq 150$ and $80 \leq y \leq 120$. *Interior:* $C_x = 50 - 0.5y = 0$ when $y = 100$; $C_y = 70 - 0.5x = 0$ when $x = 140$. This gives one point, $(140, 100)$. *Boundary segment* $y = 80$, $100 \leq x \leq 150$: $C = 115{,}600 + 10x$, $C' = 10 \neq 0$. Thus, we get only the endpoints $(100, 80)$ and $(150, 80)$. The other three boundary segments are similar, giving us only two more points, $(100, 120)$ and $(150, 120)$. Checking all of the points $(140, 100)$, $(100, 80)$, $(150, 80)$, $(100, 120)$, and $(150, 120)$, we find that that, for the minimum cost of $16,600 you should make 100 5-speeds and 80 10-speeds; and for the maximum cost of $17,400 you should make 100 5-speeds and 120 10-speeds.

21. We wish to maximize $P(x, y)$ subject to $x + y \leq 200$, $x \geq 0$, and $y \geq 0$. *Interior:* $P_x = 20 - 0.2x = 0$ when $x = 100$; $P_y = 40 - 0.2y = 0$ when $y = 200$. The point $(100, 200)$ is not in the domain. *Boundary segment* $y = 0$, $0 \leq x \leq 200$: $P = 20x - 0.1x^2$, $P' = 20 - 0.2x = 0$ when $x = 100$. With the endpoints, this gives the points $(0, 0)$, $(100, 0)$, and $(200, 0)$. *Boundary segment* $x = 0$, $0 \leq y \leq 200$: $P = 40y - 0.1y^2$, $P' = 40 - 0.2y = 0$ when $y = 200$. The only new point we get is $(0, 200)$. *Boundary segment* $x + y = 200$, $0 \leq x \leq 200$: $P = 20x + 40(200 - x) - 0.1[x^2 + (200 - x)^2] = 8000 - 20x - 0.1(2x^2 - 400x + 40{,}000) = 4000 + 20x - 0.2x^2$, $P' = 20 - 0.4x = 0$ when $x = 50$,

$y = 150$. We've already listed the endpoints, so we get one more point, $(50, 150)$. Checking the points $(0, 0)$, $(100, 0)$, $(200, 0)$, $(0, 200)$, and $(50, 150)$, we find that, for a maximum profit of $4500, you should sell 50 copies of Walls and 150 copies of Doors.

23. *Interior:* $T_x = 2x = 0$ when $x = 0$; $T_y = 4y = 0$ when $y = 0$. $(0, 0)$ is not in the interior of the square. *Boundary segment* $y = 0$, $0 \leq x \leq 1$: $T = x^2$, $T' = 2x = 0$ when $x = 0$. This gives the endpoint we already have, and we also get the endpoint $(1, 0)$. The other three boundary segments are similar, giving the other two corners. Checking the points $(0, 0)$, $(1, 0)$, $(0, 1)$, and $(1, 1)$ we find that the hottest point is $(1, 1)$ and the coldest point is $(0, 0)$.

25. *Interior:* $T_x = 2x - 1 = 0$ when $x = 1/2$; $T_y = 4y = 0$ when $y = 0$. This gives the point $(1/2, 0)$. *Boundary* $y^2 = 1 - x^2$, $-1 \leq x \leq 1$: $T = x^2 + 2(1 - x^2) - x = 2 - x^2 - x$, $T' = -2x - 1 = 0$ when $x = -1/2$, $y = \pm\sqrt{3}/2$. This gives the two points $(-1/2, \pm\sqrt{3}/2)$ and we should also check the endpoints $(\pm 1, 0)$. We find that the hottest points are $(-1/2, \pm\sqrt{3}/2)$ and the coldest point is $(1/2, 0)$.

27. Minimize $d(x, y, z) = x^2 + y^2 + z^2$ subject to $x^2 + y - 1 - z = 0$. Using Lagrange multipliers, we need to solve $2x = 2\lambda x$, $2y = \lambda$, $2z = -\lambda$, and $x^2 + y - 1 - z = 0$. From the first equation, either $x = 0$ or $\lambda = 1$. If $x = 0$, substitute $y = \lambda/2$ and $z = -\lambda/2$ in the last equation to get $\lambda/2 - 1 + \lambda/2 = 0$, $\lambda = 1$. This gives the point $(0, 1/2, -1/2)$. On the other hand, if $\lambda = 1$, then $y = 1/2$, $z = -1/2$, and $x^2 + 1/2 - 1 + 1/2 = 0$, so $x = 0$, giving the same point. Since the distance may get arbitrarily large, but can get no less than 0, there must be a minimum distance and it must be at the point we found, $(0, 1/2, -1/2)$.

29. Minimize $d(x, y, z) = (x + 1)^2 + (y - 1)^2 + (z - 3)^2$ subject to $-2x + 2y + z - 5 = 0$. Using Lagrange multipliers, we need to solve $2(x + 1) = -2\lambda$, $2(y - 1) = 2\lambda$, $2(z - 3) = \lambda$, and $-2x + 2y + z - 5 = 0$. Solve the first three equations for $x = -\lambda - 1$, $y = \lambda + 1$, and $z = \lambda/2 + 3$. Substitute these in the last equation: $-2(-\lambda - 1) + 2(\lambda + 1) + (\lambda/2 + 3) - 5 = 0$, $\frac{9}{2}\lambda + 2 = 0$, $\lambda = -4/9$. This gives $x = -5/9$, $y = 5/9$, and $z = 25/9$. Thus, the point on the plane closest to $(-1, 1, 3)$ is $(-5/9, 5/9, 25/9)$.

31. Let the length, width, and height be l, w, and h, respectively. We wish to minimize $C = 20 \times 2 \times lw + 10(2lh + 2wh) = 40lw + 20lh + 20wh$ subject to $lwh = 2$. Solve the constraint for $l = \frac{2}{wh}$ and substitute to get $C(w, h) = \frac{80}{h} + \frac{40}{w} + 20wh$, $w > 0$ and $h > 0$. $C_w = -\frac{40}{w^2} + 20h = 0$ when $h = \frac{2}{w^2}$; $C_h = -\frac{80}{h^2} + 20w = 0$ when $w = \frac{4}{h^2} = w^4$, giving $w = 1$ (recall that $w > 0$). Thus, $h = 2$ and $l = 1$. Making either of h or w small or both large makes C large, so the point we found must be a minimum. Thus, the dimensions of the box of least cost are $l \times w \times h = 1 \times 1 \times 2$.

33. Let L be the cost of lightweight cardboard and let H be the cost of heavy-duty cardboard. Minimize $C = 2Hlw + 2Llh + 2Lwh$ subject to $lwh = 2$. Solve the constraint for $l = \frac{2}{wh}$ and substitute to get $C(w, h) = \frac{4H}{h} + \frac{4L}{w} + 2Lwh$. $C_w = -\frac{4L}{w^2} + 2Lh = 0$ when $h = \frac{2}{w^2}$; $C_h =$ $-\frac{4H}{h^2} + 2Lw = 0$ when $w = \frac{2H}{Lh^2} = \frac{H}{2L}w^4$, so $w = \left(\frac{2L}{H}\right)^{1/3}$. Substituting in the expressions for l and h we find $h = 2^{1/3}\left(\frac{H}{L}\right)^{2/3}$ and $l = w = \left(\frac{2L}{H}\right)^{1/3}$.
Thus, the box of least cost has dimensions
$\left(\frac{2L}{H}\right)^{1/3} \times \left(\frac{2L}{H}\right)^{1/3} \times 2^{1/3}\left(\frac{H}{L}\right)^{2/3}$.

35. Let $l =$ length, $w =$ width, and $h =$ height. We want to maximize the volume $V = lwh$. We are told that $l + 2(w + h) \le 108$; for the largest possible volume we will want $l + 2(w + h) = 108$. In the solution for Exercise 41 in Section 8.4 we showed one way to solve this problem. Here we use the alternative, which is Lagrange multipliers. So, we need to solve $wh = \lambda$, $lh = 2\lambda$, $lw = 2\lambda$, and $l + 2w + 2h - 108 = 0$. Substitute $\lambda = wh$ in the second equation to get $lh = 2wh$, so $l = 2w$ (we certainly must have $h > 0$ for a maximum volume). If we substitute $\lambda = wh$ in the third equation we get $lw = 2wh$, so $l = 2h$, hence $w = h = \frac{1}{2}l$. Substituting now in the last equation gives $3l = 108$, so $l = 36$, $w = 18$, and $h = 18$. Thus, the largest volume package has dimensions $18 \text{ in} \times 18 \text{ in} \times 36 \text{ in}$ and volume $= 11{,}664$ cubic inches.

37. If (x, y, z) is one corner of the box (with $x > 0$, $y > 0$, and $z > 0$), the box will have dimensions $2x \times 2y \times z$. We need to maximize $V = 4xyz$ subject to $x^2 + y^2 + z - 1 = 0$. Using Lagrange multipliers we must solve $4yz = 2\lambda x$, $4xz = 2\lambda y$, $4xy = \lambda$, and $x^2 + y^2 + z - 1 = 0$. Solve the first equation for $\lambda = 2\frac{yz}{x}$ and substitute in the second and third equations: $4xz = 4\frac{y^2z}{x}$, $x^2 = y^2$; $4xy = 2\frac{yz}{x}$, $2x^2 = z$. Substituting

$y^2 = x^2$ and $z = 2x^2$ in the last equation gives $x^2 + x^2 + 2x^2 = 1$, $x^2 = \frac{1}{4}$, $x = \frac{1}{2}$; $y = \frac{1}{2}$; $z = \frac{1}{2}$. The dimensions of the box are $1 \times 1 \times \frac{1}{2}$.

39. Maximize $P(x, y) = 3x + 2y - 0.01(x^2 + y^2)$ subject to $2x + y \leq 500$, $3x + 3y \leq 900$ (or $x + y \leq 300$), $x \geq 0$, and $y \geq 0$. Note that the two boundary lines meet at (200, 100). *Interior:* $P_x = 3 - 0.02x = 0$ when $x = 150$; $P_y = 2 - 0.02y = 0$ when $y = 100$. This gives the point (150, 100). *Boundary segment* $y = 0$, $0 \leq x \leq 250$: $P = 3x - 0.01x^2$, $P' = 3 - 0.02x = 0$ when $x = 150$. With the endpoints this gives three points: (0, 0), (150, 0), and (250, 0). *Boundary segment* $x = 0$, $0 \leq y \leq 300$: $P = 2y - 0.01y^2$, $P' = 2 - 0.02y = 0$ when $y = 100$. With the endpoints this gives two more points: (0, 100) and (0, 300). *Boundary segment* $2x + y = 500$, $200 \leq x \leq 250$: $P = 3x + 2(500 - 2x) - 0.01[x^2 + (500 - 2x)^2] = -1500 + 19x - 0.05x^2$, $P' = 19 - 0.1x = 0$ when $x = 190$, which is not in the domain. However, we do get one more point, the endpoint (200, 100). *Boundary segment* $x + y = 300$, $0 \leq x \leq 200$: $P = 3x + 2(300 - x) - 0.01[x^2 + (300 - x)^2] = -300 + 7x - 0.02x^2$, $P' = 7 - 0.04x = 0$ when $x = 175$, $y = 125$. This gives one more point, (175, 125). Checking the points we've found, (150, 100), (0, 0), (150, 0), (250, 0), (0, 100), (0, 300), (200, 100), and (175, 125), we find that the maximum profit occurs when you produce 150 quarts of vanilla and 100 quarts of mocha, for a profit of $325.

41. Maximize $P(x, y) = 5,000,000(1 - e^{-0.02x - 0.01y})$ subject to $60x + 50y \leq 6000$ (or $6x + 5y \leq 600$), $x + y \leq 110$, $x \geq 0$ and $y \geq 0$. Note that the two boundary lines meet at (50, 60). *Interior:* $P_x = 100,000e^{-0.02x - 0.01y} \neq 0$ (and

similarly for P_y), so there are no critical points in the interior. *Boundary segment* $y = 0$, $0 \leq x \leq 100$: $P = 5,000,000(1 - e^{-0.02x})$, $P' = 100,000e^{-0.02x} \neq 0$, so we get only the endpoints (0, 0) and (100, 0). *Boundary segment* $x = 0$, $0 \leq y \leq 110$: $P = 5,000,000(1 - e^{-0.01y})$, $P' = 50,000e^{-0.01y} \neq 0$, so we get only the endpoint (0, 110). *Boundary segment* $6x + 5y = 600$, $50 \leq x \leq 100$: $P = 5,000,000[1 - e^{-0.02x - 0.01(120 - 1.2x)}] = 5,000,000(1 - e^{-0.008x - 1.2})$, $P' = 40,000e^{-0.008x - 1.2} \neq 0$, so we get the endpoint (50, 60). *Boundary segment* $x + y = 110$, $0 \leq x \leq 50$: $P = 5,000,000[1 - e^{-0.02x - 0.01(110 - x)}] = 5,000,000(1 - e^{-0.01x - 1.1})$, $P' = 50,000e^{-0.01x - 1.1} \neq 0$, so we get no more points. Checking the point we've found, (0, 0), (100, 0), (0, 110), and (50, 60), we find that the university should offer 100 sections of Finite Math and no sections of Applied Calculus.

43. Let x be the number of servings of cereal and let y be the number of servings of dessert. Minimize $C(x, y) = 10x + 53y$ subject to $xy \geq 10$. It's clear that the minimum must occur when $xy = 10$ (otherwise decrease either x or y to get a lower cost). There are several approaches to finding the minimum, but in the spirit of this section we'll use Lagrange multipliers. So, we must solve the equations $10 = \lambda y$, $53 = \lambda x$, and $xy - 10 = 0$. Solve the first for $\lambda = \dfrac{10}{y}$ and substitute in the second: $53 = \dfrac{10x}{y}$ or $y = \dfrac{10x}{53}$. Substituting in the last equation gives $\dfrac{10x^2}{53} = 10$, $x^2 = 53$, so $x = \sqrt{53} \approx 7.280$, $y \approx 1.374$. Use 7.280 servings of Mixed Cereal and 1.374 servings of Tropical Fruit Dessert.

Section 8.5

45. Minimize $C(x, y) = x^2 + y^2 - 200x - 200y + 320{,}000$ subject to $400x + 300y \geq 48{,}000$, $80x + 100y \geq 12{,}800$, $x \geq 0$, and $y \geq 0$. Note that two of the boundary lines cross at $(60, 80)$. *Interior:* $C_x = 2x - 200 = 0$ when $x = 100$; $C_y = 2y - 200 = 0$ when $y = 100$. This gives the critical point $(100, 100)$. *Boundary segment* $y = 0$, $x \geq 160$: $C = x^2 - 200x + 320{,}000$, $C' = 2x - 200 = 0$ when $x = 100$, which is out of the domain. However, we do have the endpoint $(160, 0)$. *Boundary segment* $x = 0$, $y \geq 160$: $C = y^2 - 200y + 320{,}000$, $C' = 2y - 200 = 0$ when $y = 100$, which is out of the domain. However, we do have the endpoint $(0, 160)$. *Boundary segment* $400x + 300y = 48{,}000$, $0 \leq x \leq 60$: $C = x^2 + \left(160 - \frac{4}{3}x\right)^2 - 200x - 200\left(160 - \frac{4}{3}x\right) + 320{,}000 = \frac{25}{9}x^2 - 360x + 313{,}600$, $C' = \frac{50}{9}x - 360 = 0$ when $x = 64.8$, which is not in the domain. So, we get only the endpoint $(60, 80)$. *Boundary segment* $80x + 100y = 12{,}800$, $60 \leq x \leq 160$. $C = x^2 + \left(128 - \frac{4}{5}x\right)^2 - 200x - 200\left(128 - \frac{4}{5}x\right) + 320{,}000 = \frac{41}{25}x^2 - \frac{1224}{5}x + 310{,}784$, $C' = \frac{82}{25}x - \frac{1224}{5} = 0$ when $x = \frac{3060}{41} \approx 74.63$; $y = \frac{2188}{41} \approx 53.37$. This gives us one more point to check: $(74.63, 53.37)$. Checking the points we've found, $(100, 100)$, $(160, 0)$, $(0, 160)$, $(60, 80)$, and $(74.63, 53.37)$, the lowest value of C occurs at $(100, 100)$. The domain is unbounded, but as x or y or both increase, the terms $x^2 + y^2$ predominate and C gets large. Thus, the minimum cost occurs when the school buys 100 of each.

47. Yes. There may be relative extrema at points on the boundary of the domain of the function. The partial derivatives of the function need not be 0 at such points.

49. If the only constraint is an equality constraint, and if it is impossible to eliminate one of the variables in the objective function by substitution (solving the constraint equation for a variable or some other method).

51. In a linear programming problem, the objective function is linear, and so the partial derivatives can never all be zero. (We are ignoring the simple case in which the objective function is constant.) It follows that the extrema cannot occur in the interior of the domain (since the partial derivatives must be zero at such points).

8.6

1. $\int_0^1 \int_0^1 (x - 2y) \, dx \, dy = \int_0^1 \left[\frac{1}{2}x^2 - 2xy\right]_{x=0}^1 dy = \int_0^1 \left(\frac{1}{2} - 2y\right) dy = \left[\frac{1}{2}y - y^2\right]_{y=0}^1 = -\frac{1}{2}$

3. $\int_0^1 \int_0^2 (ye^x - x - y) \, dx \, dy = \int_0^1 \left[ye^x - \frac{1}{2}x^2 - xy\right]_{x=0}^2 dy = \int_0^1 (e^2 y - 2 - 3y) \, dy =$

$\left[\frac{1}{2}e^2 y^2 - 2y - \frac{3}{2}y^2\right]_{y=0}^1 = \frac{1}{2}e^2 - \frac{7}{2}$

5. $\int_0^2 \int_0^3 e^{x+y} \, dx \, dy = \int_0^2 [e^{x+y}]_{x=0}^3 \, dy = \int_0^2 (e^{3+y} - e^y) \, dy = [e^{3+y} - e^y]_{y=0}^2 = e^5 - e^2 - e^3 + 1 =$

$(e^3 - 1)(e^2 - 1)$. This may also be found by writing $\int_0^2 \int_0^3 e^{x+y} \, dx \, dy = \int_0^2 \int_0^3 e^x e^y \, dx \, dy = \left(\int_0^3 e^x \, dx\right)\left(\int_0^2 e^y \, dy\right)$

as in Exercise 57.

7. $\int_0^1 \int_0^{2-y} x \, dx \, dy = \int_0^1 \left[\frac{1}{2}x^2\right]_{x=0}^{2-y} dy = \int_0^1 \frac{1}{2}(2-y)^2 \, dy = \left[-\frac{1}{6}(2-y)^3\right]_{y=0}^1 = -\frac{1}{6} + \frac{8}{6} = \frac{7}{6}$

9. $\int_{-1}^1 \int_{y-1}^{y+1} e^{x+y} \, dx \, dy = \int_{-1}^1 [e^{x+y}]_{x=y-1}^{y+1} \, dy = \int_{-1}^1 (e^{2y+1} - e^{2y-1}) \, dy = \left[\frac{1}{2}e^{2y+1} - \frac{1}{2}e^{2y-1}\right]_{y=-1}^1 =$

$\frac{1}{2}(e^3 - e - e^{-1} + e^{-3})$

11. $\int_0^1 \int_{-x^2}^{x^2} x \, dy \, dx = \int_0^1 [xy]_{y=-x^2}^{x^2} \, dx = \int_0^1 2x^3 \, dx = \left[\frac{1}{2}x^4\right]_{x=0}^1 = \frac{1}{2}$

13. $\int_0^1 \int_0^x e^{x^2} \, dy \, dx = \int_0^1 [ye^{x^2}]_{y=0}^x \, dx = \int_0^1 xe^{x^2} \, dx = \left[\frac{1}{2}e^{x^2}\right]_{x=0}^1 = \frac{1}{2}e - \frac{1}{2} = \frac{1}{2}(e - 1)$

Section 8.6

15. $\displaystyle\int_0^2\int_{1-x}^{8-x}(x+y)^{1/3}\,dy\,dx = \int_0^2\left[\tfrac{3}{4}(x+y)^{4/3}\right]_{y=1-x}^{8-x}dx = \int_0^2\left(\tfrac{3}{4}(16) - \tfrac{3}{4}\right)dx = \int_0^2\tfrac{45}{4}dx = \left[\tfrac{45}{4}x\right]_{x=0}^2 = \tfrac{45}{2}$

17. $\displaystyle\int_{-1}^1\int_0^{1-x^2}2\,dy\,dx = \int_{-1}^1[2y]_{y=0}^{1-x^2}dx = \int_{-1}^1 2(1-x^2)\,dx = \left[2x - \tfrac{2}{3}x^3\right]_{x=-1}^1 = 2 - \tfrac{2}{3} + 2 - \tfrac{2}{3} = \tfrac{8}{3}$

19. $\displaystyle\int_{-1}^1\int_0^{1-y^2}(1+y)\,dx\,dy = \int_{-1}^1[x(1+y)]_{x=0}^{1-y^2}dy = \int_{-1}^1(1-y^2)(1+y)\,dy = \int_{-1}^1(1+y-y^2-y^3)\,dy =$

$\left[y + \tfrac{1}{2}y^2 - \tfrac{1}{3}y^3 - \tfrac{1}{4}y^4\right]_{y=-1}^1 = 1 + \tfrac{1}{2} - \tfrac{1}{3} - \tfrac{1}{4} + 1 - \tfrac{1}{2} - \tfrac{1}{3} + \tfrac{1}{4} = \tfrac{4}{3}$

21. $\displaystyle\int_{-1}^1\int_{-\sqrt{1-y^2}}^{\sqrt{1-y^2}}xy^2\,dx\,dy = \int_{-1}^1\left[\tfrac{1}{2}x^2y^2\right]_{x=-\sqrt{1-y^2}}^{\sqrt{1-y^2}}dy = \int_{-1}^1 0\,dy = 0$

23. $\displaystyle\int_{-1}^0\int_{-x-1}^{x+1}(x^2+y^2)\,dy\,dx + \int_0^1\int_{x-1}^{-x+1}(x^2+y^2)\,dy\,dx = \int_{-1}^0\left[x^2y + \tfrac{1}{3}y^3\right]_{y=-x-1}^{x+1}dx + \int_0^1\left[x^2y + \tfrac{1}{3}y^3\right]_{y=x-1}^{-x+1}dx =$

$\displaystyle\int_{-1}^0\left(2x^2(x+1) + \tfrac{2}{3}(x+1)^3\right)dx + \int_0^1\left(2x^2(-x+1) + \tfrac{2}{3}(-x+1)^3\right)dx = \int_{-1}^0\left(\tfrac{8}{3}x^3 + 4x^2 + 2x + \tfrac{2}{3}\right)dx +$

$\displaystyle\int_0^1\left(-\tfrac{8}{3}x^3 + 4x^2 - 2x + \tfrac{2}{3}\right)dx = \left[\tfrac{2}{3}x^4 + \tfrac{4}{3}x^3 + x^2 + \tfrac{2}{3}x\right]_{x=-1}^0 + \left[-\tfrac{2}{3}x^4 + \tfrac{4}{3}x^3 - x^2 + \tfrac{2}{3}x\right]_{x=0}^1 = -\tfrac{2}{3} + \tfrac{4}{3} -$

$1 + \tfrac{2}{3} - \tfrac{2}{3} + \tfrac{4}{3} - 1 + \tfrac{2}{3} = \tfrac{2}{3}$

25. $\displaystyle\int_0^2\int_0^{2-x}y\,dy\,dx = \int_0^2\left[\tfrac{1}{2}y^2\right]_{y=0}^{2-x}dx = \int_0^2\tfrac{1}{2}(2-x)^2\,dx = \left[-\tfrac{1}{6}(2-x)^3\right]_{x=0}^2 = \tfrac{8}{6} = \tfrac{4}{3}.\ A = \tfrac{1}{2}(2)(2) = 2$ by the

geometric formula for the area of a triangle. So, the average value $= \dfrac{4/3}{2} = \dfrac{2}{3}$.

Section 8.6

27. $\int_0^1 \int_{y-1}^{1-y} e^y \, dx \, dy = \int_0^1 [xe^y]_{x=y-1}^{1-y} \, dy = \int_0^1 2(1-y)e^y \, dy = [2(1-y)e^y + 2e^y]_{y=0}^1 = 2e - 4 = 2(e-2)$. $A = \frac{1}{2}(2)(1) = 1$ by the formula for the area of a triangle. So, the average value $= 2(e-2)$.

29. $\int_{-2}^0 \int_{x/2+1}^{x+2} x^2 \, dy \, dx + \int_0^2 \int_{-x/2+1}^{-x+2} x^2 \, dy \, dx = \frac{4}{3}$ as calculated in Exercise 24. $A = 2 \cdot \frac{1}{2}(2)(1) = 2$, since A consists of the area of two triangles. So, the average value $= \frac{4/3}{2} = \frac{2}{3}$.

31. $\int_0^1 \int_0^{1-x} f(x, y) \, dy \, dx$

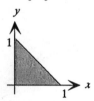

33. $\int_0^{\sqrt{2}} \int_{x^2-1}^1 f(x, y) \, dy \, dx$

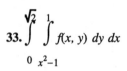

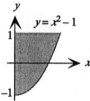

35. $\int_1^4 \int_1^{2\sqrt{y}} f(x, y) \, dx \, dy$

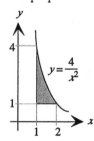

37. $V = \int_0^1 \int_0^2 (1 - x^2) \, dy \, dx = \int_0^1 [(1-x^2)y]_{y=0}^2 \, dx = \int_0^1 2(1-x^2) \, dx = \left[2x - \frac{2}{3}x^3\right]_{x=0}^1 = 2 - \frac{2}{3} = \frac{4}{3}$

202

Section 8.6

39. The top face of the tetrahedron is the graph of $z = 1 - x - y$. So, the volume is $V =$
$$\int_0^1 \int_0^{1-x} (1 - x - y)\, dy\, dx = \int_0^1 \left[(1-x)y - \tfrac{1}{2}y^2\right]_{y=0}^{1-x} dx = \int_0^1 \left[(1-x)^2 - \tfrac{1}{2}(1-x)^2\right] dx = \int_0^1 \tfrac{1}{2}(1-x)^2\, dx =$$
$$\left[-\tfrac{1}{6}(1-x)^3\right]_{x=0}^{1} = \tfrac{1}{6}.$$

41. $\int_{45}^{55}\int_{8}^{12} 10{,}000 x^{0.3} y^{0.7}\, dy\, dx = 10{,}000 \int_{45}^{55} \left[\tfrac{1}{1.7} x^{0.3} y^{1.7}\right]_{y=8}^{12} dx = \dfrac{10{,}000(12^{1.7} - 8^{1.7})}{1.7} \int_{45}^{55} x^{0.3}\, dx =$

$\dfrac{10{,}000(12^{1.7} - 8^{1.7})}{(1.7)(1.3)} [x^{1.3}]_{x=45}^{55} = \dfrac{10{,}000(12^{1.7} - 8^{1.7})(55^{1.3} - 45^{1.3})}{(1.7)(1.3)} \approx 6{,}471{,}000.$ $A = (12 - 8)(55 - 45) =$

40. So, the average is approximately $\dfrac{6{,}471{,}000}{40} \approx 162{,}000$ gadgets.

43. $R(p, q) = pq$ with $40 \le p \le 50$ and $8000 - p^2 \le q \le 10{,}000 - p^2$. We first find the extrema: *Interior:* $R_p = q \ne 0$ in the domain, so there are no critical points in the interior of the domain. *Boundary segment* $p = 40$, $6400 \le q \le 8400$: $R = 40q$, $R' = 40 \ne 0$, so we get only the endpoints $(40, 6400)$ and $(40, 8400)$. *Boundary segment* $p = 50$, $5500 \le q \le 7500$: This segment is similar, giving only the endpoints $(50, 5500)$ and $(50, 7500)$. *Boundary segment* $q = 8000 - p^2$, $40 \le p \le 50$: $R = p(8000 - p^2) = 8000p - p^3$, $R' = 8000 - 3p^2 = 0$ when $p \approx 51.6$, outside of the domain. So we get no more interesting points from this segment. *Boundary segment* $q = 10{,}000 - p^2$, $40 \le p \le 50$: $R = p(10{,}000 - p^2) = 10{,}000p - p^3$, $R' = 10{,}000 - 3p^2 = 0$ when $p \approx 57.7$, again outside the domain. Checking the four points we found, $(40, 6400)$, $(40, 8400)$, $(50, 5500)$, and $(50, 7500)$, we find that the maximum revenue is \$375,500 at $(50, 5500)$ and the minimum revenue is \$256,000 at $(40, 6400)$.

To find the average we first compute
$$\int_{40}^{50} \int_{8000-p^2}^{10{,}000-p^2} pq\, dq\, dp = \tfrac{1}{2}\int_{40}^{50} [pq^2]_{q=8000-p^2}^{10{,}000-p^2} dp =$$
$$\tfrac{1}{2}\int_{40}^{50} [p(10{,}000 - p^2)^2 - p(8000 - p^2)^2]\, dp =$$
$$\tfrac{1}{12}[-(10{,}000 - p^2)^3 + (8000 - p^2)^3]_{p=40}^{50} =$$
$6{,}255{,}000{,}000$. Next we compute $A =$
$$\int_{40}^{50} \int_{8000-p^2}^{10{,}000-p^2} dq\, dp = \int_{40}^{50} [q]_{q=8000-p^2}^{10{,}000-p^2} dp =$$
$$\int_{40}^{50} [10{,}000 - p^2 - (8000 - p^2)]\, dp =$$
$$\int_{40}^{50} 2000\, dp = [2000p]_{p=40}^{50} = 100{,}000 - 80{,}000 =$$
$20{,}000$. The average revenue is thus $\dfrac{6{,}255{,}000{,}000}{20{,}000} = \$312{,}750.$

Section 8.6

45. $R(p, q) = pq$ with $15{,}000/q \leq p \leq 20{,}000/q$ and $500 \leq q \leq 1000$. We first find the extrema: *Interior:* $R_p = q \neq 0$ in the domain, so there are no critical points in the interior of the domain. *Boundary segment* $q = 500$, $30 \leq p \leq 40$: $R = 500p$, $R' = 500 \neq 0$, so we get only the endpoints $(30, 500)$ and $(40, 500)$. *Boundary segment* $q = 1000$, $15 \leq p \leq 20$: $R = 1000p$, $R' = 1000 \neq 0$, so we get only the endpoints $(15, 1000)$ and $(20, 1000)$. *Boundary segment* $p = 15{,}000/q$, $500 \leq q \leq 1000$: $R = pq = 15{,}000$ is constant, so all points along this curve are critical. We'll concentrate on the endpoints but remember this fact. *Boundary segment* $p = 20{,}000/q$, $500 \leq q \leq 1000$: $R = pq = 20{,}000$ is constant, so again all point along this curve are critical. Checking the corner points we find that the maximum revenue is $\$20{,}000$, which is achieved at $(40, 500)$ and $(20, 1000)$ and all points along the curve $p = 20{,}000/q$, $500 \leq q \leq 1000$ connecting them. Similarly, the minimum revenue is $\$15{,}000$, achieved at $(30, 500)$ and $(15, 1000)$ and all points along the curve $p = 15{,}000/q$, $500 \leq q \leq 1000$ connecting them.

To find the average we first compute

$$\int_{500}^{1000} \int_{15{,}000/q}^{20{,}000/q} pq \, dp \, dq = \frac{1}{2}\int_{500}^{1000} [p^2 q]_{p=15{,}000/q}^{20{,}000/q} \, dq =$$

$$\frac{1}{2}\int_{500}^{1000}\left(\frac{20{,}000^2}{q} - \frac{15{,}000^2}{q}\right) dq =$$

$$\frac{175{,}000{,}000}{2}\int_{500}^{1000}\frac{1}{q}\,dq = \frac{175{,}000{,}000}{2}[\ln q]_{q=500}^{1000} =$$

$$= \frac{175{,}000{,}000}{2}(\ln 1000 - \ln 500) =$$

$$\frac{175{,}000{,}000}{2}\ln 2.\ \text{Next we compute } A =$$

$$\int_{500}^{1000}\int_{15{,}000/q}^{20{,}000/q} dp \, dq = \int_{500}^{1000}[p]_{p=15{,}000/q}^{20{,}000/q}\,dq =$$

$$5000\int_{500}^{1000}\frac{1}{q}\,dq = 5000[\ln q]_{q=500}^{1000} =$$

$5000(\ln 1000 - \ln 500) = 5000 \ln 2$. The average revenue is thus $\dfrac{175{,}000{,}000 \ln 2}{2 \cdot 5000 \ln 2} = \$17{,}500$.

47. Total population $= \displaystyle\int_0^{20}\int_0^{30} e^{-0.1(x+y)}\,dy\,dx =$

$$-10\int_0^{20}[e^{-0.1(x+y)}]_{y=0}^{30}\,dx =$$

$$-10\int_0^{20}(e^{-0.1x-3} - e^{-0.1x})\,dx =$$

$$-10(e^{-3}-1)\int_0^{20} e^{-0.1x}\,dx = 100(e^{-3}-1)[e^{-0.1x}]_{x=0}^{20}$$

$= 100(e^{-3}-1)(e^{-2}-1) \approx 82.16$ hundred people, or 8216 people.

49. $\displaystyle\int_0^1\int_0^1 (x^2 + 2y^2)\,dy\,dx = \int_0^1\left[x^2y + \frac{2}{3}y^3\right]_{y=0}^1 dx$

$$= \int_0^1\left(x^2 + \frac{2}{3}\right)dx = \left[\frac{1}{3}x^3 + \frac{2}{3}x\right]_{x=0}^1 = 1.\ A = 1,$$ the area of the square. Hence, the average temperature is $1/1 = 1$ degree.

51. The area between the curves $y = r(x)$ and $y = s(x)$ and the vertical lines $x = a$ and $x = b$ is

given by $\int_a^b \int_{r(x)}^{s(x)} dy\, dx$ assuming that $r(x) \le s(x)$
for $a \le x \le b$.

53. The first step in calculating an integral of the form $\int_a^b \int_{r(x)}^{s(x)} f(x, y)\, dy\, dx$ is to evaluate the integral $\int_{r(x)}^{s(x)} f(x, y)\, dy$, obtained by holding x constant and integrating with respect to y.

55. paintings per picasso per dali

57. The left-hand side is $\int_a^b \int_c^d f(x)g(y)\, dx\, dy =$

$\int_a^b \left(g(y) \int_c^d f(x)\, dx \right) dy$ (because $g(y)$ is treated as a constant in the inner integral) =

$\left(\int_c^d f(x)\, dx \right) \int_a^b g(y)\, dy$ (because $\int_c^d f(x)\, dx$ is a

constant and can therefore be taken outside the integral with respect to y). For example,
$\int_0^1 \int_1^2 y e^x\, dx\, dy = \frac{1}{2}(e^2 - e)$ if we compute it as an iterated integral or as $\int_1^2 e^x\, dx \int_0^1 y\, dy$.

Chapter 8 Review Test

1. (a) $g(0, 0, 0) = 0$; $g(1, 0, 0) = 1$; $g(0, 1, 0) = 0$; $g(x, x, x) = x^3 + x^2$; $g(x, y + k, z) = x(y + k)(x + y + k - z) + x^2$ **(b)** Decreases by 0.32 units; increases by 12.5 units **(c)** Reading left to right, starting at the top: 4, 0, 0, 3, 0, 1, 2, 0, 2 **(d)** Answers may vary. Two examples are $f(x, y) = 3(x - y)/2$ and $f(x, y) = 3(x - y)^3/8$.

2. (a) $f_x = 2x + y$, $f_y = x$, $f_{yy} = 0$ **(b)** $\dfrac{\partial f}{\partial x} = ye^{xy} + 6xe^{3x^2 - y^2}$, $\dfrac{\partial^2 f}{\partial x \partial y} = (xy + 1)e^{xy} - 12xye^{3x^2 - y^2}$
(c) $\dfrac{\partial f}{\partial x} = \dfrac{-x^2 + y^2 + z^2}{(x^2 + y^2 + z^2)^2}$, $\dfrac{\partial f}{\partial y} = -\dfrac{2xy}{(x^2 + y^2 + z^2)^2}$, $\dfrac{\partial f}{\partial z} = -\dfrac{2xz}{(x^2 + y^2 + z^2)^2}$, $\dfrac{\partial f}{\partial x}\bigg|_{(0,1,0)} = 1$ **(d)** $f_{xx} + f_{yy} + f_{zz} = 2 + 2 + 2 = 6$

3. (a) $f_x = 2(x - 1)$; $f_y = 2(2y - 3)$; $f_{xx} = 2$; $f_{yy} = 4$; $f_{xy} = 0$. $f_x = 0$ when $x = 1$; $f_y = 0$ when $y = 3/2$, so $(1, 3/2)$ is the only critical point. $H = 8$ and $f_{xx} > 0$, so f has an absolute minimum at $(1, 3/2)$.
(b) $g_x = 2(x - 1)$; $g_y = -6y$; $g_{xx} = 2$; $g_{yy} = -6$; $g_{xy} = 0$. $g_x = 0$ when $x = 1$; $g_y = 0$ when $y = 0$, so $(1, 0)$ is the only critical point. $H = -12 < 0$, so g has a saddle point at $(1, 0)$.
(c) $h_x = ye^{xy}$; $h_y = xe^{xy}$; $h_{xx} = y^2 e^{xy}$; $h_{yy} = x^2 e^{xy}$; $h_{xy} = (xy + 1)e^{xy}$. $h_x = 0$ when $y = 0$; $h_y = 0$ when $x = 0$, so $(0, 0)$ is the only critical point. $H(0, 0) = -1 < 0$, so h has a saddle point at $(0, 0)$.
(d) $j_x = y + 2x$; $j_y = x$; $j_{xx} = 2$; $j_{yy} = 0$; $j_{xy} = 1$. $j_x = 0$ when $y = -2x$; $j_y = 0$ when $x = 0$, so $(0, 0)$ is the only critical point. $H = -1 < 0$, so j has a saddle point at $(0, 0)$.

(e) Note that the domain of f contains every point except $(0, 0)$. $f_x = \dfrac{2x}{x^2 + y^2} - 2x$; $f_y = \dfrac{2y}{x^2 + y^2} - 2y$; $f_{xx} = \dfrac{-2x^2 + 2y^2}{(x^2 + y^2)^2} - 2$; $f_{yy} = \dfrac{2x^2 - 2y^2}{(x^2 + y^2)^2} - 2$; $f_{xy} = \dfrac{-4xy}{(x^2 + y^2)^2}$. $f_x = 0$ if either $x = 0$ or $x^2 + y^2 = 1$; $f_y = 0$ if either $y = 0$ or $x^2 + y^2 = 1$; since $(0, 0)$ is not in the domain, the critical points are all the points on the circle $x^2 + y^2 = 1$. At such a point we can substitute $y^2 = 1 - x^2$ to get $f_{xx} = (2 - 4x^2) - 2 = -4x^2$, $f_{yy} = 4x^2 - 4$; and $f_{xy} = \pm 4x(1 - x^2)^{1/2}$. Hence $H = -4x^2(4x^2 - 4) - 16x^2(1 - x^2) = 0$. To resolve what happens along the circle we graph the function:

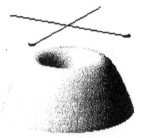

f has an absolute maximum at each point on the circle $x^2 + y^2 = 1$.

4. (a) Minimize $d(x, y, z) = x^2 + y^2 + z^2$ subject to $x^2 + 2(y - 3)^2 - z^2 = 0$ ($z \geq 0$). Using Lagrange multipliers, we need to solve $2x = 2\lambda x$, $2y = 4\lambda(y - 3)$, $2z = -2\lambda z$, and $x^2 + 2(y - 3)^2 - z^2 = 0$. From the first equation, either $x = 0$ or $\lambda = 1$. Suppose that $x = 0$. From the third equation, either $z = 0$ or $\lambda = -1$. If $z = 0$ then the final equation gives $y = 3$, but then the second equation cannot be satisfied. Therefore, $z \neq 0$ and $\lambda = -1$. From the second equation we get $2y = -4(y - 3)$, so $y = 2$. From the last equation we now get $z = \sqrt{2}$. This gives us the point $(0, 2, \sqrt{2})$. Returning to our original decision, what happens if $x \neq 0$ and $\lambda = 1$? From the second equation we get $2y = 4(y - 3)$, so $y = 6$; from the third equation we get $z = 0$. From the last equation we get $x^2 = -18$, which has no real solutions. Therefore, $(0, 2, \sqrt{2}\,)$ is the point we're looking for.

(b) $T(x, y) = (x - 1)^2 + y^2$, $0 \le x \le 3$, $0 \le y \le 1$, $x + 3y \le 3$. *Interior:* $T_x = 2(x - 1) = 0$ when $x = 1$; $T_y = 2y = 0$ when $y = 0$, but $(1, 0)$ is not an interior point. *Boundary segment* $y = 0$, $0 \le x \le 3$: $T = (x - 1)^2$, $T' = 2(x - 1) = 0$ when $x = 1$. With the endpoints this gives $(0, 0)$, $(1, 0)$, and $(3, 0)$. *Boundary segment* $x = 0$, $0 \le y \le 1$: $T = y^2$, $T' = 2y = 0$ when $y = 0$. This give only the endpoint $(0, 1)$. *Boundary segment* $x + 3y = 1$, so $x = 1 - 3y$: $T = 9y^2 + y^2 = 10y^2$; $T' = 20y = 0$ when $y = 0$. This gives no new points. Checking the points we've found, $(0, 0)$, $(1, 0)$, $(3, 0)$, $(0, 1)$, we find that T has a minimum of 0 at $(1, 0)$ and a maximum of 4 at $(3, 0)$.

(c) *Interior:* $F_x = 2x - 2 + y = 0$; $F_y = 4y - 4 + x$. The system $2x + y = 2$ and $x + 4y = 4$ has solution $(4/7, 6/7)$, which is in the interior of the domain. *Boundary segment* $y = -2$, $-1 \le x \le 1$: $F = x^2 - 4x + 16$, $F' = 2x - 4 = 0$ when $x = 2$, which is outside the domain. This give just the endpoints $(-1, -2)$ and $(1, -2)$. *Boundary segment* $y = 2$, $-1 \le x \le 1$: $F = x^2$, $F' = 2x = 0$ when $x = 0$. This gives $(-1, 2)$, $(0, 2)$, and $(1, 2)$. *Boundary segment* $x = -1$, $-2 \le y \le 2$: $F = 2y^2 - 5y + 3$, $F' = 4y - 5 = 0$ when $y = 5/4$. This gives the point $(-1, 5/4)$. *Boundary segment* $x = 1$, $-2 \le y \le 2$: $F = 2y^2 - 3y - 1$, $F' = 4y - 3 = 0$ when $y = 3/4$. This gives the point $(1, 3/4)$. Checking the points we've found, $(4/7, 6/7)$, $(-1, -2)$, $(1, -2)$, $(-1, 2)$, $(0, 2)$, $(1, 2)$, $(-1, 5/4)$ and $(1, 3/4)$, we find that F has a minimum of $-16/7$ at $(4/7, 6/7)$ and a maximum of 21 at $(-1, -2)$.

5. (a) $\displaystyle\int_0^1 \int_0^2 2xy \, dx \, dy = \int_0^1 [x^2 y]_{x=0}^2 \, dy = \int_0^1 4y \, dy$

$= [2y^2]_{y=0}^1 = 2$

(b) $\displaystyle\int_1^2 \int_0^1 xye^{x+y} \, dx \, dy = \int_1^2 [xye^{x+y} - ye^{x+y}]_{x=0}^1 \, dy =$

$\displaystyle\int_1^2 ye^y \, dy = [ye^y - e^y]_{y=1}^2 = 2e^2 - e^2 - (e - e) =$

e^2

(c) $\displaystyle\int_0^2 \int_0^{2x} \frac{1}{x^2 + 1} \, dy \, dx = \int_0^2 \left[\frac{y}{x^2 + 1}\right]_{y=0}^{2x} dy =$

$\displaystyle\int_0^2 \frac{2x}{x^2 + 1} \, dy = [\ln(x^2 + 1)]_{x=0}^2 = \ln 5$

(d) We've already computed $\displaystyle\int_1^2 \int_0^1 xye^{x+y} \, dx \, dy =$ e^2. On the other hand, $A = 1$ because the domain is a square with sides of length 1. Hence the average value is $e^2/1 = e^2$

(e) $\displaystyle\int_0^1 \int_y^{2-y} (x^2 - y^2) \, dx \, dy = \int_0^1 \left[\frac{1}{3}x^3 - xy^2\right]_{x=y}^{2-y} dy =$

$\displaystyle\int_0^1 \left[\frac{1}{3}(2 - y)^3 - (2 - y)y^2 - \frac{1}{3}y^3 + y^3\right] dy =$

$\displaystyle\int_0^1 \left(\frac{8}{3} - 4y + \frac{4}{3}y^3\right) dy = \left[\frac{8}{3}y - 2y^2 + \frac{1}{3}y^4\right]_{y=0}^1 =$

(f) $\int_{-1}^{1}\int_{0}^{1-x^2}(1-y)\,dy\,dx = \int_{-1}^{1}\left[y - \frac{1}{2}y^2\right]_{y=0}^{1-x^2} dx =$

$\int_{-1}^{1}\left[(1-x^2) - \frac{1}{2}(1-x^2)^2\right] dx = \frac{1}{2}\int_{-1}^{1}(1-x^4)\,dx$

$= \frac{1}{2}\left[x - \frac{1}{5}x^5\right]_{x=-1}^{1} = \frac{1}{2}\left(1 - \frac{1}{5} + 1 - \frac{1}{5}\right) = \frac{4}{5}$

6. (a) $h(x, y) = 5000 - 0.8x - 0.6y$ hits per day (x = number of new customers at JungleBooks.com, y = number of new customers at FarmerBooks.com) (b) Set $x = 100$ and $h = 4770$: $4770 = 5000 - 0.8(100) - 0.6y$, $y = 250$ new customers. (c) $h(x, y, z) = 5000 - 0.8x - 0.6y + 0.0001z$ (z = number of new Internet shoppers) (d) Set $x = y = 100$ and $h = 5000$: $5000 = 5000 - 0.8(100) - 0.6(100) + 0.0001z$, $z = 1.4$ million new Internet shoppers.

7. (a) $h(2000, 3000) = 2320$ hits per day (b) $h_y = 0.08 + 0.00003x$ hits (daily) per dollar spent on television advertising per month; this increases with increasing x. (c) Set $h_y = 1/5 = 0.2$: $0.2 = 0.08 + 0.00003x$, $x = \$4000$ per month (d) (A) because $h_x(x, 0) = 0.05$.

8. (a) $P(10, 1000) \approx 15{,}800$ orders per day (b) Minimize $C(x, y) = 150x + y$ subject to $1000x^{0.9}y^{0.1} = 15{,}000$. Using Lagrange multipliers we need to solve $150 = 900\lambda x^{-0.1}y^{0.1}$, $1 = 100\lambda x^{0.9}y^{-0.9}$, and $1000x^{0.9}y^{0.1} - 15{,}000 = 0$. From the first equation we get $\lambda = \frac{x^{0.1}}{6y^{0.1}}$; substituting in the second equation gives $1 = \frac{100x}{6y}$, so $y = \frac{50}{3}x$. Substituting in the last equation gives $1000\left(\frac{50}{3}\right)^{0.1} x = 15{,}000$, so $x = 15\left(\frac{3}{50}\right)^{0.1} \approx 11$.

9. Minimize and maximize $C(x, y) = 0.1x^2 + 0.2y^2 - 200x - 480y + 400{,}000$ subject to $900 \le x \le 1100$ and $1000 \le y \le 1500$. Interior: $C_x = 0.2x - 200 = 0$ if $x = 1000$; $C_y = 0.4y - 480 = 0$ if $y = 1200$, giving us (1000, 1200). Boundary segment $y = 1000$, $900 \le x \le 1100$: $C = 0.1x^2 - 200x + 120{,}000$, $C' = 0.2x - 200 = 0$ when $x = 1000$. With the endpoints this gives (900, 1000), (1000, 1000), and (1100, 1000). Boundary segment $y = 1500$, $900 \le x \le 1100$: This is similar, giving the points (900, 1500), (1000, 1500), and (1100, 1500). Boundary segment $x = 900$, $1000 \le y \le 1500$: $C = 0.2y^2 - 480y + 301{,}000$, $C' = 0.4y - 480 = 0$ when $y = 1200$. This gives us one more point, (900, 1200). Boundary segment $x = 1100$, $1000 \le y \le 1500$: This is similar, giving the point (1100, 1200). Checking the points we've found, (1000, 1200), (900, 1000), (1000, 1000), (1100, 1000), (900, 1500), (1000, 1500), (1100, 1500), (900, 1200), and (1100, 1200), we find that selling 1000 paperbacks and 1200 hardcover books costs the least, while selling 900 or 1100 paperbacks and 1500 hardcover books costs the most.

10. $\int_{1200}^{1500}\int_{1800}^{2000}(3x + 10y)\,dy\,dx =$

$\int_{1200}^{1500}\left[3xy + 5y^2\right]_{y=1800}^{2000} dx =$

$\int_{1200}^{1500}(600x + 3{,}800{,}000)\,dx = [300x^2 + 3{,}800{,}000x]_{x=1200}^{1500} = 1{,}383{,}000{,}000$. $A = (1500 - 1200)(2000 - 1800) = 60{,}000$, so the average profit is $1{,}383{,}000{,}000/60{,}000 = \$23{,}050$.

Chapter 9
9.1

1.

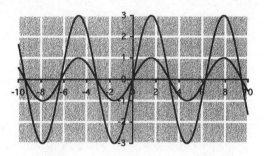

3.

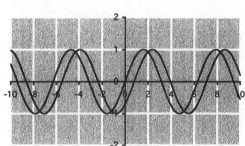

5.

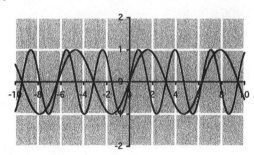

7.

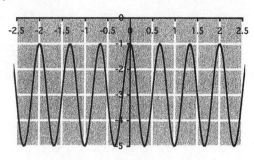

9.

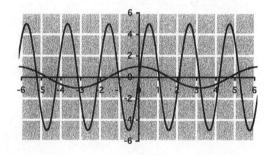

11.

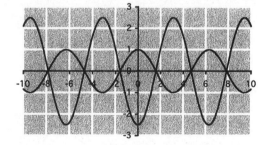

13. From the graph, $C = 1$, $A = 1$, $\alpha = 0$, and $P = 1$, so $\omega = 2\pi$. Thus, $f(x) = \sin(2\pi x) + 1$.

15. From the graph, $C = 0$, $A = 1.5$, $\alpha = 0.25$, and $P = 0.5$, so $\omega = 4\pi$. Thus, $f(x) = 1.5\sin[4\pi(x - 0.25)]$.

17. From the graph, $C = -50$, $A = 50$, $\alpha = 5$, and $P = 20$, so $\omega = \pi/10$. Thus, $f(x) = 50\sin[\pi(x - 5)/10] - 50$.

19. From the graph, $C = 0$, $A = 1$, $\alpha = 0$, and $P = 1$, so $\omega = 2\pi$. Thus, $f(x) = \cos(2\pi x)$.

21. From the graph, $C = 0$, $A = 1.5$, $\alpha = 0.375$, and $P = 0.5$, so $\omega = 4\pi$. Thus, $f(x) = 1.5\cos[4\pi(x - 0.375)]$. Alternatively, $\alpha = -0.125$, so $f(x) = 1.5\cos[4\pi(x + 0.125)]$.

23. From the graph, $C = 40$, $A = 40$, $\alpha = 10$, and $P = 20$, so $\omega = \pi/10$. Thus, $f(x) = 40\cos[\pi(x - 10)/10] + 40$.

25. $f(t) = 4.2\sin(\pi/2 - 2\pi t) + 3$

27. $g(x) = 4 - 1.3\sin[\pi/2 - 2.3(x - 4)]$

29. $\sin^2 x + \cos^2 x = 1$; $(\sin^2 x + \cos^2 x)/\cos^2 x = 1/\cos^2 x$; $\tan^2 x + 1 = \sec^2 x$.

31. $\sin(\pi/3) = \sin(\pi/6 + \pi/6) = \sin(\pi/6)\cos(\pi/6) + \cos(\pi/6)\sin(\pi/6) = (1/2)(\sqrt{3}/2) + (\sqrt{3}/2)(1/2) = \sqrt{3}/2$

33. $\sin(t + \pi/2) = \sin t \cos(\pi/2) + \cos t \sin(\pi/2) = \cos t$ because $\cos(\pi/2) = 0$ and $\sin(\pi/2) = 1$.

35. $\sin(\pi - x) = \sin \pi \cos x - \cos \pi \sin x = \sin x$ because $\sin \pi = 0$ and $\cos \pi = -1$.

37. $\tan(x + \pi) = \dfrac{\sin(x + \pi)}{\cos(x + \pi)} = \dfrac{\sin x \cos \pi + \cos x \sin \pi}{\cos x \cos \pi - \sin x \sin \pi} = \dfrac{-\sin x}{-\cos x} = \tan x$

39. (a) $P = 2\pi/0.602 \approx 10.4$ years.

(b) Maximum: $58.8 + 57.7 = 116.5 \approx 117$; minimum: $58.8 - 57.7 = 1.1 \approx 1$ **(c)** $1.43 + P/4 + P = 1.43 + 13.05 \approx 14.5$ years, or midway through 2011

41. (a)

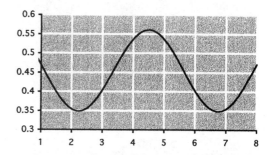

Maximum sales occurred when $t \approx 4.5$ (during the first quarter of 1996). Minimum sales occurred when $t \approx 2.2$ (during the third quarter of 1995) and $t \approx 6.8$ (during the third quarter of 1996). **(b)** Maximum quarterly revenues were $0.561 billion; minimum quarterly revenues were $0.349 billion. **(c)** The maximum and minimum values are $C \pm A$, so the maximum is $0.455 + 0.106 = 0.561$ and the minimum is $0.455 - 0.106 = 0.349$.

43. Amplitude = 0.106, vertical offset = 0.455, phase shift = $-1.61/1.39 \approx -1.16$, angular frequency = 1.39, period = 4.52. In 1995 and 1996, quarterly revenue from the sale of computers at Computer City fluctuated in cycles of 4.52 quarters about a baseline of $0.455 billion. Every cycle, quarterly revenue peaked at $0.561 billion ($0.106 above the baseline) and dipped to a low of $0.349 billion. Revenue peaked in the middle of the first quarter of 1996 (at $t = -1.16 + (5/4) \times 4.52 = 4.49$).

45. $C = 12.5$, $A = 7.5$, period = 52 weeks, so $\omega = \pi/26$, $\alpha = 52/4 = 13$. So, $P(t) =$

Section 9.1

$7.5\sin[\pi(t-13)/26] + 12.5$.

47. $C = 87.5$, $A = 7.5$, period = 12 months, so $\omega = \pi/6$, $\alpha = 6 + 12/4 = 9$. So, $s(t) = 7.5\sin[\pi(t-9)/6] + 87.5$.

49. $C = 87.5$, $A = 7.5$, period = 12 months, so $\omega = \pi/6$, $\alpha = 0$. So, $s(t) = 7.5\cos(\pi t/6) + 87.5$.

51. $C = 10$, $A = 5$, period = 13.5 hours, so $\omega = 2\pi/13.5$, $\alpha = 5 - 13.5/4 = 1.625$. So, $d(t) = 5\sin[2\pi(t - 1.625)/13.5] + 10$.

53. (a) $C = 7.5$, $A = 2.5$, period = 1 year, so $\omega = 2\pi$, $\alpha = 0.75$. So, $u(t) = 2.5\sin[2\pi(t-0.75)] + 7.5$. **(b)** $c(t) = 1.04^t\{2.5\sin[2\pi(t-0.75)] + 7.5\}$.

55. (a) $P \approx 8$, $C \approx 6$, $A \approx 2$, $\alpha \approx 8$ (Answers may vary)

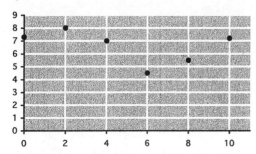

(b) $C(t) = 1.755\sin[0.636(t - 9.161)] + 6.437$

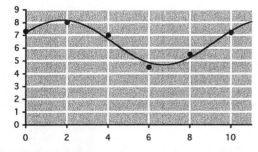

(c) 9.9 [$\approx 2\pi/0.636$], 4.7% [$\approx 6.437 - 1.755$], 8.2% [$\approx 6.437 + 1.755$]

57. (a)

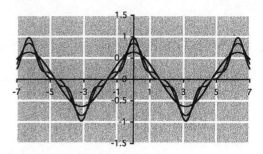

(b) $y_{11} = \dfrac{2}{\pi}\cos x + \dfrac{2}{3\pi}\cos 3x + \dfrac{2}{5\pi}\cos 5x + \dfrac{2}{7\pi}\cos 7x + \dfrac{2}{9\pi}\cos 9x + \dfrac{2}{11\pi}\cos 11x$

Graph of all four functions:

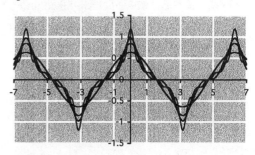

Graph of y_{11} alone:

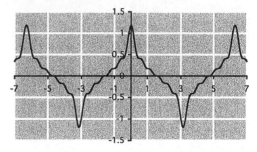

(c) Multiply the amplitudes by 3 and change ω to 1/2: $y_{11} = \dfrac{6}{\pi}\cos\dfrac{x}{2} + \dfrac{6}{3\pi}\cos\dfrac{3x}{2} + \dfrac{6}{5\pi}\cos\dfrac{5x}{2} + \dfrac{6}{7\pi}\cos\dfrac{7x}{2} + \dfrac{6}{9\pi}\cos\dfrac{9x}{2} + \dfrac{6}{11\pi}\cos\dfrac{11x}{2}$

59.

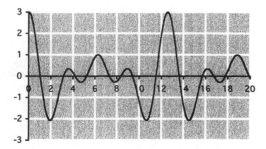

The period is approximately 12.6 units.

61. Lows: $B - A$; Highs: $B + A$

63. He is correct. The other trig functions can be obtained from the sine function by first using the formula $\cos x = \sin(x + \pi/2)$ to obtain cosine, and then using the formulas $\tan x = \dfrac{\sin x}{\cos x}$, $\cot x = \dfrac{\cos x}{\sin x}$, $\sec x = \dfrac{1}{\cos x}$, $\csc x = \dfrac{1}{\sin x}$ to obtain the rest.

65. The largest B can be is A. Otherwise, if B is larger than A, the low figure for sales would have the negative value of $A - B$.

Section 9.2

9.2

1. $f'(x) = \cos x + \sin x$

3. $g'(x) = \cos x \tan x + \sin x \sec^2 x = \sin x (1 + \sec^2 x)$

5. $h'(x) = -2\csc x \cot x - \sec x \tan x + 3$

7. $r'(x) = \cos x - x \sin x + 2x$

9. $s'(x) = (2x - 1)\tan x + (x^2 - x + 1)\sec^2 x$

11. $t'(x) = -[\csc^2 x (1 + \sec x) + \cot x \sec x \tan x]/(1 + \sec x)^2$

13. $k'(x) = -2 \cos x \sin x$

15. $j'(x) = 2 \sec^2 x \tan x$

17. $p'(x) = \pi \cos\left[\dfrac{\pi}{5}(x - 4)\right]$

19. $u'(x) = -(2x - 1)\sin(x^2 - x)$

21. $v'(x) = (2.2x^{1.2} + 1.2)\sec(x^{2.2} + 1.2x - 1) \cdot \tan(x^{2.2} + 1.2x - 1)$

23. $w'(x) = \sec x \tan x \tan(x^2 - 1) + 2x \sec x \sec^2(x^2 - 1)$

25. $y'(x) = e^x[-\sin(e^x)] + e^x \cos x - e^x \sin x = e^x[-\sin(e^x) + \cos x - \sin x]$

27. $z'(x) = \dfrac{\sec x \tan x + \sec^2 x}{\sec x + \tan x} = \dfrac{\sec x (\tan x + \sec x)}{\sec x + \tan x} = \sec x$

29. $\dfrac{d}{dx} \sec x = \dfrac{d}{dx}\left[\dfrac{1}{\cos x}\right] = \dfrac{\sin x}{\cos^2 x} = \dfrac{1}{\cos x} \dfrac{\sin x}{\cos x} = \sec x \tan x$

31. $\dfrac{d}{dx} \csc x = \dfrac{d}{dx}\left[\dfrac{1}{\sin x}\right] = -\dfrac{\cos x}{\sin^2 x} = -\dfrac{1}{\sin x} \dfrac{\cos x}{\sin x} = -\csc x \cot x$

33. $\dfrac{d}{dx} [e^{-2x} \sin(3\pi x)] = -2e^{-2x} \sin(3\pi x) + 3\pi e^{-2x} \cos(3\pi x) = e^{-2x}[-2\sin(3\pi x) + 3\pi \cos(3\pi x)]$

35. $\dfrac{d}{dx} [\sin(3x)]^{0.5} = 0.5[\sin(3x)]^{-0.5} \, 3\cos(3x) = 1.5[\sin(3x)]^{-0.5} \cos(3x)$

37. $\dfrac{d}{dx} \sec\left(\dfrac{x^3}{x^2 - 1}\right) = \dfrac{3x^2(x^2 - 1) - 2x^4}{(x^2 - 1)^2} \sec\left(\dfrac{x^3}{x^2 - 1}\right) \tan\left(\dfrac{x^3}{x^2 - 1}\right) = \dfrac{x^4 - 3x^2}{(x^2 - 1)^2} \sec\left(\dfrac{x^3}{x^2 - 1}\right) \tan\left(\dfrac{x^3}{x^2 - 1}\right)$

39. $\dfrac{d}{dx} [\ln |x| \cot(2x - 1)] = \dfrac{\cot(2x - 1)}{x} - 2 \ln |x| \csc^2(2x - 1)$

41. (a) Not differentiable at 0:
$$\lim_{h \to 0^+} \dfrac{|\sin h| - |\sin 0|}{h} = \lim_{h \to 0^+} \dfrac{\sin h}{h} = 1 \text{ but}$$
$$\lim_{h \to 0^-} \dfrac{|\sin h| - |\sin 0|}{h} = \lim_{h \to 0^-} \dfrac{-\sin h}{h} = -1.$$
(b) $\sin(1) \approx 0.84 > 0$ so $|\sin x| = \sin x$ for x near 1; hence $f'(1)$ exists, and $f'(1) \approx \dfrac{\sin 1.0001 - \sin 0.9999}{0.0002} \approx 0.5403$.

213

Section 9.2

43. 0: Write $f(x) = (\sin^2 x)/x$.

x	$f(x)$
−0.1	−0.0997
−0.01	−0.01
−0.001	−0.001
−0.0001	−0.0001
0	
0.0001	0.0001
0.001	0.001
0.01	0.01
0.1	0.0997

By L'Hospital's rule: $\lim\limits_{x \to 0} \dfrac{\sin^2 x}{x} =$

$\lim\limits_{x \to 0} \dfrac{2 \sin x \cos x}{1} = 0.$

45. 2: Write $f(x) = (\sin 2x)/x$.

x	$f(x)$
−0.1	1.98669331
−0.01	1.99986667
−0.001	1.99999867
−0.0001	1.99999999
0	
0.0001	1.99999999
0.001	1.99999867
0.01	1.99986667
0.1	1.98669331

By L'Hospital's rule: $\lim\limits_{x \to 0} \dfrac{\sin 2x}{x} =$

$\lim\limits_{x \to 0} \dfrac{2 \cos 2x}{1} = 2.$

47. Does not exist: Write $f(x) = (\cos x - 1)/x^3$.

x	$f(x)$
−0.1	4.99583472
−0.01	49.9995833
−0.001	499.999958
−0.0001	4999.99997
0	
0.0001	−5000
0.001	−499.99996
0.01	−49.999583
0.1	−4.9958347

By L'Hospital's rule: $\lim\limits_{x \to 0} \dfrac{\cos x - 1}{x^3} =$

$\lim\limits_{x \to 0} \dfrac{-\sin x}{3x^2} = \lim\limits_{x \to 0} \dfrac{-\cos x}{6x}$ does not exist.

49. $1 = (\sec^2 y) \dfrac{dy}{dx}$, so $\dfrac{dy}{dx} = 1/\sec^2 y$

51. $1 + \dfrac{dy}{dx} + \left(y + x \dfrac{dy}{dx}\right) \cos(xy) = 0,$

$[1 + x \cos(xy)] \dfrac{dy}{dx} = -[1 + y \cos(xy)]$, so $\dfrac{dy}{dx} = -[1 + y \cos(xy)]/[1 + x \cos(xy)]$

53. $c'(t) = 7\pi \cos[2\pi(t - 0.75)]$; $c'(0.75) \approx$ 21.99 per $year \approx \$0.42$ per week.

55. $N'(t) = 34.7354 \cos[0.602(t - 1.43)]$, $N'(6) \approx -32.12$. On January 1, 2003, the number of sunspots was decreasing at a rate of 32.12 sunspots per year.

57. $c'(t) = 1.035^t[\ln(1.035)(0.8 \sin(2\pi t) + 10.2) + 1.6\pi \cos(2\pi t)]$; $c'(1) \approx \$5.57$ per year, or $0.11 per week.

59. (a) $d(t) = 5\cos(2\pi t/13.5) + 10$ (b) $d'(t) = -(10\pi/13.5)\sin(2\pi t/13.5)$; $d'(7) \approx 0.270$. At noon, the tide was rising at a rate of 0.270 feet per hour.

Section 9.2

61. (a) (C): From what we are told we must have $\alpha = -500$ and $P = 40$, so $\omega = 2\pi/40$. **(b)** $A'(t) = -(\pi/20)\sin[2\pi(t+500)/40]$, $A'(-150) \approx 0.157$. The tilt was increasing at a rate of 0.157 degrees per thousand years.

63. -6; 6

65. Answers will vary. Examples: $f(x) = \sin x$; $f(x) = \cos x$

67. Answers will vary. Examples: $f(x) = e^{-x}$; $f(x) = -2e^{-x}$

69. The graph of $\cos x$ slopes down over the interval $(0, \pi)$, so that its derivative is negative over that interval. The function $-\sin x$, and not $\sin x$, has this property.

71. The derivative of $\sin x$ is $\cos x$. When $x = 0$, this is $\cos(0) = 1$. Thus, the tangent to the graph of $\sin x$ at the point $(0, 0)$ has slope 1, which means it slopes upward at $45°$.

Section 9.3

9.3

1. $\int (\sin x - 2\cos x)\, dx = -\cos x - 2\sin x + C$

3. $\int (2\cos x - 4.3\sin x - 9.33)\, dx = 2\sin x + 4.3\cos x - 9.33x + C$

5. $\int \left(3.4\sec^2 x + \dfrac{\cos x}{1.3} - 3.2e^x\right) = 3.4\tan x + \dfrac{\sin x}{1.3} - 3.2e^x + C$

7. $\int 7.6\cos(3x-4)\, dx = \dfrac{7.6}{3}\sin(3x-4) + C$

9. $\int x\sin(3x^2 - 4)\, dx = -\dfrac{1}{6}\cos(3x^2 - 4) + C$:

Substitute $u = 3x^2 - 4$.

11. $\int (4x + 2)\sin(x^2 + x)\, dx = -2\cos(x^2 + x) + C$: Substitute $u = x^2 + x$.

13. $\int (x + x^2)\sec^2(3x^2 + 2x^3)\, dx = \dfrac{1}{6}\tan(3x^2 + 2x^3) + C$: Substitute $u = 3x^2 + 2x^3$.

15. $\int x^2\tan(2x^3)\, dx = -\dfrac{1}{6}\ln|\cos(2x^3)| + C$:

Substitute $u = 2x^3$.

17. $\int 6\sec(2x - 4)\, dx = 3\ln|\sec(2x-4) + \tan(2x-4)| + C$: Substitute $u = 2x - 4$.

19. $\int e^{2x}\cos(e^{2x} + 1)\, dx = \dfrac{1}{2}\sin(e^{2x} + 1) + C$:

Substitute $u = e^{2x} + 1$.

21. $\int_{-\pi}^{0} \sin x\, dx = [-\cos x]_{-\pi}^{0} = -\cos(0) + \cos(-\pi)$

$= -1 - 1 = -2$

23. $\int_{0}^{\pi/3} \tan x\, dx = -[\ln|\cos x|]_0^{\pi/3} =$

$-\ln|\cos(\pi/3)| + \ln|\cos(0)| = -\ln(1/2) + \ln(1) = \ln(2)$

25. $\int_{1}^{\sqrt{\pi+1}} x\cos(x^2 - 1)\, dx = \int_{0}^{\pi} \dfrac{1}{2}\cos u\, du$

(substitute $u = x^2 - 1$) $= \dfrac{1}{2}[\sin u]_0^{\pi} = \dfrac{1}{2}[\sin(\pi) - \sin(0)] = 0$

27. $\int_{1/\pi}^{2/\pi} \dfrac{\sin(1/x)}{x^2}\, dx = -\int_{\pi}^{\pi/2} \sin u\, du$ (substitute $u = 1/x$) $= [\cos u]_{\pi}^{\pi/2} = \cos(\pi/2) - \cos(\pi) = 1$

29. $\int \cos(ax + b)\, dx = \int \dfrac{1}{a}\cos u\, du$ (substitute $u = ax + b$) $= \dfrac{1}{a}\sin u + C = \dfrac{1}{a}\sin(ax + b) + C$

31. $\int \cot x\, dx = \int \dfrac{\cos x}{\sin x}\, dx = \int \dfrac{1}{u}\, du$ (substitute $u = \sin x$) $= \ln|u| + C = \ln|\sin x| + C$

Section 9.3

33. $\int \sin(4x)\, dx = -\frac{1}{4}\cos(4x) + C$

35. $\int \cos(-x+1)\, dx = -\sin(-x+1) + C$

37. $\int \sin(-1.1x - 1)\, dx = \frac{1}{1.1}\cos(-1.1x - 1) + C$

39. $\int \cot(-4x)\, dx = -\frac{1}{4}\ln|\sin(-4x)| + C$

41. $\int_{-\pi/2}^{\pi/2} \sin x\, dx = 0$ because, by symmetry, there is as much area above the x axis as below.

43. $\int_{0}^{2\pi} (1 + \sin x)\, dx = 2\pi$ because, by symmetry, the average value of $1 + \sin x$ over $[0, 2\pi]$ is 1, hence the area under the curve is the same as the area under the line of height 1, which is 2π.

45.

	D	I
+	x	$\sin x$
−	1	$-\cos x$
+$\int$	0 →	$-\sin x$

$\int x \sin x\, dx = -x\cos x + \sin x + C$

47.

	D	I
+	x^2	$\cos(2x)$
−	$2x$	$\frac{1}{2}\sin(2x)$
+	2	$-\frac{1}{4}\cos(2x)$
−$\int$	0 →	$-\frac{1}{8}\sin(2x)$

$\int x^2 \cos(2x)\, dx = \left(\frac{x^2}{2} - \frac{1}{4}\right)\sin(2x) + \frac{x}{2}\cos(2x) + C$

49.

	D	I
+	$\sin x$	e^{-x}
−	$\cos x$	$-e^{-x}$
+$\int$	$-\sin x$ →	e^{-x}

$\int e^{-x} \sin x\, dx = -e^{-x}\sin x - e^{-x}\cos x - \int e^{-x}\sin x\, dx$, so $2\int e^{-x}\sin x\, dx = -e^{-x}\sin x - e^{-x}\cos x + C$; $\int e^{-x}\sin x\, dx = -\frac{1}{2}e^{-x}\sin x - \frac{1}{2}e^{-x}\cos x + C$

51.

	D	I
+	x^2	$\sin x$
−	$2x$	$-\cos x$
+	2	$-\sin x$
−∫	0	$\cos x$

$$\int_0^\pi x^2 \sin x \, dx = [-x^2 \cos x + 2x \sin x + 2 \cos x]_0^\pi$$
$$= \pi^2 - 4$$

53. Average $= \dfrac{1}{\pi} \displaystyle\int_0^\pi \sin x \, dx = \dfrac{1}{\pi}[-\cos x]_0^\pi = \dfrac{2}{\pi}$

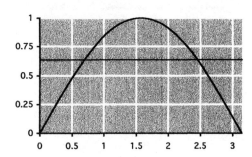

55. $\displaystyle\int_0^{+\infty} \sin x \, dx = \lim_{M \to +\infty} \int_0^M \sin x \, dx =$

$\displaystyle\lim_{M \to +\infty} [-\cos x]_0^M = \lim_{M \to +\infty} (1 - \cos M)$

diverges

57.

	D	I
+	$\cos x$	e^{-x}
−	$-\sin x$	$-e^{-x}$
+∫	$-\cos x$	e^{-x}

$\displaystyle\int e^{-x} \cos x \, dx = -e^{-x} \cos x + e^{-x} \sin x -$

$\displaystyle\int e^{-x} \cos x \, dx$, so $2\displaystyle\int e^{-x} \cos x \, dx = -e^{-x} \cos x +$

$e^{-x} \sin x + C$; $\displaystyle\int e^{-x} \cos x \, dx = -\dfrac{1}{2} e^{-x} \cos x +$

$\dfrac{1}{2} e^{-x} \sin x + C$; $\displaystyle\int_0^{+\infty} e^{-x} \cos x \, dx =$

$\displaystyle\lim_{M \to +\infty} \int_0^M e^{-x} \cos x \, dx = \lim_{M \to +\infty} [-\dfrac{1}{2} e^{-x} \cos x +$

$\dfrac{1}{2} e^{-x} \sin x]_0^M = \displaystyle\lim_{M \to +\infty} (-\dfrac{1}{2} e^{-M} \cos M +$

$\dfrac{1}{2} e^{-M} \sin M + \dfrac{1}{2}) = \dfrac{1}{2}$; converges

59. $C(t) = \displaystyle\int \left\{0.04 - 0.1 \sin\left[\dfrac{\pi}{26}(t - 25)\right]\right\} dt =$

$0.04t + \dfrac{2.6}{\pi} \cos\left[\dfrac{\pi}{26}(t - 25)\right] + K$; $C(12) = 1.50$

gives $K = 1.02$, so $C(t) = 0.04t +$
$\dfrac{2.6}{\pi} \cos\left[\dfrac{\pi}{26}(t - 25)\right] + 1.02$

61. Position $= \displaystyle\int_0^{10} 3\pi \cos\left[\dfrac{\pi}{2}(t - 1)\right] dt =$

$\left[6 \sin\left[\dfrac{\pi}{2}(t - 1)\right]\right]_0^{10} = 6 \sin \dfrac{9\pi}{2} - 6 \sin\left(-\dfrac{\pi}{2}\right) =$

12 feet

63. Average =

$$\frac{1}{2}\int_5^7 \{57.7\sin[0.602(t-1.43)] + 58.8\}\,dt =$$

$$\frac{1}{2}[-95.847\cos[0.602(t-1.43)] + 58.8t]_5^7 \approx 79$$

sunspots

65. $P(t) = 7.5\sin[\pi(t-13)/26] + 12.5$ (see Exercise 45 in Section 9.1). Average =

$$\frac{1}{13}\int_0^{13} \{7.5\sin[\pi(t-13)/26] + 12.5\}\,dt =$$

$$\frac{1}{13}[-62.07\cos[\pi(t-13)/26] + 12.5t]_0^{13} \approx 7.7\%$$

67. (a) Average voltage over [0, 1/6] is

$$6\int_0^{1/6} 165\cos(120\pi t)\,dt = 2.63[\sin(120\pi t)]_0^{1/6} = 0.$$

In one second the voltage goes through 60 periods of the cosine wave, hence reaches its maximum 60 times; hence the electricity has a frequency of 60 cycles per second.

(b)

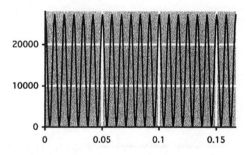

(c) We can use technology to estimate the average $\bar{S} \approx 13{,}612.5$, hence the RMS voltage is approximately $\sqrt{13{,}612.5} \approx 116.673$ volts. Or, we can notice that the graph in (b) appears to be a sinusoid with an average value of $\bar{S} = 165^2/2 = 13{,}612.5$, so that the RMS voltage is $165/\sqrt{2} \approx 116.673$ volts.

69. $TV = \int_0^1 [50{,}000 + 2000\pi\sin(2\pi t)]\,dt =$

$[50{,}000t - 1000\cos(2\pi t)]_0^1 = \$50{,}000$

71. The integral over a whole number of periods is always zero, by symmetry.

73. 1, because the average value of $2\cos x$ will be approximately 0 over a large interval.

75. Integrate twice to get $s = -\dfrac{K}{\omega^2}\sin(\omega t - \alpha) + Lt + M$ for constants L and M.

Chapter 9 Review Test

1. (a) $C = 1$, $A = 2$, $\alpha = 0$, and $P = 2\pi$, so $\omega = 1$. Thus, $f(x) = 2 \sin x + 1$. **(b)** $C = 0$, $A = 10$, $\alpha = -\pi/4$, and $P = \pi$, so $\omega = 2$. Thus, $f(x) = 10 \sin[2(x + \pi/4)] = 10 \sin(2x + \pi/2)$ **(c)** $C = 2$, $A = 2$, $\alpha = 1$, and $P = 2$, so $\omega = \pi$. Thus, $f(x) = 2 \sin[\pi(x - 1)] + 2$. We could also take $\alpha = -1$, getting $f(x) = 2 \sin[\pi(x + 1)] + 2$ **(d)** $C = -2$, $A = 3$, $\alpha = 0.25$, and $P = 0.5$, so $\omega = 4\pi$. Thus, $f(x) = 3 \sin[4\pi(x - 0.25)] - 2$. We could also take $\alpha = 0.25$, getting $f(x) = 3 \sin[4\pi(x + 0.25)] - 2$.

2. In each case, substitute $\sin z = \cos(z - \pi/2)$ in the answer from Question 1 and simplify as necessary. **(a)** $f(x) = 2 \cos(x - \pi/2) + 1$ **(b)** $f(x) = 10 \cos(2x)$ **(c)** $f(x) = 2 \cos[\pi(x - 3/2)] + 2 = 2 \cos[\pi(x + 1/2)] + 2$ **(d)** $f(x) = 3 \cos[4\pi(x - 0.375)] - 2 = 3 \cos[4\pi(x + 0.125)] - 2$

3. (a) $-2x \sin(x^2 - 1)$
(b) $2x[\cos(x^2 + 1) \cos(x^2 - 1) - \sin(x^2 + 1) \sin(x^2 - 1)]$ **(c)** $2e^x \sec^2(2e^x - 1)$
(d) $\dfrac{(2x - 1)\sec\sqrt{x^2 - x}\tan\sqrt{x^2 - x}}{2\sqrt{x^2 - x}}$
(e) $4x \sin(x^2) \cos(x^2)$
(f) $4 \cos(2x) \cos[1 - \sin(2x)] \sin[1 - \sin(2x)]$

4. (a) $\displaystyle\int 4 \cos(2x - 1) \, dx = 2 \sin(2x - 1) + C$

(substitute $u = 2x - 1$)

(b) $\displaystyle\int (x - 1)\sin(x^2 - 2x + 1) \, dx =$

$-\dfrac{1}{2}\cos(x^2 - 2x + 1) + C$ (substitute $u = x^2 -$

$2x + 1$) **(c)** $\displaystyle\int x \tan(x^2 + 1) \, dx =$

$-\dfrac{1}{2}\ln|(\cos(x^2 + 1)| + C$ (substitute $u = x^2 + 1$)

(d) $\displaystyle\int_0^\pi \cos(x + \pi/2) \, dx = [\sin(x + \pi/2)]_0^\pi =$

$\sin(3\pi/2) - \sin(\pi/2) = -2$ **(e)** $\displaystyle\int_{\ln(\pi/2)}^{\ln(\pi)} e^x \sin(e^x) \, dx =$

$\displaystyle\int_{\pi/2}^{\pi} \sin u \, du$ (substitute $u = e^x$) $= [-\cos u]_{\pi/2}^{\pi} =$

$-\cos \pi + \cos(\pi/2) = 1$ **(f)** $\displaystyle\int_\pi^{2\pi} \tan(x/6) \, dx =$

$[-6 \ln|\cos(x/6)|]_\pi^{2\pi} = -6 \ln|\cos(\pi/3)| +$
$6 \ln|\cos(\pi/6)| = -6 \ln(1/2) + 6 \ln(\sqrt{3}/2) = 3 \ln 3$

5. (a)

D	I
$+\quad x^2$	$\sin x$
$-\quad 2x$	$-\cos x$
$+\quad 2$	$-\sin x$
$-\int\quad 0$	$\cos x$

$\displaystyle\int x^2 \sin x \, dx = -x^2 \cos x + 2x \sin x + 2\cos x + C$

(b)

D	I
$+\quad \sin 2x$	e^x
$-\quad 2 \cos 2x$	e^x
$+\int\quad -4 \sin 2x$	e^x

Chapter 9 Review Test

$\int e^x \sin 2x \, dx = e^x \sin 2x - 2e^x \cos 2x -$

$4\int e^x \sin 2x \, dx$, so $5\int e^x \sin 2x \, dx = e^x \sin 2x -$

$2e^x \cos 2x + C$, $\int e^x \sin 2x \, dx = \frac{1}{5} e^x \sin 2x -$

$\frac{2}{5} e^x \cos 2x + C$

6. $C = 10{,}500$, $A = 1500$, $\alpha = 52/2 = 26$, and $P = 52$, so $\omega = 2\pi/52$. Thus, $s(t) = 1500 \sin[(2\pi/52)(t - 26)] + 10{,}500 \approx 1500 \sin(0.12083t - 3.14159) + 10{,}500$

7. $R'(t) = 15{,}000\{-0.12e^{-0.12t} \cos[(\pi/6)(t - 4)] - (\pi/6)e^{-0.12t} \sin[(\pi/6)(t - 4)]\}$; $R'(10) \approx \$542$ per month

8. $\int_0^{10} \{20{,}000 + 15{,}000e^{-0.12t} \cos[(\pi/6)(t - 4)]\} \, dt = [20{,}000t]_0^{10} + 15{,}000 \int_0^{10} e^{-0.12t} \cos[(\pi/6)(t - 4)] \, dt =$

$200{,}000 + 15{,}000 \int_0^{10} e^{-0.12t} \cos[(\pi/6)(t - 4)] \, dt$. Evaluate the latter integral using integration by parts:

	D	I
+	$\cos[(\pi/6)(t - 4)]$	$e^{-0.12t}$
−	$-(\pi/6) \sin[(\pi/6)(t - 4)]$	$-\frac{1}{0.12} e^{-0.12t}$
+∫	$-(\pi/6)^2 \cos[(\pi/6)(t - 4)]$	$\frac{1}{0.12^2} e^{-0.12t}$

$\int e^{-0.12t} \cos[(\pi/6)(t - 4)] \, dt = -\frac{1}{0.12} e^{-0.12t} \cos[(\pi/6)(t - 4)] + \frac{\pi/6}{0.12^2} e^{-0.12t} \sin[(\pi/6)(t - 4)] -$

$\left(\frac{\pi/6}{0.12}\right)^2 \int e^{-0.12t} \cos[(\pi/6)(t - 4)] \, dt$; $\int_0^{10} e^{-0.12t} \cos[(\pi/6)(t - 4)] \, dt =$

$\left[1 + \left(\frac{\pi/6}{0.12}\right)^2\right]^{-1} \left[-\frac{1}{0.12} e^{-0.12t} \cos[(\pi/6)(t - 4)] + \frac{\pi/6}{0.12^2} e^{-0.12t} \sin[(\pi/6)(t - 4)]\right]_0^{10} \approx 1.489$. Hence,

$\int_0^{10} \{20{,}000 + 15{,}000e^{-0.12t} \cos[(\pi/6)(t - 4)]\} \, dt \approx 200{,}000 + 15{,}000(1.489) \approx \$222{,}300$

9. Total consumption $= \int_0^t \{150 + 50 \sin[(\pi/2)(x - 1)]\} \, dx = [150x - (100/\pi) \cos[(\pi/2)(x - 1)]]_0^t =$

$150t - \frac{100}{\pi} \cos\left[\frac{\pi}{2}(t - 1)\right]$ grams

Chapter 9 Review Test

Section A.1

Appendix A
A.1

1. $2(4 + (-1))(2 \cdot -4)$
$= 2(3)(-8) = (6)(-8) = -48$

3. $20/(3*4) - 1$
$= \dfrac{20}{12} - 1 = \dfrac{5}{3} - 1 = \dfrac{2}{3}$

5. $\dfrac{3 + ([3 + (-5)])}{3 - 2\times 2} = \dfrac{3 + (-2)}{3 - 4}$
$= \dfrac{1}{-1} = -1$

7. $(2-5*(-1))/1 - 2*(-1)$
$= \dfrac{2 - 5\cdot(-1)}{1} - 2\cdot(-1)$
$= \dfrac{2+5}{1} + 2 = 7 + 2 = 9$

9. $2\cdot(-1)^2/2 = \dfrac{2\times(-1)^2}{2} = \dfrac{2\times 1}{2} = \dfrac{2}{2} = 1$

11. $2\cdot 4^2 + 1 = 2\times 16 + 1 = 32 + 1 = 33$

13. $3\wedge 2 + 2\wedge 2 + 1$
$= 3^2 + 2^2 + 1 = 9 + 4 + 1 = 14$

15. $\dfrac{3 - 2(-3)^2}{-6(4-1)^2} = \dfrac{3 - 2\times 9}{-6(3)^2} = \dfrac{3 - 18}{6\times 9}$
$= \dfrac{-15}{54} = \dfrac{5}{18}$

17. $10*(1+1/10)\wedge 3$
$= 10\left(1 + \dfrac{1}{10}\right)^3 = 10(1.1)^3$
$= 10\times 1.331 = 13.31$

19. $3\left[\dfrac{-2\cdot 3^2}{-(4-1)^2}\right] = 3\left[\dfrac{-2\times 9}{-3^2}\right] = 3\left[\dfrac{-18}{-9}\right]$
$= 3\times 2 = 6$

21. $3\left[1 - \left(-\dfrac{1}{2}\right)^2\right]^2 + 1 = 3\left[1 - \dfrac{1}{4}\right]^2 + 1$
$= 3\left[\dfrac{3}{4}\right]^2 = 3\left[\dfrac{9}{16}\right] + 1 = \dfrac{27}{16} + 1 = \dfrac{43}{16}$

23. $(1/2)\wedge 2 - 1/2\wedge 2$
$= \left[\dfrac{1}{2}\right]^2 - \dfrac{1}{2^2} = \dfrac{1}{4} - \dfrac{1}{4} = 0$

25. $3\times(2-5) = 3*(2-5)$

27. $\dfrac{3}{2-5} = 3/(2-5)$
Note $3/2-5$ is wrong, since it corresponds to $\dfrac{3}{2} - 5$.

29. $\dfrac{3-1}{8+6} = (3-1)/(8+6)$
Note $3-1/8-6$ is wrong, since it corresponds to $3 - \dfrac{1}{8} - 6$.

31. $3 - \dfrac{4+7}{8} = 3 - (4+7)/8$

33. $\dfrac{2}{3+x} - xy^2 = 2/(3+x) - x*y\wedge 2$

35. $3.1x^3 - 4x^{-2} - \dfrac{60}{x^2-1}$
$= 3.1x\wedge 3 - 4x\wedge(-2) - 60/(x\wedge 2 - 1)$

37. $\dfrac{\left[\dfrac{2}{3}\right]}{5} = (2/3)/5$

Note that we use only (round) parentheses in technology formulas, and not brackets.

39. $3^{4-5}\times 6 = 3\wedge(4-5)*6$
Note that the entire exponent is in parentheses.

41. $3\left[1 + \dfrac{4}{100}\right]^{-3} = 3*(1+4/100)\wedge(-3)$

Note that we use only (round) parentheses in technology formulas, and not brackets.

43. $3^{2x-1} + 4^x - 1 = 3\wedge(2*x-1) + 4\wedge x - 1$
Note that the entire exponent of 3 is in parentheses.

45. $2^{2x^2-x+1} = 2\wedge(2x\wedge 2 - x + 1)$

223

Section A.1

Note that the entire exponent is in parentheses.

47. $\dfrac{4e^{-2x}}{2-3e^{-2x}}$

```
  = 4*e^(-2*x)/(2-3e^(-2*x))
or 4(*e^(-2*x))/(2-3e^(-2*x))
or (4*e^(-2*x))/(2-3e^(-2*x))
```

49. $3\left[1-\left(-\dfrac{1}{2}\right)^2\right]^2 + 1$

```
  = 3(1-(-1/2)^2)^2+1
```
Note that we use only (round) parentheses in technology formulas, and not brackets.

Section A.2

A.2

1. $3^3 = 27$

3. $-(2 \cdot 3)^2 = -(2^2 \cdot 3^2) = -(4 \cdot 9) = -36$ or
$-(2 \cdot 3)^2 = -(6^2) = -36$

5. $\left(\dfrac{-2}{3}\right)^2 = \dfrac{(-2)^2}{3^2} = \dfrac{4}{9}$

7. $(-2)^{-3} = \dfrac{1}{(-2)^3} = \dfrac{1}{-8} = -\dfrac{1}{8}$

9. $\left(\dfrac{1}{4}\right)^{-2} = \dfrac{1}{(1/4)^2} = \dfrac{1}{1/4^2} = \dfrac{1}{1/16} = 16$

11. $2 \cdot 3^0 = 2 \cdot 1 = 2$

13. $2^3\, 2^2 = 2^{3+2} = 2^5 = 32$ or $2^3\, 2^2 = 8 \cdot 4 = 32$

15. $2^2\, 2^{-1}\, 2^4\, 2^{-4} = 2^{2-1+4-4} = 2^1 = 2$

17. $x^3 x^2 = x^{3+2} = x^5$

19. $-x^2 x^{-3} y = -x^{2-3} y = -x^{-1} y = -\dfrac{y}{x}$

21. $\dfrac{x^3}{x^4} = x^{3-4} = x^{-1} = \dfrac{1}{x}$

23. $\dfrac{x^2 y^2}{x^{-1} y} = x^{2-(-1)} y^{2-1} = x^3 y$

25. $\dfrac{(xy^{-1}z^3)^2}{x^2 yz^2} = \dfrac{x^2(y^{-1})^2(z^3)^2}{x^2 yz^2} = x^{2-2} y^{-2-1} z^{6-2} =$
$y^{-3} z^4 = \dfrac{z^4}{y^3}$

27. $\left(\dfrac{xy^{-2}z}{x^{-1}z}\right)^3 = \dfrac{(xy^{-2}z)^3}{(x^{-1}z)^3} = \dfrac{x^3 y^{-6} z^3}{x^{-3} z^3} = x^{3-(-3)} y^{-6} z^{3-3} =$
$x^6 y^{-6} = \dfrac{x^6}{y^6}$

29. $\left(\dfrac{x^{-1} y^{-2} z^2}{xy}\right)^{-2} = (x^{-1-1} y^{-2-1} z^2)^{-2} = (x^{-2} y^{-3} z^2)^{-2} =$
$x^4 y^6 z^{-4} = \dfrac{x^4 y^6}{z^4}$

31. $3x^{-4} = \dfrac{3}{x^4}$

33. $\dfrac{3}{4} x^{-2/3} = \dfrac{3}{4x^{2/3}}$

35. $1 - \dfrac{0.3}{x^{-2}} - \dfrac{6}{5} x^{-1} = 1 - 0.3x^2 - \dfrac{6}{5x}$

37. $\sqrt{4} = 2$

39. $\sqrt{\dfrac{1}{4}} = \dfrac{\sqrt{1}}{\sqrt{4}} = \dfrac{1}{2}$

41. $\sqrt{\dfrac{16}{9}} = \dfrac{\sqrt{16}}{\sqrt{9}} = \dfrac{4}{3}$

43. $\dfrac{\sqrt{4}}{5} = \dfrac{2}{5}$

45. $\sqrt{9} + \sqrt{16} = 3 + 4 = 7$

47. $\sqrt{9 + 16} = \sqrt{25} = 5$

49. $\sqrt[3]{8 - 27} = \sqrt[3]{-19} \approx -2.668$

51. $\sqrt[3]{\dfrac{27}{8}} = \dfrac{\sqrt[3]{27}}{\sqrt[3]{8}} = \dfrac{3}{2}$

53. $\sqrt{(-2)^2} = \sqrt{4} = 2$

Section A.2

55. $\sqrt{\frac{1}{4}(1 + 15)} = \sqrt{\frac{16}{4}} = \frac{\sqrt{16}}{\sqrt{4}} = \frac{4}{2} = 2$

57. $\sqrt{a^2 b^2} = \sqrt{a}\sqrt{b} = ab$

59. $\sqrt{(x+9)^2} = x + 9$ ($x + 9 > 0$ because x is positive)

61. $\sqrt[3]{x^3(a^3 + b^3)} = \sqrt[3]{x^3}\sqrt[3]{a^3 + b^3} = x\sqrt[3]{a^3 + b^3}$
(Notice: Not $x(a + b)$.)

63. $\sqrt{\frac{4xy^3}{x^2 y}} = \sqrt{\frac{4y^2}{x}} = \frac{\sqrt{4}\sqrt{y^2}}{\sqrt{x}} = \frac{2y}{\sqrt{x}}$

65. $\sqrt{3} = 3^{1/2}$

67. $\sqrt{x^3} = x^{3/2}$

69. $\sqrt[3]{xy^2} = (xy^2)^{1/3}$

71. $\frac{x^2}{\sqrt{x}} = \frac{x^2}{x^{1/2}} = x^{2-1/2} = x^{3/2}$

73. $\frac{3}{5x^2} = \frac{3}{5}x^{-2}$

75. $\frac{3x^{-1.2}}{2} - \frac{1}{3x^{2.1}} = \frac{3}{2}x^{-1.2} - \frac{1}{3}x^{-2.1}$

77. $\frac{2x}{3} - \frac{x^{0.1}}{2} + \frac{4}{3x^{1.1}} = \frac{2}{3}x - \frac{1}{2}x^{0.1} + \frac{4}{3}x^{-1.1}$

79. $\frac{1}{(x^2 + 1)^3} - \frac{3}{4\sqrt[3]{(x^2 + 1)}} =$
$\frac{1}{(x^2 + 1)^3} - \frac{3}{4(x^2 + 1)^{1/3}} =$
$(x^2 + 1)^{-3} - \frac{3}{4}(x^2 + 1)^{-1/3}$

81. $2^{2/3} = \sqrt[3]{2^2}$

83. $x^{4/3} = \sqrt[3]{x^4}$

85. $(x^{1/2} y^{1/3})^{1/5} = \sqrt[5]{\sqrt{x}\sqrt[3]{y}}$

87. $-\frac{3}{2}x^{-1/4} = -\frac{3}{2x^{1/4}} = -\frac{3}{2\sqrt[4]{x}}$

89. $0.2x^{-2/3} + \frac{3}{7x^{-1/2}} = \frac{0.2}{x^{2/3}} + \frac{3x^{1/2}}{7} = \frac{0.2}{\sqrt[3]{x^2}} + \frac{3\sqrt{x}}{7}$

91. $\frac{3}{4(1-x)^{5/2}} = \frac{3}{4\sqrt{(1-x)^5}}$

93. $4^{-1/2} 4^{7/2} = 4^{-1/2+7/2} = 4^3 = 64$

95. $3^{2/3} 3^{-1/6} = 3^{2/3-1/6} = 3^{1/2} = \sqrt{3}$

97. $\frac{x^{3/2}}{x^{5/2}} = x^{3/2-5/2} = x^{-1} = \frac{1}{x}$

99. $\frac{x^{1/2} y^2}{x^{-1/2} y} = x^{1/2+1/2} y^{2-1} = xy$

101. $\left(\frac{x}{y}\right)^{1/3} \left(\frac{y}{x}\right)^{2/3} = \left(\frac{y}{x}\right)^{-1/3} \left(\frac{y}{x}\right)^{2/3} = \left(\frac{y}{x}\right)^{1/3}$

103. $x^2 - 16 = 0$, $x^2 = 16$, $x = \pm\sqrt{16} = \pm 4$

105. $x^2 - \frac{4}{9} = 0$, $x^2 = \frac{4}{9}$, $x = \pm\sqrt{\frac{4}{9}} = \pm\frac{2}{3}$

107. $x^2 - (1 + 2x)^2 = 0$, $x^2 = (1 + 2x)^2$,

Section A.2

$x = \pm(1 + 2x)$; if $x = 1 + 2x$ then $-x = 1$, $x = -1$; if $x = -(1 + 2x)$ then $3x = -1$, $x = -1/3$. So, $x = -1$ or $-1/3$.

109. $x^5 + 32 = 0$, $x^5 = -32$, $x = \sqrt[5]{-32} = -2$

111. $x^{1/2} - 4 = 0$, $x^{1/2} = 4$, $x = 4^2 = 16$

113. $1 - \frac{1}{x^2} = 0$, $1 = \frac{1}{x^2}$, $x^2 = 1$, $x = \pm\sqrt{1} = \pm 1$

115. $(x - 4)^{-1/3} = 2$, $x - 4 = 2^{-3} = \frac{1}{8}$, $x = 4 + \frac{1}{8} = \frac{33}{8}$

Section A.3

A.3

1. $x(4x + 6) = 4x^2 + 6x$

3. $(2x - y)y = 2xy - y^2$

5. $(x + 1)(x - 3) = x^2 + x - 3x - 3 = x^2 - 2x - 3$

7. $(2y + 3)(y + 5) = 2y^2 + 3y + 10y + 15 = 2y^2 + 13y + 15$

9. $(2x - 3)^2 = 4x^2 - 12x + 9$

11. $\left(x + \dfrac{1}{x}\right)^2 = x^2 + 2 + \dfrac{1}{x^2}$

13. $(2x - 3)(2x + 3) = (2x)^2 - 3^2 = 4x^2 - 9$

15. $\left(y - \dfrac{1}{y}\right)\left(y + \dfrac{1}{y}\right) = y^2 - \left(\dfrac{1}{y}\right)^2 = y^2 - \dfrac{1}{y^2}$

17. $(x^2 + x - 1)(2x + 4) = (x^2 + x - 1)2x + (x^2 + x - 1)4 = 2x^3 + 2x^2 - 2x + 4x^2 + 4x - 4 = 2x^3 + 6x^2 + 2x - 4$

19. $(x^2 - 2x + 1)^2 = (x^2 - 2x + 1)(x^2 - 2x + 1)$
$= x^2(x^2 - 2x + 1) - 2x(x^2 - 2x + 1) + (x^2 - 2x + 1) = x^4 - 2x^3 + x^2 - 2x^3 + 4x^2 - 2x + x^2 - 2x + 1 = x^4 - 4x^3 + 6x^2 - 4x + 1$

21. $(y^3 + 2y^2 + y)(y^2 + 2y - 1) = y^3(y^2 + 2y - 1) + 2y^2(y^2 + 2y - 1) + y(y^2 + 2y - 1) = y^5 + 2y^4 - y^3 + 2y^4 + 4y^3 - 2y^2 + y^3 + 2y^2 - y = y^5 + 4y^4 + 4y^3 - y$

23. $(x + 1)(x + 2) + (x + 1)(x + 3) = (x + 1)(x + 2 + x + 3) = (x + 1)(2x + 5)$

25. $(x^2 + 1)^5(x + 3)^4 + (x^2 + 1)^6(x + 3)^3 = (x^2 + 1)^5(x + 3)^3(x + 3 + x^2 + 1) =$

228

$(x^2 + 1)^5(x + 3)^3(x^2 + x + 4)$

27. $(x^3 + 1)\sqrt{x + 1} - (x^3 + 1)^2\sqrt{x + 1} = (x^3 + 1)\sqrt{x + 1}\,[1 - (x^3 + 1)] = -x^3(x^3 + 1)\sqrt{x + 1}$

29. $\sqrt{(x + 1)^3} + \sqrt{(x + 1)^5} = \sqrt{(x + 1)^3} \cdot [1 + \sqrt{(x + 1)^2}\,] = \sqrt{(x + 1)^3}\,(1 + x + 1) = (x + 2)\sqrt{(x + 1)^3}$

31. (a) $2x + 3x^2 = x(2 + 3x)$ (b) $x(2 + 3x) = 0$; $x = 0$ or $2 + 3x = 0$; $x = 0$ or $-2/3$

33. (a) $6x^3 - 2x^2 = 2x^2(3x - 1)$
(b) $2x^2(3x - 1) = 0$; $x = 0$ or $3x - 1 = 0$; $x = 0$ or $1/3$

35. (a) $x^2 - 8x + 7 = (x - 1)(x - 7)$
(b) $(x - 1)(x - 7) = 0$; $x - 1 = 0$ or $x - 7 = 0$; $x = 1$ or 7

37. (a) $x^2 + x - 12 = (x - 3)(x + 4)$
(b) $(x - 3)(x + 4) = 0$; $x - 3 = 0$ or $x + 4 = 0$; $x = 3$ or -4

39. (a) $2x^2 - 3x - 2 = (2x + 1)(x - 2)$
(b) $(2x + 1)(x - 2) = 0$; $2x + 1 = 0$ or $x - 2 = 0$; $x = -1/2$ or 2

41. (a) $6x^2 + 13x + 6 = (2x + 3)(3x + 2)$
(b) $(2x + 3)(3x + 2) = 0$; $2x + 3 = 0$ or $3x + 2 = 0$; $x = -3/2$ or $-2/3$

43. (a) $12x^2 + x - 6 = (3x - 2)(4x + 3)$
(b) $(3x - 2)(4x + 3) = 0$; $3x - 2 = 0$ or $4x + 3 = 0$; $x = 2/3$ or $-3/4$

45. (a) $x^2 + 4xy + 4y^2 = (x + 2y)^2$
(b) $(x + 2y)^2 = 0$; $x + 2y = 0$; $x = -2y$

Section A.3

47. (a) $x^4 - 5x^2 + 4 = (x^2 - 1)(x^2 - 4) = (x - 1)(x + 1)(x - 2)(x + 2)$
(b) $(x - 1)(x + 1)(x - 2)(x + 2) = 0$; $x - 1 = 0$ or $x + 1 = 0$ or $x - 2 = 0$ or $x + 2 = 0$; $x = \pm 1$ or ± 2

A.4

1. $\dfrac{x-4}{x+1} \cdot \dfrac{2x+1}{x-1} = \dfrac{(x-4)(2x+1)}{(x+1)(x-1)} = \dfrac{2x^2-7x-4}{x^2-1}$

3. $\dfrac{x-4}{x+1} + \dfrac{2x+1}{x-1} = \dfrac{(x-4)(x-1)+(x+1)(2x+1)}{(x+1)(x-1)} = \dfrac{3x^2-2x+5}{x^2-1}$

5. $\dfrac{x^2}{x+1} - \dfrac{x-1}{x+1} = \dfrac{x^2-(x-1)}{x+1} = \dfrac{x^2-x+1}{x+1}$

7. $\dfrac{1}{\left(\dfrac{x}{x-1}\right)} + x - 1 = \dfrac{x-1}{x} + x - 1 = \dfrac{x-1+x(x-1)}{x} = \dfrac{x^2-1}{x}$

9. $\dfrac{1}{x}\left(\dfrac{x-3}{xy}+\dfrac{1}{y}\right) = \dfrac{1}{x}\left(\dfrac{x-3+x}{xy}\right) = \dfrac{2x-3}{x^2 y}$

11. $\dfrac{(x+1)^2(x+2)^3 - (x+1)^3(x+2)^2}{(x+2)^6} = \dfrac{(x+1)^2(x+2)^2[(x+2)-(x+1)]}{(x+2)^6} = \dfrac{(x+1)^2}{(x+2)^4}$

13. $\dfrac{(x^2-1)\sqrt{x^2+1} - \dfrac{x^4}{\sqrt{x^2+1}}}{x^2+1} = \dfrac{(x^2-1)(x^2+1) - x^4}{(x^2+1)\sqrt{x^2+1}} = \dfrac{-1}{\sqrt{(x^2+1)^3}}$

15. $\dfrac{\dfrac{1}{(x+y)^2} - \dfrac{1}{x^2}}{y} = \dfrac{x^2-(x+y)^2}{yx^2(x+y)^2} = \dfrac{x^2-x^2-2xy-y^2}{yx^2(x+y)^2} = \dfrac{-y(2x+y)}{yx^2(x+y)^2} = \dfrac{-(2x+y)}{x^2(x+y)^2}$

A.5

1. $x + 1 = 0$, $x = -1$

3. $-x + 5 = 0$, $x = 5$

5. $4x - 5 = 8$, $4x = 13$, $x = 13/4$

7. $7x + 55 = 98$, $7x = 43$, $x = 43/7$

9. $x + 1 = 2x + 2$, $-x = 1$, $x = -1$

11. $ax + b = c$, $ax = c - b$, $x = (c - b)/a$

13. $2x^2 + 7x - 4 = 0$, $(2x - 1)(x + 4) = 0$, $x = -4, \frac{1}{2}$

15. $x^2 - x + 1 = 0$, $\Delta = -3 < 0$, so this equation has no real solutions

17. $2x^2 - 5 = 0$, $x^2 = \frac{5}{2}$, $x = \pm\sqrt{\frac{5}{2}}$

19. $-x^2 - 2x - 1 = 0$, $-(x + 1)^2 = 0$, $x = -1$

21. $\frac{1}{2}x^2 - x - \frac{3}{2} = 0$, $x^2 - 2x - 3 = 0$, $(x + 1)(x - 3) = 0$, $x = -1, 3$

23. $x^2 - x = 1$, $x^2 - x - 1 = 0$, $x = \frac{1 \pm \sqrt{5}}{2}$ by the quadratic formula

25. $x = 2 - \frac{1}{x}$, $x^2 = 2x - 1$, $x^2 - 2x + 1 = 0$, $(x - 1)^2 = 0$, $x = 1$

27. $x^4 - 10x^2 + 9 = 0$, $(x^2 - 1)(x^2 - 9) = 0$, $x^2 = 1$ or $x^2 = 0$, $x = \pm 1, \pm 3$

29. $x^4 + x^2 - 1 = 0$, $x^2 = \frac{-1 \pm \sqrt{5}}{2}$ by the quadratic formula, $x = \pm\sqrt{\frac{-1 \pm \sqrt{5}}{2}}$

31. $x^3 + 6x^2 + 11x + 6 = 0$, $(x + 1)(x + 2)(x + 3) = 0$, $x = -1, -2, -3$

33. $x^3 + 4x^2 + 4x + 3 = 0$, $(x + 3)(x^2 + x + 1) = 0$, $x = -3$ (For $x^2 + x + 1 = 0$, $\Delta = -3 < 0$, so there are no real solutions to this quadratic equation.)

35. $x^3 - 1 = 0$, $x^3 = 1$, $x = \sqrt[3]{1} = 1$

37. $y^3 + 3y^2 + 3y + 2 = 0$, $(y + 2)(y^2 + y + 1) = 0$, $y = -2$ (For $y^2 + y + 1 = 0$, $\Delta = -3 < 0$, so there are no real solutions to this quadratic equation.)

39. $x^3 - x^2 - 5x + 5 = 0$, $(x - 1)(x^2 - 5) = 0$, $x = 1, \pm\sqrt{5}$

41. $2x^6 - x^4 - 2x^2 + 1 = 0$, $(2x^2 - 1)(x^4 - 1) = 0$, [or $(2x^2 - 1)(x^2 - 1)(x^2 + 1) = 0$; in any case, think of the cubic you get by substituting y for x^2], $x = \pm 1, \pm\frac{1}{\sqrt{2}}$

43. $(x^2 + 3x + 2)(x^2 - 5x + 6) = 0$, $(x + 2)(x + 1)(x - 2)(x - 3) = 0$, $x = -2, -1, 2, 3$

A.6

1. $x^4 - 3x^3 = 0$, $x^3(x - 3) = 0$, $x = 0, 3$

3. $x^4 - 4x^2 = -4$, $x^4 - 4x^2 + 4 = 0$, $(x^2 - 2)^2 = 0$, $x = \pm\sqrt{2}$

5. $(x + 1)(x + 2) + (x + 1)(x + 3) = 0$, $(x + 1)(x + 2 + x + 3) = 0$, $(x + 1)(2x + 5) = 0$, $x = -1, -5/2$

7. $(x^2 + 1)^5(x + 3)^4 + (x^2 + 1)^6(x + 3)^3 = 0$,
$(x^2 + 1)^5(x + 3)^3(x + 3 + x^2 + 1) = 0$,
$(x^2 + 1)^5(x + 3)^3(x^2 + x + 4) = 0$, $x = -3$
(Neither $x^2 + 1 = 0$ nor $x^2 + x + 4 = 0$ has a real solution.)

9. $(x^3 + 1)\sqrt{x + 1} - (x^3 + 1)^2\sqrt{x + 1} = 0$,
$(x^3 + 1)\sqrt{x + 1}\,[1 - (x^3 + 1)] = 0$,
$-x^3(x^3 + 1)\sqrt{x + 1} = 0$, $x = 0, -1$

11. $\sqrt{(x + 1)^3} + \sqrt{(x + 1)^5} = 0$, $\sqrt{(x + 1)^3}\,(1 + x + 1) = 0$, $(x + 2)\sqrt{(x + 1)^3} = 0$, $x = -1$ ($x = -2$ is not a solution because $\sqrt{(x + 1)^3}$ is not defined for $x = -2$.)

13. $(x + 1)^2(2x + 3) - (x + 1)(2x + 3)^2 = 0$,
$(x + 1)(2x + 3)(x + 1 - 2x - 3) = 0$,
$(x + 1)(2x + 3)(-x - 2) = 0$, $x = -2, -3/2, -1$

15. $\dfrac{(x + 1)^2(x + 2)^3 - (x + 1)^3(x + 2)^2}{(x + 2)^6} = 0$,
$\dfrac{(x + 1)^2(x + 2)^2[(x + 2) - (x + 1)]}{(x + 2)^6} = 0$,
$\dfrac{(x + 1)^2}{(x + 2)^4} = 0$, $(x + 1)^2 = 0$, $x = -1$

17. $\dfrac{2(x^2 - 1)\sqrt{x^2 + 1} - \dfrac{x^4}{\sqrt{x^2 + 1}}}{x^2 + 1} = 0$,
$\dfrac{2(x^2 - 1)(x^2 + 1) - x^4}{(x^2 + 1)\sqrt{x^2 + 1}} = 0$, $\dfrac{x^4 - 2}{(x^2 + 1)\sqrt{x^2 + 1}} = 0$,
$x^4 - 2 = 0$, $x = \pm\sqrt[4]{2}$

19. $x - \dfrac{1}{x} = 0$, $x^2 - 1 = 0$, $x = \pm 1$

21. $\dfrac{1}{x} - \dfrac{9}{x^3} = 0$, $x^2 - 9 = 0$, $x = \pm 3$

23. $\dfrac{x - 4}{x + 1} - \dfrac{x}{x - 1} = 0$,
$\dfrac{(x - 4)(x - 1) - x(x + 1)}{(x + 1)(x - 1)} = 0$,
$\dfrac{-6x + 4}{(x + 1)(x - 1)} = 0$, $-6x + 4 = 0$, $x = 2/3$

25. $\dfrac{x + 4}{x + 1} + \dfrac{x + 4}{3x} = 0$,
$\dfrac{3x(x + 4) + (x + 1)(x + 4)}{3x(x + 1)} = 0$,
$\dfrac{(x + 4)(3x + x + 1)}{3x(x + 1)} = 0$, $\dfrac{(x + 4)(4x + 1)}{3x(x + 1)} = 0$,
$(x + 4)(4x + 1) = 0$, $x = -4, -1/4$